FAMILY	SUBFAMILY	GENUS

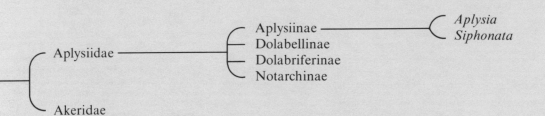

Aplysiidae

Aplysiinae — *Aplysia* / *Siphonata*
Dolabellinae
Dolabriferinae
Notarchinae

Akeridae

Behavioral Biology of *Aplysia*

For Denise, Paul, and Minouche

The Aplisa
by Minouche

An aplisa is like a squishy snail.

In rain in snow, in sleet, in hail.

When it is angry, it shoots out ink.
The ink is purple, its not pink.

An aplisa cannot live on land.

It doesn't have feets so it can't stand.
It has a very funny mouth
And in winter it goes to the south

Minouche Kandel
1972 (age 7)

Behavioral Biology of *Aplysia*

A CONTRIBUTION TO
THE COMPARATIVE STUDY OF OPISTHOBRANCH MOLLUSCS

Eric R. Kandel
COLUMBIA UNIVERSITY

W. H. Freeman and Company
SAN FRANCISCO

A Series of Books in Psychology

EDITORS: *Richard C. Atkinson*
Jonathan Freedman
Gardner Lindzey
Richard F. Thompson

Library of Congress Cataloging in Publication Data

Kandel, Eric R
Behavioral biology of Aplysia.

(A Series of books in psychology)
Bibliography: p.
Includes index.
1. Aplysia—Behavior. 2. Aplysia—Physiology.
3. Mollusks—Behavior. 4. Mollusks—Physiology.
I. Title.
QL430.5.A66K35   593'.3   78–18226
ISBN 0-7167-0021-2
ISBN 0-7167-1070-6 pbk.

Life drawings by Darwen Hennings and Vally Hennings.

Printed in the United States of America

9 8 7 6 5 4 3 2 1

Contents

Preface   xiii

Introduction: The Comparative Study of Behavior   1

1   The Comparative Biology of Molluscs   5

Body Plan   5
Evolution of Molluscs   8
The Hypothetical Proto-Mollusc   11
Classification   12
Monoplacophora   12
Aplacophora   14
Polyplacophora   14
Scaphopoda   14
Bivalvia (Pelecypoda)   14
Cephalopoda   15
Gastropoda   16
Torsion and Detorison in Gastropods   18
Ontogenetic Origins of Torsion   20
Phylogenetic Origins and Adaptive Advantage of Torsion   22
Detorsion   23
Summary and Perspective   24
Selected Reading   26

2   The Comparative Biology of *Aplysia*: Systematics and Distribution   27

The Opisthobranchs   29
Evolutionary Relationships Among Opisthobranch Orders   31
The Anaspids   32
Distribution   38
The Shore Zones and Their *Aplysia* Populations   40
Relationship of Littoral and Sublittoral Populations   43

Summary and Perspective 45
Selected Reading 45

3 *Aplysia* Among the Molluscs I: Sensory Capabilities and Effector
Physiology 47

Body Plan 47
The Head 47
The Sense Organs 50
   The Eyes 50
   The Tentacles: Chemosensation 56
   The Osphradium: Osmoreception 58
   The Statocysts: Equilibrium Reception 61
   Mechanoreception 65
The Foot and Parapodia: The Somatic Motor Apparatus 66
The Mantle Cavity: The Respiratory Space 69
   The Gill: The Respiratory Organ 72
   Muscle Groups of the Gill 75
The Visceral Mass: Internal Organ Systems 76
   Cardiorespiratory System 76
      Circulation 76
      The Heart 79
      Pressure Gradients and Cardiac Filling 80
      Regulation of Cardiac Output 82
      Blood and Respiratory Pigments 83
   Urinary System: Water Balance 83
      Urine Formation 83
      Osmotic Regulation 86
   Digestive System 88
      Foregut 91
      Midgut 96
      Hindgut 96
   Reproductive System 97
Innervation of Effector Structures 104
Summary and Perspective 105
Selected Reading 106

4 *Aplysia* Among the Molluscs II: The Nervous System 107

The Molluscan Central Nervous Systems 108
Aplysia Nervous System 119
   Head Ganglia 123
      Cerebral Ganglia 123
      Buccal Ganglia 127
      Pedal Ganglia 129
      Pleural Ganglia 132
   Abdominal Ganglion 134
Variations in the *Aplysia* Central Nervous System 138
Variations in the Anaspid Nervous System 140
Neuroendocrine Systems of Molluscs 142
Summary and Perspective 145
Selected Reading 147

5   The Abdominal Ganglion of *Aplysia*: A Central Ganglion   149

    Types of Invertebrate Ganglia   149
    General Structure of a Molluscan Ganglion   150
        Connective Tissue Sheath   150
        Cell-Body Region   153
    Glial Cells   156
        Connectives and Peripheral Nerves   156
        Cell-Body Region and Neuropil   157
    Neurons   157
        Fine Structure of the Neuron   157
        Fine Structure of the Nucleus   160
        Pigment Granules in the Neuronal Cell Body   163
    The Neuropil Region   165
        Three-Dimensional Architecture at the Light-Microscopic Level   165
        Fine Structure   170
            Electron-Microscopic Labels   170
            Morphological Basis of Synaptic Action   175
    Relationship Between Cell Body and Axon Terminals   181
    Summary and Perspective   183
    Selected Reading   185

6   Interaction Between the Peripheral and Central Nervous Systems
   in *Aplysia*   187

    Three Patterns of Innervation in a Common Effector System   188
        Relayed Reflexes: The Gill-Withdrawal Reflex   188
        Local Responses: The Pinnule Response   196
        Conjoint Responses: Central and Peripheral Mediation of Siphon
          Withdrawal   199
    Peripheral Motor Cells in the Siphon of *Aplysia*   203
    Peripheral-Central Interactions in the Siphon of *Spisula*   208
    Innervation of the Parapodia and Foot of *Aplysia*   209
    Central Innervation Without Peripheral Motor Cells   211
    Capability for Habituation in Central and Peripheral Pathways   211
    Summary and Perspective   213
    Selected Reading   215

7   Development of *Aplysia* and Related Opisthobranchs   217

    Phases in the Life Cycle of *Aplysia californica*   218
        Sperm, Egg, and Fertilization   220
        Embryonic Development   223
            Cleavage   223
            Late Embryonic Development   230
        Larval Development   235
        Metamorphic Development   238
        Juvenile Development   239
    Development of the Nervous System   240
        Stages in Development   240
        Torsion and Detorsion   245
        Postmetamorphic Development   249
        Neuronal Differentiation and Development of Peripheral Organs   249

Development of Behavior 250
   Locomotion 250
   Feeding 250
   Defensive Withdrawal 252
Interaction Between Environment and Development 254
Developmental Malformations 254
Types of Development in Opisthobranchs 256
Metamorphosis and Postmetamorphic Development in Opisthobranchs 259
Summary and Perspective 260
Selected Reading 263

8   Species-Specific Behavior of Opisthobranchs 265

Classification of Behavior 267
Reflex and Fixed Acts 269
   Defensive Reflexes 269
   Defensive Fixed Act: Inking 272
   Effects of Environment on Elementary Behavior 274
Complex Behavior 274
   Autonomic Regulation 274
      Respiratory Pumping Movements in *Aplysia* 274
      Circulatory Adjustments in *Aplysia* 276
   Locomotion 277
      Crawling in *Aplysia* 277
      Circadian Periodicity in Crawling 284
      Crawling in Pulmonates 287
      Burrowing in *Aplysia* 288
      Swimming in *Aplysia* 290
      Swimming in Other Opisthobranchs 292
   Escape and Predator-Prey Recognition 295
   Feeding 298
      Feeding Behavior in *Aplysia* 299
      Nutrition in *Aplysia* 301
      Orienting, Food-Seeking, and Ingestion in *Aplysia* 304
      Feeding Behavior of Other Opisthobranchs 310
      Feeding Behavior of Pulmonates 318
Higher-Order Behavior 321
   Mating Patterns in *Aplysia* 321
   Mating Patterns in Other Opisthobranchs 322
   Spawn Production and Egg-Laying in *Aplysia* 323
Hormonal Modulation of Behavior 325
Response Hierarchies and Behavioral Choice 325
Summary and Perspective 327
Selected Reading 329

9   Learning, Arousal, and Motivation in Opisthobranchs 331

Learning Capabilities of Opisthobranchs and Pulmonates 332
Nonassociative Learning 333
   Habituation and Sensitization of Elementary Behavior in
      Aplysia 333
   Cellular Analysis of Short-Term Habituation and Sensitization 337
   Cellular Analysis of Long-Term Habituation 345

Effects of Different Environments on Habituation of Defensive
  Reflexes in *Aplysia*   346
Habituation of Complex Behavior in *Tritonia*   348
Sensitization of Complex Behavior in *Aplysia*   348
Arousal and Sensitization   351
  The Multivariant Nature of Arousal in *Aplysia*   354
    Appetitive Stimuli   354
    Noxious Stimuli   356
  Independent Positive Sensitizing Systems   356
Motivational State   358
  Goal-Specific Effects   360
  Nonspecific Effects   364
Relationships Between Sensitization, Arousal, and Motivation   365
Motivational State and the Hierarchy of Behaviors   368
Motivation, Homeostasis, and Neurobiological Explanations   370
Associative Learning   371
  Early Attempts to Produce Classical Conditioning in
    Gastropods   372
  Attempts to Produce Operant Conditioning in *Aplysia*   372
  Sensory Interactions in *Hermissenda*   374
  Modification of Feeding in *Aplysia*   377
  Modification of Feeding in Other Gastropods   379
    Avoidance Training and Habituation of Feeding in
      Navanax   379
    Classical Conditioning of *Pleurobranchaea*   379
    Food-Avoidance Learning in *Limax*   382
Summary and Perspective   383
Selected Reading   385

10 Epilog: Cellular and Behavioral Homologies, Divergence and
    Speciation   387

Homologies Between the Central Ganglia of Opisthobranchs   388
Behavioral Homologies Between Closely Related Opisthobranchs   394
The Adaptive Function of Behavior   399
The Role of Behavior in Speciation   400
The Cellular Basis of Speciation   403
Selected Reading   405

References   407

Name Index   443

Subject Index   449

Preface

This book is a companion volume to *Cellular Basis of Behavior* (1976), my earlier attempt to write an introduction to the neurobiology of behavior based upon cellular studies of behavior and learning in invertebrates, particularly *Aplysia* and other opisthobranchs. In writing this book I have had two purposes in mind. My primary purpose has been to write a handbook on the opisthobranch mollusc *Aplysia*, one that brings together the extensive literature that has appeared since Nellie B. Eales published her classic monograph on the genus in 1921. Before that monograph there was little interest in *Aplysia* as an experimental animal. It was primarily being studied by a few molluscan biologists who saw it as an interesting transitional form between the torted prosobranchs and the untorted nudibranch opisthobranchs. Since 1955 interest in *Aplysia* has grown considerably. The large identifiable neurons of its central nervous system have offered experimental opportunities to neurobiologists interested in cell and molecular biological studies of neuronal functioning and to psychologists interested in relating nerve cells to behavior. In addition, the ability to culture certain species in the laboratory makes *Aplysia* interesting to developmental biologists concerned with the differentiation of specific identified nerve cells.

These advantages of *Aplysia* are also present in other opisthobranchs. My second purpose, therefore, has been to relate recent neurobiological and behaviorial studies of *Aplysia* to those of other opisthobranchs so as to establish a background for, and encourage interest in, a comparative study of the behavioral biology of opisthobranchs, a study for which they are particularly suited.

The two purposes are interrelated, since reviews of other opisthobranchs are as much needed as a review of *Aplysia*. There are few discussions of the opisthobranchs comparable to those of the prosobranchs and pulmonates (e.g., Fretter and Graham, 1962; Fretter and Peake, 1975), although the systematic classification of opisthobranchs, long controversial, now appears to be well worked out (Thompson, 1976). Yet the molluscan species that currently interest neurobiologists, and those most suitable for comparative analysis (*Aplysia*, *Navanax*, *Pleurobranchaea*, *Tritonia*, *Hermissenda*, *Melibe*, *Anisodoris*), belong to this subclass. Although I do not review other opisthobranchs in the detail that I devote to *Aplysia*, I have tried to draw attention to some important points of comparison. I also refer briefly to studies of certain pulmonates for which information on cellular physiological interactions is available (*Helix*, *Helisoma*, and *Limax*).

I have also tried to bridge the research in neurobiology and mulluscan biology. With few exceptions, the main thrust of molluscan biology is surprisingly isolated from recent developments in neurobiology, even though these developments owe a great deal to studies using molluscs such as squid and *Aplysia*. Conversely, many neurobiologists working with molluscs often have little familiarity with molluscan biology. This was certainly true of me; I first approached the readings summarized in this book for my own education. I now hope this review will prove useful to others.

I have benefited greatly by comments on this book from several colleagues, particularly Robin Brace, Thomas Carew, Vincent Castellucci, James Blankenship, Alan Gelperin, Irving Kupfermann, Richard Lee, Eveline Marcus, and Dennis Willows. Selected chapters were read by Nellie B. Eales, Michael Ghiselin, Robert Hawkins, Behrus Jahan–Parwar, John Koester, Arnold Kriegstein, David Leibowitz, James H. Schwartz, Elizabeth Thompson, and Jeffrey Wine. I am grateful to each of them for guiding me to additional literature and for pointing out errors of fact and interpretation. I am solely responsible for errors that remain.

I am again deeply indebted to Kathrin Hilten, who prepared the preliminary versions of the illustrations for this volume, as she did for its companion; to Darwen Hennings, who did the final drawings with marvelous skill; to Sally Muir and Sally Fields, who edited earlier drafts and checked the bibliography; and to Howard Beckman, who, with his usual diligence and quick wit, edited the final manuscript and prepared it for publication.

My own research has been supported by a Research Scientist Award from the National Institute of Mental Health and by research grants MH-19795 from the National Institute of Mental Health and NS-09361 from the National Institute of Neurological and Communicative Disorders and Stroke. I am grateful for this continued support. I am also indebted, along with so many others working on *Aplysia*, to Dr. Rimmon C. Fay of the Pacific Biomarine Laboratories, who for over a decade and a half has supplied *Aplysia* and other opisthobranch specimens to scientists throughout the world.

August, 1978   *Eric R. Kandel*
                                                         WELLFLEET, MASSACHUSETTS

Behavioral Biology of *Aplysia*

Introduction: The Comparative Study of Behavior

The study of behavior, like that of other areas of biology, can be approached from two overlapping perspectives: a general or functional perspective and a comparative or evolutionary perspective. From a general perspective one asks, What principles underlie behavior and how do these principles operate? That is the question I asked in a companion volume to this book, entitled *Cellular Basis of Behavior* (W. H. Freeman and Co., 1976). There I focused on general behavioral questions and searched for general cellular solutions. The discussion centered on the opisthobranch mollusc *Aplysia,* but similar principles would have emerged from a review of any of several other invertebrates, such as the leeches, locusts, or crayfish. In the functional study of behavior a specific animal is only a means toward an end. In this context *Aplysia* is a convenient (and somewhat arbitrary) model, but the animal *per se* is not of primary concern. Even less a concern, from the general perspective, are the specific behaviors of an animal. The gill-withdrawal reflex of *Aplysia,* for example, serves as a proto-type, a model, of a simple defensive (flexion) reflex. One is tempted to add (in hushed tones) that this reflex may lead to a fuller understanding of the defen-sive reflexes common to all animals, including man. For in general biology one is interested in behavioral problems that are, in principle, applicable to all ani-mals, including humans. Indeed, the assumption of general biology is that throughout phylogeny similar central problems share similar solutions. A particular solution may not be the only mechanism for a given biological process, but it is likely to be a mechanism that will be repeatedly encountered throughout phylogeny. The *particular* experimental animal used is therefore a convenient and necessary substitute for all animals—even for people.

Both biology and behavior can, however, also be examined from a comparative or evolutionary perspective, where the focus is on the species, on its naturally occurring (species-specific) responses, and on differences between species in homologous behavior. Indeed, modern ethologists and evolutionary biologists have argued effectively that one cannot fully understand a behavior or a form of learning until one knows its function *in the animal's natural environment* and is in a position to suggest how it evolved (Wilson, 1975; Manning, 1976). In evolutionary biology one is concerned less with functional mechanisms of behavior or *proximal causation,* the factors that operate during the lifespan of an organism, than with *ultimate causation,* the factors that produce evolutionary change—weather, predators, new sources of food, accessibility of mates, etc. Species respond to these environmental influences by genetic evolution through natural selection, which consequently shapes the structure and behavior of individual animals. These factors operate over a long time span covering generations of animals, and from a biological perspective they are the ultimate causes of which the proximal causes are a by-product (Wilson, 1975).

During the past 25 years functional studies in biology and behavior have been in the ascendancy. The cellular physiological and molecular techniques for studying cells and organs and the conditioning paradigms for studying learning have produced insights that have provoked widespread interest. By contrast, the interpretation of comparative studies often has been less powerful scientifically and therefore less compelling. Phylogenetic studies in particular often lack rigor because they are historical and reconstructive rather than predictive and mechanistic. Like all historical research, the reconstruction of the phylogenetic past is based on inferences that cannot be experimentally tested. The problem is particularly severe when it comes to behavior. Evolutionary studies of behavior must rely on comparisons between fossil forms and extant organisms. Since fossils are rarely useful for behavioral studies (Atz, 1970), one must usually rely almost exclusively on comparisons between extant forms. Because of these difficulties, comparative biology and specifically the comparative study of behavior have not been as actively pursued as functional studies of behavioral mechanisms. But two recent developments promise to enliven the evolutionary study of behavior. One comes from studies of new forms of associative learning, such as bait-shyness (Garcia, Hankins, and Rusiniak, 1974). These studies have shown that in order to study learning effectively one must know not only the general rules of animal behavior but also the particular evolutionary history and environment of the particular animal, its sensory and motor capabilities, its species-specific behavior, and its prey–predator relationships. The need to know more about behavior in an ecological context has stimulated comparative studies of the processes of adaptation in related species—the relationships of structure to function (including behavior), and their implications for the adaptation of the species.

The second development stems from the use of identified cells to study behavioral homologies. The greatest problem in the comparative study of behavior is the lack of rigorous criteria for homologies of behavior. Homologies between morphological entities or between functional entities are said to exist when they can be traced to a common ancestor (for discussion see Boyden, 1943; Haas and Simpson, 1946; Hodos and Campbell, 1969). Although the establishment of homologies need not require that entities be structurally similar, such similarities are nonetheless the most useful criteria and those upon which it is easiest to obtain agreement. Three of the four commonly listed criteria for homology (fossil record, minuteness of similarity, multiplicity of similarity, and ontogenetic similarity) depend upon structural similarities. In contrast to the measurable (and readily comparable) morphological characters used by comparative anatomists, such as the bones of the hand, ethologists usually have had to base their comparisons on the sum of a variety of qualitative behavioral characteristics. Despite this limitation, ethologists have made a beginning. But as Atz (1970: 68f.) has pointed out:

> The extent to which [homologous behaviors can be established] is directly correlated with the degree to which [behavior] can be conceived or abstracted in morphological terms. Nevertheless, no morphological correlates have ever been found, either in the nervous system or peripheral structures, by which the homology of behavior can be established. On the other hand, to deny that homologous behavior exists would seem to deny that behavior is a characteristic of animals that is subject to evolutionary change. What is the basis of this paradoxical situation and can it be resolved? Its resolution rests on the recognition that the idea of homology is essentially a morphological concept. It is operationally impossible to conceive of two homologous entities without some kind of structure being involved. Therefore, when one thinks of homologous behavior, one is perforce thinking of behavior in morphological terms (fixed-action patterns and the like) or as closely associated with some structure (neural or otherwise). The difficulties in homologizing behavior that have arisen are almost all the result of the lack of morphological correlates in behavior. Until the time that behavior, like more and more physiological functions, can critically be associated with structure, the application of the idea of homology to behavior is operationally unsound and fraught with danger, since the history of the study of animal behavior shows that to think of behavior *as* structure has led to the most pernicious kind of oversimplification.

The newly developed ability to describe the neural architecture of behavior, using identified cells that are found in every member of a species (and even across species), places behavior on a new functional-morphological basis and provides the first opportunity for establishing rigorous homologies in behavior. It now has become possible (and interesting) to see how different patterns of adaptation are reflected in different patterns of neural architecture in closely related species. Such studies, although comparative in nature, can reveal information pertinent to general evolutionary processes, e.g., the behavioral and neuronal mechanisms underlying divergence and speciation.

The problem of adaptation is particularly opportune in the field of opisthobranch biology. Much is being learned about the neural basis of behavior in different opisthobranch molluscs, but there are few detailed comparisons between these animals, of the sort ethologists have carried out with birds, fishes, and insects (various *Drosophila* and cricket species have been studied particularly well). Yet opisthobranchs are a particularly useful group for comparative studies. Most opisthobranch orders include extended series of forms ranging from primitive ones—which resemble their possible prosobranch ancestors—to advanced forms. In addition, opisthobranchs are perhaps unique among animal groups in having many extant intermediate forms that may represent possible ancestral connecting links between orders. Most important, the behavior of a number of different opisthobranchs is now being studied at the cellular level. There has been particularly good progress in analyzing defensive behavior (withdrawal, inking, and escape locomotion), feeding behavior, and simple forms of learning. As will become apparent in the subsequent chapters, the necessary comparative data for meaningful analysis are often not yet at hand. Nonetheless, they are often readily obtainable, so that the time seems ripe for a systematic comparison of certain aspects of behavior and learning in the opisthobranchs. My purpose in emphasizing the interest and importance of comparative studies, therefore, is less to outline the results that have already been achieved than to encourage new studies.

CHAPTER 1

The Comparative Biology of Molluscs

This book concentrates on the various species of the molluscan genus *Aplysia,* using them as a reference point for comparisons of neurobiologically interesting molluscs, particularly opisthobranchs and pulmonates. But to understand any genus, one must understand the whole phylum, at least superficially. The purpose of this chapter is therefore to provide an overview of molluscan biology.

BODY PLAN

Molluscs constitute the second largest phylum in the animal kingdom, after the arthropods (see front endpapers.)[1] They are built on an apparently simple and highly adaptable body plan that is thought to have contributed greatly to their ecological and behavioral range. In fact, molluscs show more instances of

[1]The term "mollusc" (Latin *mollis,* soft) was introduced by Jonston in 1650 (Hyman, 1967). A full description of molluscs as a group was not achieved until the end of the nineteenth century. This description was the cumulative work of several comparative anatomists, beginning with Georges Cuvier in the early part of the century (1817) and ending in 1883 with E. Ray Lankester's chapter on "Mollusca" in the ninth edition of the *Encyclopaedia Britannica.* In that article Lankester finally distinguished the molluscs from the tunicates and brachiopods, thereby providing the basis for modern classification of the phylum.

evolutionary parallelism than any other phylum (Wilbur and Yonge, 1964).[2] This extensive parallelism contributes to the interest that molluscs hold for students of comparative behavior.

There are about 100,000 living molluscan species (Winckworth, 1949; Abbott, 1958;), but only a handful of genera are commonly studied by neurobiologists: *Anisodoris, Aplysia, Hermissenda, Pleurobranchaea, Tritonia, Helisoma, Helix, Loligo, Sepia,* and *Octopus.* Molluscs are soft bodied and include some of the most beautifully colored invertebrates. Adult animals in the phylum range in size from minute snails, less than a millimeter long, to giant squid, more than 16 meters long.

In spite of this heterogeneity, all modern molluscs are built according to one basic plan that is thought to derive from an ancestral type. Inferences about the structure of this ancestral mollusc have been made from comparisons of living species, from developmental evidence, and from study of the fossil record, which at present contains about 35,000 molluscan species. From these studies specialists in molluscan biology (malacologists) gradually gained the impression that certain structural characteristics are so basic that they are likely to have derived from ancestral forms. Based on these characteristics, malacologists have reconstructed the body plan of a hypothetical ancestral type (Figure 1-1), which has served also as a morphotype, or ideal, of modern molluscs and contains the defining characters of the phylum (Lankester, 1883; Pelseneer, 1906; Yonge, 1947; Knight, 1952; Knight and Yochelson, 1958; Fretter and Graham, 1962).

The four key features of the body plan are:

1. A head, with a pair of eyes, tentacles, and mouth.

---

[2]*Evolutionary parallelism* is the independent development of similar structures (and functions) in related groups in response to similar ecological conditions. Related groups that undergo similar adaptive changes in response to comparable environmental pressures thus maintain certain essential similarities (Simpson, 1949). For example, the nervous systems of different orders of opisthobranchs evolved separately but parallel with each other so that they share a common organizing feature: concentrations of nerve cells into ganglia, which make up the central nervous system (see Chapter 4). Because the groups are related, there is likely to be a common structural basis for parallel changes, and such changes are likely to have a similar genetic basis in each group. *Evolutionary convergence* is the independent development of similar structures and functions in unrelated groups. Convergent evolutionary changes (such as the development of wings and flight in insects, flying reptiles, birds, and bats) do not arise from a common ancestor with the function in question; they develop completely independently, sometimes in nonhomologous organs. There is no fundamental difference in the processes of parallel and convergent evolution. Later (Chapter 10) we shall also consider *evolutionary divergence,* which is the independent development of different structures and functions in closely related groups of animals as a result of different evolutionary opportunities (Simpson, 1949).

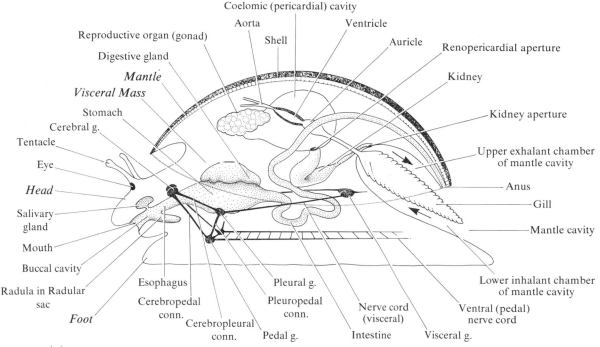

FIGURE 1-1.

The common body plan or morphotype of molluscs. Although modern molluscs comprise a diverse range of organisms, all share a common body plan, which also describes the proto-mollusc, the theoretical ancestral animal. The body plan consists of four main regions: (1) the head, containing eyes and tentacles; (2) foot; (3) mantle and mantle (respiratory) cavity; and (4) visceral mass. [After Meglitsch, 1972.]

2. A foot. In many groups the head and foot are confluent ("head–foot") and are best treated together.

3. A protective covering for the visceral mass, called the pallium or *mantle,* which secretes the shell and which is peripherally thickened to form the mantle edge. The space between the mantle and visceral mass, called the *mantle cavity,* is a *respiratory space* that contains the respiratory organs, sense organs (the osphradia) and the gills (usually ctenidium), and into which the reproductive and renal systems and the rectum open.

4. A visceral mass that contains the kidneys, heart, reproductive organs, and digestive gland.

The morphotype has been used to extrapolate the evolution of molluscs as well as to understand the general plan of existing groups. (For a discussion of the concept of morphotypes, see Nelson, 1969, 1970; Hodos, 1970, 1974.)

Evolution of Molluscs

Because the embryological development of living molluscs resembles that of segmented worms (annelids) and turbellarians (see below and Chapter 7), molluscs are included in the protostome division of invertebrates (Figure 1-2).[3] However, unlike annelids and arthropods, the body of molluscs is not segmented (Pelseneer, 1906). Nonetheless, the early stages of molluscan development are remarkably similar to those of annelids. This makes it likely that both share a common ancestor. But it is not clear whether molluscs and annelids arose together from a common nonsegmented animal, or whether molluscs arose later via the segmented annelid line and went on to lose their segmentation. Early students of molluscan phylogeny (Pelseneer, 1906; Naef, 1926; Garstang, 1928) thought that molluscs were derived from segmented annelid worms and lost segmentation through specialization (divergent evolution). More recently, Graham (1953; and Fretter and Graham, 1962) and Morton and Yonge (1964) have suggested that molluscs evolved directly from a nonsegmented turbellarian (flatworm) or nemertean (ribbon worm) ancestor. Part of the argument for the origin of the phylum prior to the appearance of segmentation rests on the fact that segmentation has not been observed in the larvae of any molluscs (Morton and Yonge, 1964).

However, the exciting discovery in 1952 of a living, apparently segmented primitive monoplacophoran (one-shelled) mollusc *Neopilina* revived interest in a possible ancestral relationship between molluscs and the segmented members of the annelid–arthropod line (Lemche, 1957, 1958; Lemche and Wingstrand, 1959).[4] *Neopilina galatheae* was found at a depth of 3,500 m off the Pacific coast of Mexico by the Danish malacologist Henning Lemche. The organs of *Neopilina galatheae* are arranged symmetrically along the longitudinal axis. There are 11 pairs of muscles, six pairs of kidneys, five pairs of

[3]Molluscs, annelids, and arthropods are coelomate animals. In molluscs the coelom is small and restricted to the pericardium and the kidney and gonadal chambers. The main body cavity is not a true coelom but a hemocoele (see Chapter 3 for further discussion). The presence or absence of a coelom is of great taxonomic significance and separates higher invertebrates (echinoderms, annelids, molluscs, and arthropods) from lower acoelomate invertebrates (Figures 1-1, 1-2). Coelomates are in turn divided into two major groups, the *protostomes* (flatworms, annelids, arthropods, and molluscs) and *deuterostomes* (echinoderms, hemichordates, and chordates).

[4]Prior to Lemche's discovery, the monoplacophorans were thought to be extinct and were known only from Cambrian and Devonian fossil remains (Wenz, 1938; Morton and Yonge, 1964). But, remarkably, Wenz (1938) had inferred their symmetry and untorted nervous system from his study of their fossil records.

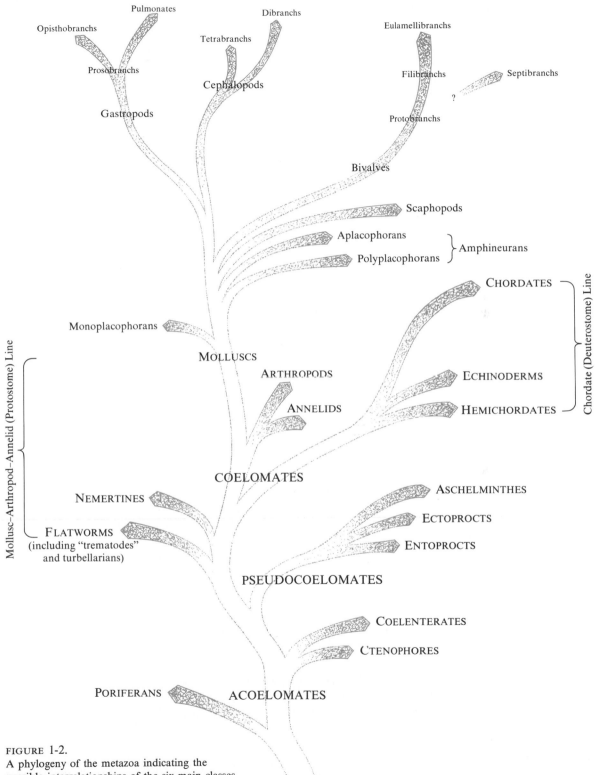

Pulmonates

Opisthobranchs

Dibranchs

Eulamellibranchs

Tetrabranchs

Prosobranchs

Filibranchs

Septibranchs

Cephalopods

?

Gastropods

Protobranchs

Bivalves

Scaphopods

Aplacophorans

Amphineurans

Polyplacophorans

CHORDATES

Chordate (Deuterostome) Line

Monoplacophorans

MOLLUSCS

ARTHROPODS

ECHINODERMS

ANNELIDS

HEMICHORDATES

Mollusc-Arthropod-Annelid (Protostome) Line

COELOMATES

NEMERTINES

ASCHELMINTHES

ECTOPROCTS

FLATWORMS
(including "trematodes"
and turbellarians)

ENTOPROCTS

PSEUDOCOELOMATES

COELENTERATES

CTENOPHORES

PORIFERANS

ACOELOMATES

FIGURE 1-2.
A phylogeny of the metazoa indicating the
possible interrelationships of the six main classes
of molluscs and some of the major orders.

PROTOZOANS

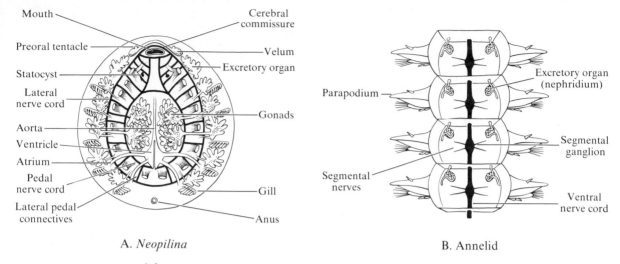

Mouth
Preoral tentacle
Statocyst
Lateral nerve cord
Aorta
Ventricle
Atrium
Pedal nerve cord
Lateral pedal connectives

Cerebral commissure
Velum
Excretory organ
Gonads
Gill
Anus

Parapodium
Segmental nerves

Excretory organ (nephridium)
Segmental ganglion
Ventral nerve cord

A. *Neopilina*

B. Annelid

FIGURE 1-3.

Comparison of the reduplication of the organs of *Neopilina* to the true segmentation of annelids

A. Serial repetition of various organ systems in *Neopilina galatheae.* Whereas in annelids there is a regular repetition of parts (figure **B**), in *Neopilina* the repetition of parts differs in the different organ systems; moreover, segmentation does not involve the body wall and the nervous system. The muscular system is repeated in 11 segments: eight body segments and three anterior segments in the vicinity of the head (not shown). The kidneys are present in six body segments; gonads in two body segments (their gonad ducts communicate to the exterior via the third and fourth nephridia), and gills in five body segments. Only the outline of the nervous system is indicated. It consists of two cords (a lateral or palliovisceral and a ventral pedal cord) that run on each side of the body. These nerve cords are linked together by a regular series of 10 lateropedal connectives. The lateral nerve cords swell anteriorly to form the cerebral ganglia. [After Lemche and Wingstrand, 1959.]

B. Schematic illustration of the highly repetitious annelid body plan. [After Buchsbaum, 1965.]

gills, and two pairs of gonads (Figure 1-3). A second, very similar species, *Neopilina (Vema) ewingi,* was discovered in 1959 at a depth of 5,700 m off the coast of Peru (Clarke and Menzies, 1959). Several additional species in the subgenera *Neopilina* and *Vema* have now also been found in the eastern tropical Pacific (Filatova, Sokolova, and Levenstein, 1968; Rokop, 1972).

Based on his studies of *Neopilina,* which appeared to him bilaterally symmetrical and containing a number of body segments, Lemche (1958) suggested that the ancestral molluscs were segmented. He argues that *Neopilina* and, to a lesser degree, some polyplacophorans (chitons) retained traces of this segmentation because of their ecological isolation, while the other less isolated groups of modern molluscs have lost all signs of segmentation.

Most scholars, however, disagree with Lemche's view and believe that the original molluscs were not segmented but descended from free-living flatworms (turbellarian–nemertine) that had no connection with the stock that gave rise to the annelid–arthropod phyla (see for example, Morton, 1963; Morton and Yonge, 1964; Stasek, 1972; Runnegar and Pojeta, 1974). In fact, it is not at all clear that *Neopilina* is truly segmented. Chitons and tetrabranch cephalopods (e.g., *Nautilus*) also have more than one pair of gills and genital ducts, which gives the impression of a segmented body organization, but they are not truly segmented. The duplication of organs represents a secondary multiplication. For example, the shell of chitons is composed of eight plates, but the plates seem to be secreted by a single shell gland. As is the case with chitons and tetrabranch cephalopods, *Neopilina* lacks the classic repetitive body plan of truly segmented animals (see Figure 1-3). But this issue probably cannot be fully settled until the development of *Neopilina* has been studied.

Whether or not the earliest molluscs were segmented, there is general agreement that the major molluscan classes are traced most easily from primitive nonsegmented ancestors.

The Hypothetical Proto-Mollusc

The hypothetical proto-mollusc (see Figure 1-1) is thought to have inhabited the shallow shore waters of pre-Cambrian oceans more than 600 million years ago (Runnegar and Pojeta, 1974). Most scholars believe that this animal was flat bodied, bilaterally symmetrical, with a well-defined head with eyes and tentacles. Its ventral surface was flat, and the thick muscular covering on this surface formed a foot for creeping. The dorsal surface was covered by an oval, convex shield-like shell that protected it against waves and predators. The shell was originally a tough cuticle secreted by the underlying epithelium—the mantle. Later this cuticle became reinforced by deposits of calcium carbonate and developed into a calcareous exoskeleton. A pair of retractor muscles, attached to the inner surface of the shell and inserted into each side of the foot, enabled the body, including the foot, to be retracted into the shell. Anteriorly, the body was slightly overhung by the periphery of the shell; posteriorly, the overhang was greater, giving rise to the mantle cavity, a chamber that contained the gills, the anus, and the opening for the kidneys.

The symmetrical pair of gills divided the mantle cavity into upper and lower chambers. Water entered the lower, inhalant chambers, passed upwards and radially through the gill into the upper chamber, and then moved posteriorly out of the cavity again. The anus, the genital ducts, the kidneys, and the hypobranchial gland all discharged into the exhalant chamber. Paired osphradia (sense organs that are thought to sample the chemical and osmotic properties of the incoming water current) lay in the path of the inhalant current. The animal presumably ate minute plant materials that were raked in by the scraping action of the radula, its food-gathering organ.

A key step in the evolution of this primitive mollusc from flatworm ancestors is thought to have been the development of a protective dorsal cuticle that formed the substrate for calcium deposits and gradually evolved into a hard shell exoskeleton (Stasek, 1972). The characteristic shell attachment muscles of the molluscan foot are thought to have developed from the dorsoventral muscles of the ancestral flatworm. As a result of the development of the exoskeleton dorsally (increasing the thickness of the body ventrally) and the enlargement of the viscera, the epithelial respiration of flatworms became inadequate. This evolutionary pressure is thought to have led to the development of the respiratory organs (gills) and the respiratory space (the mantle cavity). The development of gills in turn led to the formation of a more effective circulatory system for distributing blood and nutrients and a heart for pumping the blood (Fretter and Graham, 1962). A contractile heart required a space in which to beat, and this led to the development of the pericardial coelomic cavity, which seems to have arisen by enlargement of the gonadal spaces.

The morphotype also serves as a general model for considering the common structural features of the seven classes of modern molluscs.

CLASSIFICATION

Three of the seven classes of the phylum Mollusca are considered major: Bivalvia, Gastropoda, and Cephalopoda. The remaining four (Monoplacophora, Aplacophora, Polyplacophora, and Scaphopoda) are considered minor. The minor classes are relatively more primitive than the major classes. However, all classes have one or more primitive members that display several basic characteristics of the archetypal form (Figure 1-4), and even the more advanced members of each class exhibit one or more features of the archetype (for discussion see Morton, 1958a; Morton and Yonge, 1964; Hyman, 1967).

Monoplacophora

The class Monoplacophora ("single-shelled") was originally created by paleontologists to describe a group of extinct limpets discovered from fossil evidence and characterized by a series of paired muscle scars. Only later was *Neopilina,* the first living monoplacophoran, discovered. Monoplacophorans are untorted (see below), almost bilaterally symmetrical, segmented molluscs

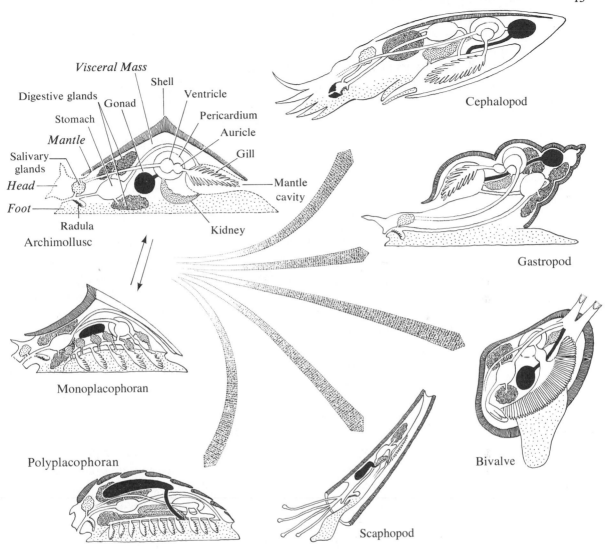

FIGURE 1-4.
Adaptive radiation in molluscs. Monoplacophorans, of which there is only one living genus, *Neopilina*, constitute the most primitive class. They have hard parts that are thought to most resemble the theoretical archimollusc. The six other molluscan classes (of which five are illustrated) can also be derived from this hypothetical archimollusc. Homologous parts are identified by similar shading. [After Wells, 1968.]

(Figures 1-3 and 1-4). Their primitive nervous system has small central ganglia, an anterior esophageal ring, and two long longitudinal cords—a lateral (palliovisceral) cord and a ventral (pedal) cord—that have 10 runglike interconnections.

Aplacophora

Aplacophora ("lacking shells") are small wormlike animals without shells. They are characterized by a cuticle studded with calcareous spicules and a rudimentary head that lacks tentacles and eyes. Their primitive nervous system has two paired longitudinal cords (two palliovisceral and two pedal) and runglike interconnections. There are discrete pedal and pleural ganglia. Two orders are distinguished: Neomenioidea (*Paramenia*) and Chaetodermatoidea (*Chaetoderma*).

Polyplacophora

Polyplacophora ("many-shelled"), commonly called chitons, are small, flattened animals covered with eight arched transverse articulated plates (Figure 1-4). The nervous system is double corded like that of the monoplacophorans, but it is even more primitive with little tendency for ganglion formation.

Until recently Aplacophora and Polyplacophora were grouped together as subclasses under the name Amphineura. However, Thompson (1960a), Stasek (1972), Runnegar and Pojeta (1974), and others have argued effectively that the differences in embryonic development, structure, and habitat between the two groups justify a class designation for each.

Scaphopoda

Scaphopoda ("boat-footed") are elongated, bilaterally symmetrical molluscs without eyes or gills (Figure 1-4). They are enclosed in a tusklike shell. Their nervous system consists of cerebral, buccal, pleural, pedal, and visceral ganglia. This minor class is typified by *Dentalium* (elephant's tooth shell).

Bivalvia (Pelecypoda)

Bivavlia ("two valves") are largely bilaterally symmetrical, flattened molluscs that lack a head (Figure 1-4). The mantle may bear tentacles and eyes, as in *Pecten* (scallop). Bivalves are enclosed in a shell with two lateral valves, hinged together mid-dorsally. They have a correspondingly bilobed mantle, containing a pair of large gills used for filter feeding. Their nervous system consists of fused cerebropleural and visceral ganglia.

Bivalvia comprises 420 genera and is the second largest class of molluscs. Among the molluscs it presents perhaps the greatest problems in classification, in large part because of the extensive parallel and convergent evolution that has produced many independent but similar forms. The most commonly used

cords into a figure eight. This condition of the nervous system is called *strepto-neuran*. In some orders there is a subsequent detorsion that results in the uncoiling of the nervous system, called *euthyneuran*. The effects of torsion and detorsion are the basis of gastropod classification. This classification is based on the adult position and character of the mantle cavity and of the gill or lung. Thus, gastropods are divided into three subclasses: (1) Prosobranchia (forward gill), which have a mantle cavity and gills that are rotated 180° and therefore lie above the heart, and a nervous system that is twisted into a figure eight; (2) Opisthobranchia (rearward gill), which have a mantle cavity and gill that are detorted to varying degrees, so that they lie behind the heart; and (3) Pulmonata (carrying lung), which have a lung instead of a mantle cavity and gill; and an untorted nervous system.

Subclass Prosobranchia. Fully torted, shelled snails. These are the most numerous and varied of gastropods, and are divided into three *orders:*
Archeogastropoda: top shells, limpets, and abalone (*Haliotis, Puncturella*)
Mesogastropoda: periwinkles, *Littorina*
Neogastropoda: whelks, oyster drills, and conchs (*Busycon, Buccinum*)

Subclass Opisthobranchia.[5] Marine gastropods with shells reduced or absent and detorsion. There are eight *orders:*
Cephalaspidea: *Bulla, Acteon, Navanax*
Thecosomata: *Limacina*
Acholidiacea: *Acochlidium*
Anaspidea: *Akera, Aplysia*
Gymnosomata: *Clione*
Sacoglossa: *Oxynoe*
Notaspidea: *Pleurobranchaea*
Nudibranchia [Acoela]

Subclass Pulmonata. Mantle vascularized to form a lung. Visceral nerve shortened due to concentration of ganglia in the head region (cephalization). Pulmonates are divided into two *orders* based upon eye position:
Basommatophora (eye at the base of the tentacles):
freshwater and marine limpets, pond snails (*Lymnea, Planorbis, Helisoma*)
Stylommatophora (eye on the pedicle): *Helix, Polygra* (land snails); *Limax* (land slug); *Onchidium*

[5]See Chapter 2 for alternative classifications.

TORSION AND DETORSION IN GASTROPODS

The molluscs that we shall consider in later chapters belong to the class Gastropoda. Although the gastropods conform to the general molluscan plan outlined above (Figure 1-1), they have all undergone torsion. As a result, Gastropoda is the only class of molluscs—indeed, the only large group in the animal kingdom—that exhibits consistent and sometimes striking asymmetry (Eales, 1950b). Much research has been carried out on the neural control of the mantle and the visceral mass by the abdominal ganglion of *Aplysia* (for review see Kandel, 1976). To appreciate how these effector systems and the abdominal ganglia that innervate them vary among the gastropods it is important that the processes of torsion and detorsion be understood.

The typical mollusc is an elongated, bilaterally symmetrical animal; the middle third of the body consists of a dorsal hump that contains the visceral mass, the mantle, and the mantle cavity. In its most schematic form torsion is a 180° rotation of this dorsal hump with the head and foot region remaining fixed (Figure 1-5). Torsion alters the orientation of many organs. The left gill, hypobranchial gland, osphradium, auricle, and kidney are shifted to the right and in many groups they tend to decrease in size and finally to disappear. The organs originally on the right become those on the left. The anus and the opening for the kidney are brought from a posterior to an anterior position and discharge over the animal's head (Figure 1-6A).

Torsion was discovered in 1881 by the German neuroanatomist J. W. Spengel while studying the innervation of the osphradium (the "organ of Spengel"). He realized that torsion could explain the anatomical asymmetries of molluscs, including the peculiar figure-eight configuration of the visceral nerve cords of prosobranchs (see also Boutan, 1899, 1902). "Prosobranchs and opisthobranchs are descended from a common primitive form (phylogenetically older species). The prosobranchs originate from it with a 180° rotation of the perianal organ complex. . . . On the other hand the opisthobranchs (and above all the tectibranchs) and the pulmonates originate from the primitive molluscs without rotation of the circumanal organs" (Spengel, 1881, my translation).

According to Spengel, the archetypal gastropod nervous system was symmetrical, consisting of paired ganglia that were connected to each other by commissures, to other ganglia by connectives, and to the periphery by peripheral nerves (Figure 1-6B). The paired cerebral ganglia were interconnected by commissures and each ganglion innervated the eyes and tentacles. Three pairs of cerebral connectives led to the buccal ganglia rostrally and to the pleural and pedal ganglia caudally. The buccal ganglia innervated the muscles of the radula and the mouth parts and the pedal ganglia innervated the foot.

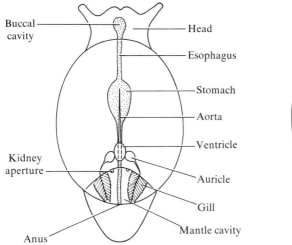

Hypothetical Pre-torsion
Mollusc

Post-torsion Mollusc

A

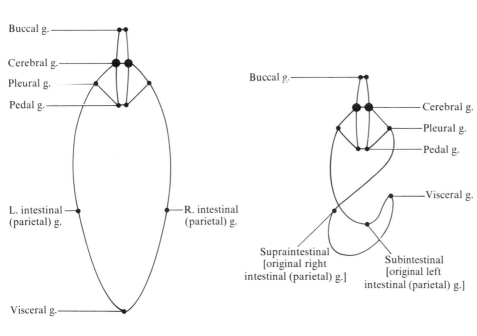

Hypothetical Pre-torsion
Nervous System

Post-torsion
Nervous System

B

FIGURE 1-6.
Dorsal view of the consequences of torsion in a hypothetical adult ancestral gastropod.

A. Body and external organs. After torsion the anus, gills, and heart are brought anteriorly, as in prosobranchs, and the position of the gills is reversed (the original left gill becomes the right one). [After Barnes, 1968.]

B. Nervous system. After torsion the nervous system shows a figure eight configuration of the connectives typical of modern prosobranchs. [After Spengel, 1881; Barnes, 1968.]

On each side the pleural ganglion sent a connective to the intestinal (parietal) ganglion on the same side, which in turn sent a connective to the usually unpaired visceral ganglion. The visceral ganglion sent peripheral nerves to the internal organs of the visceral mass as well as to the mantle cavity and its organs.

Following torsion, the pleurovisceral connectives became twisted into a figure eight so that the original left intestinal ganglion ended up below and on the right as the subintestinal ganglion, and the original right intestinal ganglion ended up above and on the left as the supraintestinal ganglion. The remainder of the nervous system was not affected by torsion (Figure 1-6B). Spengel thought that torsion occurred during larval life as an adaptation for life during the veliger stage.

Ontogenetic Origins of Torsion

In 1928 Garstang postulated that torsion resulted from a mutation in the ancestral mollusc that affected the symmetrical balance between the two larval velar retractor muscles, so that the right pedal retractor had a more forward shell attachment. Garstang's hypothesis recognized that larvae can evolve independently of the adult form and that evolutionary changes in the larva can persist into adult life (Fretter, 1969). The embryological work of Smith (1935) and Crofts (1937, 1955) has now supported Garstang's basic idea, although it has proved wrong some details in Garstang's speculations about how torsion is achieved.

Studies of the development of archeogastropods (*Calliostoma, Haliotis, Patella,* and *Patina*) have shown that there is delayed development of the left pretorsional member of a pair of larval retractor muscles, which delay allows the right retractor (whose fibers insert on both sides of the head and foot) to act in isolation. Contraction of the unbalanced retractor muscle rotates the mantle complex and the visceral hump in a counterclockwise direction relative to the head–foot. This rotation (Figure 1-7A) enables the foot and velum to be partially retracted for protective purposes (Smith, 1935; Crofts, 1937, 1955).

Garstang (1928) assumed that in the ancestral form full larval rotation of 180° took place in one step. This does not occur in contemporary forms. Torsion takes several hours to several weeks and in a number of prosobranchs it occurs in two discrete 90° stages, the first rapid and the second slow. In *Calliostoma* the first step takes four hours, the second 32 hours; in *Haliotis* the first step takes up to six hours, the second 200 hours; in *Patella* and *Patina* the first step takes 15 to 20 hours and the second 26 to 36 hours. The two stages

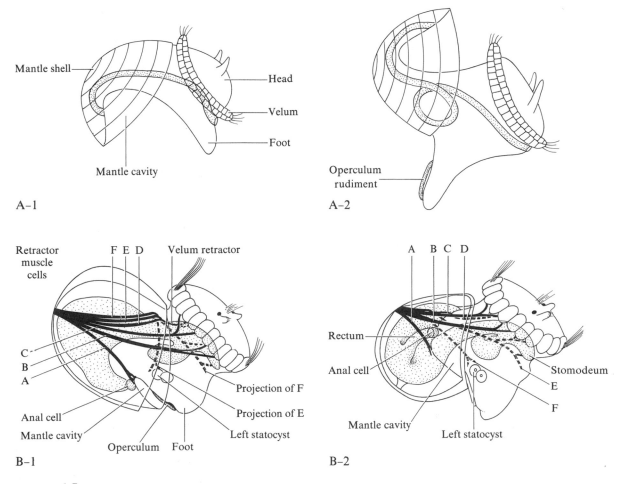

Mantle shell

Head

Velum

Foot

Mantle cavity

A–1

Operculum rudiment

A–2

Retractor muscle cells

F E D

Velum retractor

C

B

A

Anal cell

Mantle cavity

Operculum Foot

Projection of F

Projection of E

Left statocyst

B–1

A B C D

Rectum

Anal cell

Stomodeum

E

F

Mantle cavity

Left statocyst

B–2

FIGURE 1-7.
Larval torsion in prosobranch gastropods.

A. Diagram of possible larval advantages following torsion. **(1)** Before torsion evolved the head and velum of the veliger were the last structures to be retracted in response to a noxious stimulus. **(2)** After torsion evolved the foot became the last to be retracted. [After Wilmoth, 1967.]

B. Drawings of torsion in *Patella vulgata*. **(1)** Reconstruction of larva at 70 hours, ready for torsion. The view is from the right side showing unopposed right larval retractor muscle cell inserts on both sides of the head–foot (A to F). **(2)** Reconstruction at 76 hours following a 90° torsion due to the contraction of the unopposed right-side muscles. The site of torsion is the narrow neck behind the head–foot region. [After Crofts, 1955.]

of torsion are controlled by different mechanisms. As indicated above, the first stage is initiated by the contraction of the single velar retractor muscle (Figure 1-7B) and is assisted mechanically by the operculum, which develops immediately before torsion and acts as a wedge pushing the mantle and shelf region away from the foot (Crofts, 1937). The second stage results from a supplementary mechanism, the differential growth of larval structures, particularly the retractor muscle of the foot and its migration (Crofts, 1937).

Not all instances of torsion involve two stages, however, and not all require muscular asymmetries. These exceptions are interesting because they illustrate the opportunism of evolutionary processes that lead to parallel development so characteristic of molluscs. In some specialized prosobranchs, such as *Pomatias,* which has no veliger larval stage, torsion begins before muscles develop and is completed by differential growth in 31 days (Creek, 1951). In the absence of a free larval stage the velar rectractor muscles have no adaptive value. As a result, their size and power decrease and their role in torsion is taken up by differential growth of larval structures, a mechanism that originally was only supplementary to muscular contraction (Creek, 1951). Similar processes determine torsion in *Littorina, Crepidula, Ocenebra,* and *Viviparus* (Fretter and Graham, 1962). Finally, in nudibranchs (*Adalaria* and *Tritonia*) and perhaps in other opisthobranchs, torsion does not involve an actual movement of the visceral mass. Differential growth occurs before the viscera are formed, so that when the organs first appear they are already in a posttorsional position (Thompson, 1958, 1962).

Phylogenetic Origins and Adaptive Advantage of Torsion

Torsion presumably offers a selective advantage since all gastropods undergo torsion to some degree. The nature of this advantage is not known, but there are many speculations (for a critical evaluation of theories of torsion see Ghiselin, 1965; Underwood, 1972). Garstang (1928) argued that torsion is advantageous to the veliger and protects it from predators. Before torsion evolved, when the veliger retreated into its shell, the posterior mantle cavity could receive the head and velum only after the foot was already inside. After torsion evolved the mantle cavity came to the front, so that the delicate head and velum could be accommodated first with subsequent entry of the foot (Figure 1-7A). Although popular, this theory is not well supported, particularly since the main predators, fish, swallow the veliger whole (for criticisms, see Ghiselin, 1965; Thompson, 1967; Underwood, 1972). Some researchers (Lang, 1900; Naef, 1911; Morton, 1958b; Ghiselin, 1965) have argued that torsion is advantageous for the adult snail or for the larva at settlement because it aids locomotion and facilitates the flow of a respiratory current through the mantle cavity. This is consistent with Croft's (1937) finding that in *Haliotis,* a primitive archeogastropod (whose development perhaps most closely resembles the type of animal in which torsion first occurred), only 90° of torsion is undertaken during the larval stage. The remaining 90° is undertaken after settlement, which would suggest that its value is primarily for the adult. In more advanced prosobranchs torsion may have larval as well as adult advantages. Advanced prosobranchs have a long planktonic period and torsion is completed during this phase (Fretter, 1969). The resulting anteriorly facing mantle cavity protects the large velum from damage by allowing it to retract.

Detorsion

A counter movement, often called detorsion, is found in all opisthobranchs. Brace (1977a) suggests that this movement is related to the evolution of the burrowing mode of life commonly found in primitive forms of cephalaspid and anaspid opisthobranchs. He argues that to streamline the body the neck region was reduced, thereby producing a slug-like form and necessarily restricting space anteriorly for the mantle complex. Because of lack of space the mantle cavity has rotated to the right and subsequently posteriorly.

Detorsion again places the mantle cavity and anus in a posterior position, and the body assumes bilateral symmetry. The shell and mantle cavity become reduced or wholly lost, and the body reorganizes along bilateral lines in a slug-like shape. In *Aplysia,* for example, the mantle cavity is reduced to a narrow triangular cleft that opens on the right side. It is almost filled by the gill and contains a small osphradium. The edge of the mantle shelf carries the purple gland (Blochmann's gland), possibly a derivative of the hypobranchial gland of more primitive molluscs. The mantle cavity is covered by a vestigial shell plate located within the mantle shelf. The cavity is also protected by parapodia on either side. The mantle cavity and its organs are therefore centrally located and do not disrupt body symmetry.

Although Spengel realized that the twisting of the connectives in prosobranchs resulted from torsion of a primitive symmetrical ancestor, he failed to appreciate that opisthobranchs were not simply untorted molluscs but detorted ones. Detorsion was discovered by Pelseneer (1894) in the course of studying the larval development of *Philine.* "The torsion which occurs during development of Streptoneura also occurs at the beginning of the embryonic life of Euthyneura; but toward the end of that life, this torsion is diminished (and mostly destroyed) by a movement in the opposite direction which I would call 'detorsion'."

Although the pulmonates are grouped with the opisthobranchs as being euthyneurans, they are not detorted forms but achieve a secondary symmetry of their nervous system by a shortening of the pleurovisceral connectives, loss of the supraintestinal ganglion, and ganglionic fusion (see Chapter 4, p 115). However, an examination of several primitive forms (*Chilina, Amphibola*) indicates that, like opisthobranchs, a certain amount of detorsion took place before shortening occurred. This again can be related to a burrowing mode of life (Brace, 1977a).

Thus, whereas the larval torsion of many forms appears similar, in opisthobranchs the secondary compensations for torsion are varied and represent parallel evolution (for a detailed discussion of detorsion and secondary symmetry see Eales, 1950a, b). All forms of detorsion involve reduction of the shell, the mantle, the mantle cavity, and the mantle organs. Reduction of the shell in turn renders these gastropods less suitable for life in the littoral zone, but, as a result, larger and more varied species have evolved that successfully live in sublittoral zones (see Chapter 2 for discussion of the subdivision of the coastal environment).

Detorsion is generally less rapid than the original torsion and is achieved by postlarval growth (Pelseneer, 1906; Eales, 1950b; Thompson, 1958). In some forms, as in the nudibranch *Adalaria,* detorsion is brought about in two stages, one rapid, the other slow (Thompson, 1958).

SUMMARY AND PERSPECTIVE

In spite of their heterogeneity, molluscs form a unified group that shares a common plan or morphotype. The morphotype is both a conjectural model of the ancestral mollusc and an idealized form of modern molluscs. The morphotypical molluscan body can be divided into four main parts: (1) the *head* (lacking in bivalves), which contains the mouth, tentacles, and eyes; (2) the *foot,* a ventral muscular surface for locomotion, often fused with the head; (3) the *visceral mass,* a domelike structure containing the heart, digestive gland, kidney, and reproductive organs; and (4) the *mantle,* a wide skirt that serves as a protective covering for the visceral mass and secretes the shell. The space between the mantle and the visceral mass forms a sheltered external cavity, the *mantle* (or *pallial*) *cavity.* Primarily a respiratory chamber containing the gill, this cavity also houses the terminations of the digestive, renal, and reproductive systems, which discharge into it. In some gastropods a fold of the mantle forms a siphon that channels water into or out of the mantle cavity. The mantle cavity and its organs have a significant influence on the appearance of molluscs and on their mode of life. In addition to respiration and excretion, the mantle cavity can also participate in food collection, incubation of early larval stages, and locomotion.

The ancestral mollusc that presumably inhabited the pre-Cambrian oceans more than 600 million years ago is thought to have derived from the protostome division of invertebrates that includes segmented worms (annelids) and arthropods. Both molluscs and annelids are believed to originate from common, unsegmented flatworm (turbellarian–nemertine) stock. Whether molluscs arose before or after the annelid branch became segmented is not completely clear, in view of the recent discovery of *Neopilina,* a primitive monoplacophoran mollusc thought by some workers to have features of segmentation. But most students of molluscan phylogeny favor the view that molluscs arose from an unsegmented ancestor.

Molluscs are classified into four relatively primitive, minor classes, Monoplacophora, Aplacophora, Polyplacophora, and Scaphopoda, and three more advanced, major classes, Bivalvia, Gastropoda, and Cephalopoda. *Aplysia* and related molluscs belong to the class Gastropoda. This class comprises the subclasses Prosobranchia, Opisthobranchia (e.g., *Aplysia, Navanax, Pleurobranchaea, Tritonia,* and *Anisodoris*), and Pulmonata (e.g., *Helix, Helisoma,* and *Limax*).

Gastropods have undergone extensive adaptive divergence. Among the 35,000 modern species, there are marine species that have invaded fresh water and species that have become land dwellers. Although gastropods fit the general molluscan plan, most have undergone torsion, a characteristic 180° horizontal rotation of the visceral mass and mantle with the head–foot region remaining fixed. Torsion altered the orientation of many organs, including the nervous system. The anus and kidney opening were brought from a posterior to an anterior position, where they discharge over the animal's head. The organs originally on the right became those on the left. The left gill, hypobranchial gland, osphradium, auricle, and kidney at first became those of the right (as is evident in tectibranchs). These organs then tended to decrease and finally (in nudibranchs) to disappear.

The archetypal gastropod nervous system is thought to have been symmetrical, consisting of paired cerebral, buccal, pleural, and pedal ganglia. On each side the pleural ganglion sent a connective to the intestinal ganglion on the same side, which in turn sent a connective to the unpaired visceral ganglion. The visceral ganglion sent peripheral nerves to the internal organs of the visceral mass as well as to the mantle cavity and its organs. Following torsion, the pleurovisceral connectives became twisted into a figure eight so that the right intestinal ganglion ended up on the left side as the supraintestinal ganglion and the left intestinal ganglion ended up below and on the right as the subintestinal ganglion. The remainder of the nervous system was not affected by torsion.

Torsion may have arisen in an ancestral gastropod as an adaptation to a larval need, but it was probably of continued use to the adult for either respiration or locomotion, or both. Embryological studies on archeogastropods have shown that as a result of the early development of only one of two retractor muscles (the right pretorsional), the visceral hump develops asymmetrically and bulges to the pretorsional left. The contraction of this single unbalanced retractor rotates the head–foot mass dorsally and the mantle cavity and the visceral hump in a counterclockwise direction. The time needed for torsion to be completed varies from several hours to several weeks. In a number of prosobranch species it takes place in two discrete 90° stages, the first being rapid and the second slow. In opisthobranchs torsion is followed by detorsion. This returns the mantle cavity and anus to the posterior end of the animal, and the body assumes a secondarily derived bilateral symmetry.

SELECTED READING

Fretter, V., and A. Graham. 1962. *British Prosobranch Molluscs: Their Functional Anatomy and Ecology.* London: Ray Society. The definitive book on the prosobranch molluscs. Contains an excellent chapter on development.

Fretter, V., and A. Graham. 1976. *A Functional Anatomy of Invertebrates.* New York: Academic Press. A well-illustrated introduction to the structure of invertebrates.

Fretter, V., and J. Peake, eds. 1975. *Pulmonates, Vol. I, Functional Anatomy and Physiology.* New York: Academic Press. An interesting collection of essays on various aspects of pulmonate physiology.

Hyman, L. H. 1967. *The Invertebrates, Vol. 6, Mollusca I.* New York: McGraw-Hill. The most detailed general source on molluscs in English.

Meglitsch, P. A. 1972. *Invertebrate Zoology,* 2nd ed. London: Oxford University Press. A scholarly and clearly written introduction to invertebrate biology.

Morton, J. E. 1958a. *Molluscs.* London: Hutchinson University Library. A small and selective introduction to the biology of molluscs.

Runnegar, B., and J. Pojeta, Jr. 1974. Molluscan phylogeny: The paleontological viewpoint. *Science* (Wash., D.C.), **186**:311–317. A recent review of the origins of the modern classes of molluscs.

Thompson, T. E. 1976. *Biology of Opisthobranch Molluscs,* Vol. 1. London: The Ray Society. A modern classification and review of the opisthobranchs by one of the most able students of opisthobranch biology.

Wilbur, K. M., and C. M. Yonge. 1966. *Physiology of Mollusca,* Vol. 2. New York: Academic Press. A useful handbook on molluscan biology.

CHAPTER 2

The Comparative Biology of *Aplysia*: Systematics and Distribution

Aplysia was possibly the first opisthobranch genus described in the zoological literature. A large marine snail, probably an *Aplysia*, is mentioned by Pliny the Elder in *Naturalis Historia*, written during the first century A.D., and again in the writings of Galen and Claudius Aelian in the second century. The ancient scholars called it *Lepus marinus*, "sea hare," because its posterior tentacles are thick and long and the posterior body is flabby, so that it resembles a rabbit when sitting still and contracted (Blochmann, 1884; Eales, 1921, 1960). Because of the sea hare's tendency to ink profusely when disturbed, many ancient naturalists thought it was poisonous or holy or both.

The sea hare is next mentioned in the late Renaissance by the French naturalists Pierre Belon (1553) and G. Rondelet (1554); its internal anatomy was described by F. Redi (1684) and J. B. Bohadsch (1761). In the ninth edition of *Systema Naturae* (1756) Linneaus referred to the sea hare as *Lernea*, the name used by Rondelet and Bohadsch. In the tenth edition (1758) he changed the name to *Tethys limacina*, and in the twelfth edition (1766) to *Laplysia depilans*. The generic name *Aplysia* ("that which does not wash") was finally given to sea hares by Gmelin in 1789 (Eales, 1921).

The first anatomical study of the nervous system of *Aplysia* was by Georges Cuvier in 1803. This was republished as part of his classic work *Mémoires pour Servir à l'Histoire et à l'Anatomie des Mollusques* (1817). A brilliant comparative anatomist and student of molluscan structure, Cuvier distinguished *A. punctata, A. depilans,* and *A. fasciata* as separate species. Cuvier's superb lithographs and woodcuts of *Aplysia* and its internal structure, as well as those by Rang (1828), the first student of the natural history of *Aplysia,* are still delightful to examine (Figure 2-1).

Relaxed

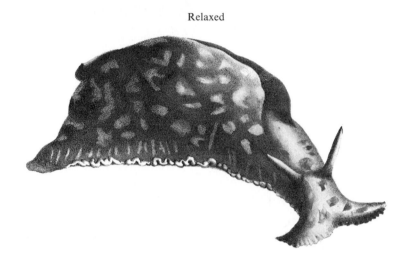

Contracted

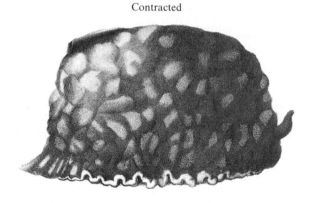

FIGURE 2-1.

A woodcut from the first monograph on *Aplysia, Histoire naturelle des Aplysiens,* by S. Rang, published in 1828. This publication followed the appearance of works by Cuvier in 1803 and 1817, which also included tinted lithographs and woodcuts. The illustration is of *A. (Aplysia) depilans* in relaxed and contracted positions.

Thirty-five species of the genus *Aplysia* have been identified (Eales, 1960). The European species were studied by several of the leading nineteenth-century biologists, including S. delle Chiaje (1828), Henri Milne-Edwards (1849), Felix Blochmann (1884), G. Mazzarelli (1893), Henri de Lacaze-Duthiers (1898), F. Bottazzi (1899), and Jules Guiart (1901). Modern scholarship on *Aplysia* started with Mazzarelli's monograph (1893) and two important papers by Nellie B. Eales (1921, 1960). Research on the five North American species began in 1863 with J. G. Cooper's description of the most common species, *A. californica;* a second California species, *A. vaccaria,* was described in 1955 by L. R. Winkler (for two other proposed American species, see Winkler, 1959; Beeman, 1963). Three other species, *A. brasiliana, A. willcoxi,* and *A. dactylomela* are found in the Gulf of Mexico off the shores of Texas and Florida. In this chapter I will consider the systematics, distribution, and shore environments of *Aplysia.*

THE OPISTHOBRANCHS

The genus *Aplysia* belongs to the subclass Opisthobranchia (hind-gilled snails). Opisthobranchs are thought to derive from the order Mesogastropoda of the more primitive subclass Prosobranchia (fore-gilled snails). They differ from the prosobranchs in having developed detorsion, which carried the mantle cavity backward along the right side of the body and ultimately led to the reduction (and even, in nudibranchs, to the loss) of the mantle cavity and its organs, as well as of the shell (Chapter 1). In opisthobranchs the gill (if present) is posterior to the heart, and the mantle cavity is usually completely open, whereas in prosobranchs the gill is anterior to the heart and to the rest of the visceral mass.

Three classifications of opisthobranchs have been proposed (Table 2-1). Pelseneer (1906) divided the opisthobranchs into two graded and somewhat overlapping *orders* on the basis of the presence of shell and gill.

1. *Order* Tectibranchia [Opisthobranchia palliata]. Shelled opisthobranchs, with a gill covered by an overhanging mantle and shell. Comprises the *suborders* Bullomorpha, Aplysiomorpha, and Pleurobranchomorpha.

2. *Order* Nudibranchia [Opisthobranchia nonpalliata]. No shell and the body is naked. Comprises the *suborders* Tritoniomorpha, Doridomorpha, Eolidomorpha, and Elysiomorpha.

TABLE 2-1.

Three schemes of classification of the subclass Opisthobranchia. Names in small capitals are orders. The braces and solid lines indicate the derivation of Thiele's groupings from those of Pelseneer. The dashed lines indicate the derivation of Odhner's groups. (After Purchon, 1968.)

Pelseneer (1906)	Thiele (1931, 1935)	Odhner (1932, 1939)*

1. TECTIBRANCHIA
   a. Bullomorpha
      Families 1–12 ———————————— a. Cephalaspidea ————1. CEPHALASPIDEA [Bullomorpha]

      Families 13–16 (= thechosomatous pteropods ————— b. Anaspidea ————

   b. Aplysiomorpha ———————————————— 2. THECOSOMATA
      2. PTEROPODA ———— 3. ACOCHILIDIACEA
      Families 2–7 (= gymnosomatous 4. ANASPIDEA [Aplysiomorpha]
      pteropods) —————— 5. GYMNOSOMATA
      Aplysiidae

   c. Pleurobranchomorpha ——————————
2. NUDIBRANCHIA 3. SACOGLOSSA ————— 6. SACOGLOSSA
   a. Tritonomorpha 4. ACOELA 7. NOTASPIDEA [Pleuro-branchomorpha]

   b. Doridomorpha 8. NUDIBRANCHIA [Acoela]
   c. Eolidomorpha
   d. Elysiomorpha

*As developed by Morton and Yonge (1964), Hyman (1967), and Thompson (1976).

Pelseneer's classification emphasized two important variations in opisthobranchs, but it ignored others. Although it is still used, most modern malacologists prefer either the slightly more precise classification by Thiele (1931, 1935) or the more detailed one originally proposed by Odhner (1932, 1939) and developed by Morton and Yonge (1964). In Thiele's system the 3,000 living opisthobranch species are divided into four orders (Table 2-1), whereas in the system based on Odhner's scheme, which is the most common one today (Hyman, 1967; Thompson, 1976), they are divided into eight orders:

1. *Order* Cephalaspidea [or Bullomorpha] (*Bulla, Acteon, Navanax*). Snails with cephalic shields, usually having moderately well-developed shell and mantle cavity, although both may be absent; prominent lateral parapodia; often burrow; sometimes actively carnivorous.

2. *Order* Thecosomata (*Limacina, Creseis, Diacria*). Planktonic snails with parapodial fins, well-developed mantle cavity; spirally coiled shell or nonspiral "pseudoconch."

3. *Order* Acochlidiacea (*Acochlidium, Hedylopsis*). Minute animals living in coarse sand; naked visceral sac; feed on microorganisms.

4. *Order* Anaspidea [Aplysiomorpha] (*Akera, Aplysia, Dolabella, Dolabrifera, Notarchus*). Lack cephalic shield. Shell reduced and internal; mantle cavity small and on the right side; prominent parapodia; often burrow; herbivorous.

5. *Order* Gymnosomata (*Clione*). Naked planktonic carnivorous snails with small parapodial fins; no mantle cavity.

6. *Order* Sacoglossa (*Oxynoe, Cylindrobulla, Elysia*). Herbivorous slugs; buccal complex modified for sucking; primitively shelled; with or without gill.

7. *Order* Notaspidea [Pleurobranchomorpha] (*Tylodina, Pleurobranchaea*). Slug-like and flattened; shell external, reduced and internal, or absent; no mantle cavity; naked gill; carnivorous.

8. *Order* Nudibranchia [Acoela]. Colorful naked carnivorous slugs; almost bilaterally symmetrical; no shell, mantle cavity, or gill; dorsal surface bears respiratory branchiae in rows or circlet surrounding anus. Includes four suborders: Doridacea [Doridomorpha] (*Anisodoris*); Dendronotacea [Tritoniomorpha] (*Tritonia*); Arminacea; and Eolidacea [Eolidiomorpha]. (In Odhner's original scheme the four suborders of Nudibranchia were classified as orders, giving a total of 11 orders.)

Evolutionary Relationships Among Opisthobranch Orders

The initial difficulty in devising a natural classification of the opisthobranchs has suggested to some workers (Purchon, 1968; Thompson, 1976) that the various opisthobranch orders may not have had a single prosobranch source. That is, two or more prosobranch lines may independently have given rise to what are now recognized as separate orders of opisthobranchs. A more attractive idea, advanced by others (Boettger, 1955; Morton, 1958a; Fretter, 1969; Brace, 1977a), is that opisthobranchs descended from a common prosobranch ancestor but diverged early in their evolution. This is supported by the finding that several modern orders of opisthobranchs include primitive representatives that resemble the mesogastropod prosobranchs in retaining a coiled shell and mantle cavity and showing only partial detorsion. This argument holds especially for the primitive cephalaspid *Acteon*, but also for the anaspid *Akera*, the sacoglossan *Cylindrobulla*, the thecosomatous family Limacinidae, and to a lesser degree for the notaspid genus *Pleurobranchaea* (Figure 2-2). According to the hypothesis of a common ancestor, each of these opisthobranch groups diverged early from the ancestral line and experienced what Morton (1958a) calls *programmed* (i.e., parallel) *evolution*.

The most primitive existing opisthobranch, the cephalaspid *Acteon*, bears the greatest resemblance to the mesogastropod prosobranchs and is thought to serve as an important bridging (annectant) form. The cephalaspids, both the carnivorous group (*Scaphander, Philine, Navanax*) and the herbivorous group (*Bulla, Haminoea*), the anaspids (through *Akera* and *Aplysia*), and the sacoglossans (through *Cylindrobulla*) can all be derived from *Acteon*. To some degree, and with less confidence, the notaspids and dorid nudibranchs can be derived from a form similar to *Acteon*, while the other nudibranch suborders can be derived from the dorid nudibranchs and notaspids. Finally, *Acteon* strongly resembles *Chilina*, the most primitive member of the pulmonate class (Figure 2-3; Morton, 1958a, b; Brace, 1977a).

THE ANASPIDS

The opisthobranch order *Anaspidea* ("without shield," referring to the absence of a cephalic shield) consists of two families, Akeridae and Aplysiidae (Figures 2-4, 2-5). The family Akeridae contains a single genus, the primitive form *Akera*.[1] The family Aplysiidae has four subfamilies and 10 genera (Eales, 1944, 1960). One of the four subfamilies of Aplysiidae, Aplysiinae, includes the genera *Aplysia* and *Siphonata*. The genus *Aplysia* can be subdivided into five subgenera (see back end papers). The main characteristics of the five subgenera are listed in Table 2-2, and some of the features are illustrated in Figures 2-6 and 2-7.

---

[1]*Akera* (also *Acera*) is usually included in Anaspidea because its opaline gland, digestive tract, reproductive tract, sperm, and nervous system resemble those of Aplysiidae (Pelseneer, 1894; Guiart, 1901; Minichev, 1963 [cited in Brace, 1977a]; Morton and Yonge, 1964). However, some authors (see Thompson, 1976) classify *Akera* in Cephalaspidea because of its cephalic shield and large external shell. Other authors (Boettger, 1955; Zulch, 1954, 1960) have argued for maintaining the order intact but changing its name to Aplysiacea so that the name of the order does not indicate that lack of a head shield is a major character.

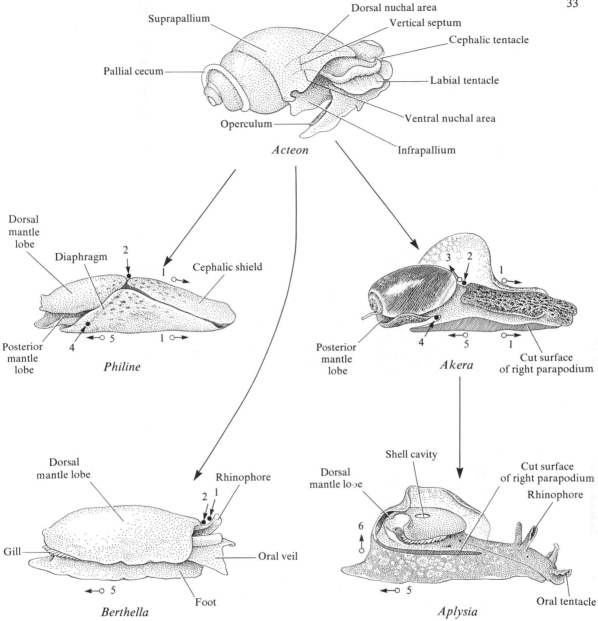

FIGURE 2-2.

Diagrams of connecting species illustrating the changes in gross anatomy incurred in the transition from the primitive cephalaspid *Acteon tornatilis* to the advanced cephalaspid *Philine aperta*, the notaspid *Berthella plumula*, the primitive anaspid *Akera bullata*, and subsequently to the more advanced anaspid *Aplysia punctata*. Empty and filled circles denote evolutionary elongation and shortening, respectively. Areas referred to are: (1) anterior portion of the head–foot; (2,3) dorsal nuchal area; (4) ventral nuchal area; (5) mesopodium; and (6) posterior portion of the head–foot. [After Brace, 1977a.]

34

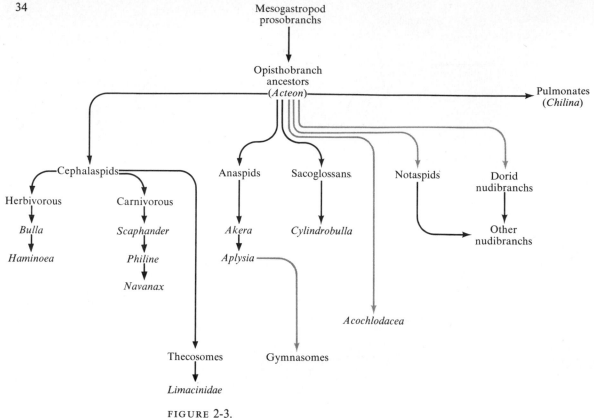

FIGURE 2-3.
A speculative outline of opisthobranch evolution. [After Brace, 1977a,b.]

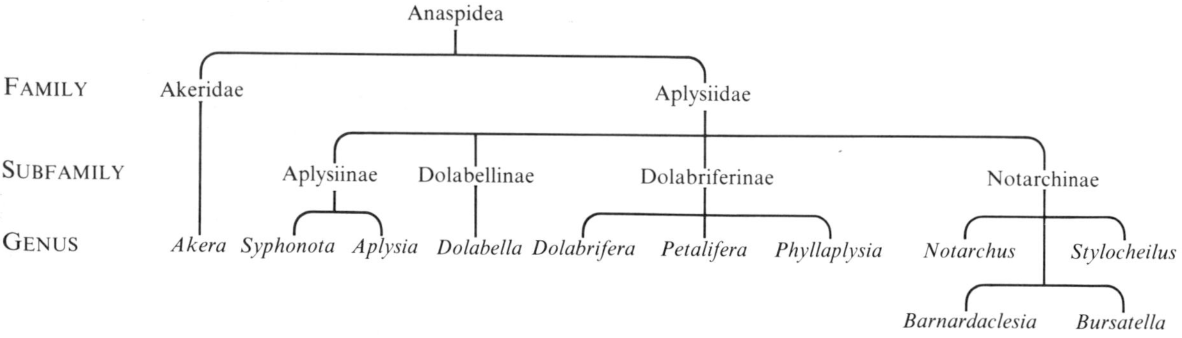

FIGURE 2-4.
Phylogeny of the opisthobranch order Anaspidea. Representative genera
are illustrated in Figure 2-5. [After Eales, 1960.]

TABLE 2-2.
Characteristics of the subgenera of Aplysia (see the back endpapers for a list of species in each subgenus). (After Eales, 1960.)

	Pruvotaplysia	Neoaplysia	Varria	Aplysia	Phycophyla
Size	Medium (60 mm to 200 mm)	Giant (to 1000 mm)	Med-Large (to 400 mm)	Giant (to 1000 mm)	Small (20 mm)
Body	Long, narrow	High, plump	High, narrow	Low, flat, bulky	Slender
Foot	Narrow	Broad	Usually narrow but may be medium or broad	Broad, often has a posterior sucking cup	Narrow
Tail	Pointed	Long, slender	Long, slender	Short	Long, tapering
Parapodia	Joined high posteriorly; closing mantle cavity in behind	Joined low posteriorly	Joined low posteriorly, mobile, used in swimming	Small, joined high posteriorly, used by some species for swimming	Freely mobile, joined low down on tail
Shell	Markedly concave, unrayed mantle aperture	Flattened calcareous apex, mantle aperture minute and closed	Concave and shallow	Flat	Mantle small, shell sac closed
Purple gland	Present	Present	Present	Not present; mantle gland usually secretes white fluid	Not described
Opaline gland	Simple	Large	Simple or complex	Simple	Simple
Radula	Small, 40 rows (18.1.18)	Well developed, 80 rows	Well developed, 80 rows (50.1.50)	Simple, 80 rows (40.1.40)	Not described
Penis	Small and spoon-shaped	Filiform-spatulate	Filiform or spatulate	Big, black, and warty	Filiform
Central nervous system	Distinct cerebrals, distinct abdominals	Fused cerebrals, fused abdominals	Fused cerebrals, fused abdominals	Fused cerebrals, fused abdominals	Not described
Distribution	A. parvula circumtropical A. punctata north-temperate and Arctic Atlantic	East North Pacific	All oceans except Arctic and Antarctic	Circumglobal, tropics and subtropics	Western Pacific
Type species	A. parvula	A. californica	A. dactylomela	A. depilans	A. euchlora

Akeridae

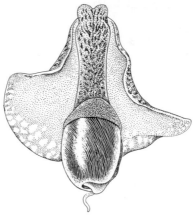

Akera

Aplysiidae

 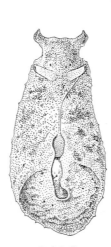 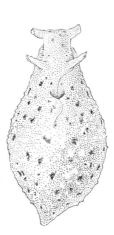 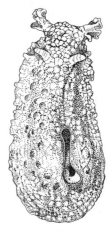

| *Aplysia* | *Dolabella* | *Notarchus* | *Dolabrifera* |
| Aplysiinae | Dolabellinae | Notarchinae | Dolabriferinae |

FIGURE 2-5.
Representative genera of the anaspid families Akeridae and Aplysiidae. See Figure 2-4 for the phylogeny of the anaspid order. [After Guiart, 1901; Marcus, 1972.]

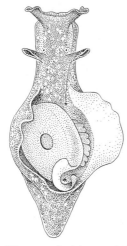

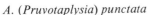

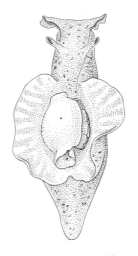

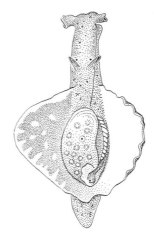

A. (Pruvotaplysia) punctata   *A. (Neoaplysia) californica*   *A. (Varria) extraordinaria*

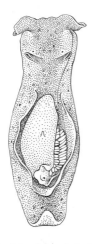

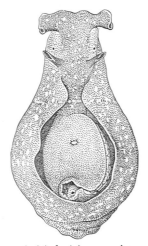

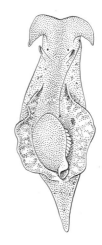

A. (Varria) rehderi   *A. (Aplysia) vaccaria*   *A. (Phycophilia) euchlora*

FIGURE 2-6.

Variations in the external features of several species of *Aplysia* from five subgenera. [After Eales, 1960.]

A. (Pruvotaplysia) punctata. Moderate-size animals found in Eastern North Atlantic. Head and neck long and narrow; elongated rhinophores and narrow foot with pointed tail; thin parapodia.

A. (Neoaplysia) californica. Large animals found on the California coast. Head small and broad; foot moderately broad and long with slender tail.

A. (Varria) extraordinaria. The sinuous edge of the parapodium is shown on the right, the pattern of the inner side on the left, with its light rim. The right mantle edge has been reflected to show the gill and the purple gland. A small papilla on the mantle is ringed, and the mantle is blotched with brown and white, but is lined near the edge.

A. (Varria) rehderi. Broad head and narrowed neck region; narrow foot and rounded tail; small parapodia, large prominent mantle.

A. (Aplysia) vaccaria. The largest of the various *Aplysia* species. Bulky body with short parapodia. Members of the subgenus *Aplysia* have no purple gland.

A. (Phycophylia) euchlora. Short neck, pointed tail.

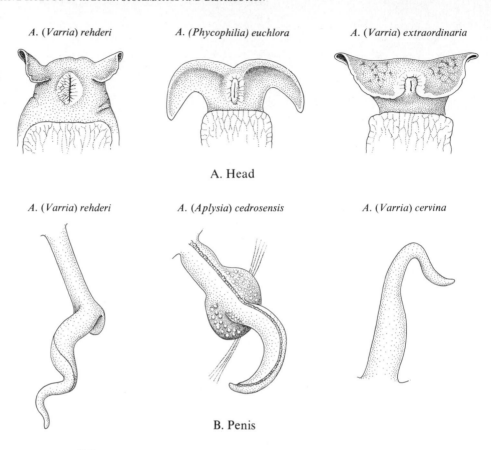

A. (Varria) rehderi   *A. (Phycophilia) euchlora*   *A. (Varria) extraordinaria*

A. Head

A. (Varria) rehderi   *A. (Aplysia) cedrosensis*   *A. (Varria) cervina*

B. Penis

FIGURE 2-7.
Variations among various *Aplysia* species in configuration of the oral veil, mouth shape, and penis. [After Eales, 1960.]

DISTRIBUTION

The 35 positively identified species of *Aplysia* generally inhabit the warm coasts throughout the world (Eales, 1960). Approximately two-thirds of the known species occupy subtropical or tropical waters; only one species, *A. punctata,* has been found within the Arctic Circle (Eales, 1960). Some species (e.g., *A. dactylomela, A. juliana,* and *A. parvula*) have an extended range, whereas others (e.g., *A. vaccaria*) have a limited range; some species with a limited range (e.g., *A. dactylomela*) can coexist with more widely distributed species. The distribution of *Aplysia* based on Eales' division of the world into 10 oceanic regions is shown in Table 2-3.

TABLE 2-3.
Distribution of the Aplysia *species according to 10 oceanic regions. (From Eales, 1960.)*

	N.Atl. W.	N.Atl. E.	N.Atl. S.E.	S.Atl. W.	S.Atl. E.	Ind. W.	Pac. N.W.	Pac. N.E.	S.Pac. W.	S.Pac. E.
brasiliana	×		×	×	×					
californica								×		
cedrosensis								×		
cervina	×			×						
cornigera						×	×			
cornullae									×	
dactylomela	×		×	×		×	×	×	×	
denisoni						×			×	
depilans		×	×							
dura					×				×	
euchlora							×			
extraordinaria									×	
fasciata		×	×		×	×				
gigantea									×	
gracilis						×				
inca										×
juliana	×	×	×	×		×	×	×	×	×
keraudreni									×	×
kurodai							×			
maculata						×				
morio	×									
nigra				×					×	×
oculifera						×	×			
parvula	×		×	×		×	×	×	×	×
pulmonica							×		×	
punctata		×	×							
rehderi								×		
reticulata									×	
robertsi								×		
sagamiana							×			
sowerbyi									×	
sydneyensis									×	
vaccaria								×		
willcoxi	×									
winneba			×							

The species of *Aplysia* most extensively investigated by neural scientists come from the Northeast Atlantic, the Mediterranean coasts of France, Monaco, and Italy, and the Northeast Pacific, specifically the coast of California. In the Mediterranean the three best-known species are *A. depilans, A. punctata,* and *A. fasciata.* In the Northeast Pacific the important species are *A. californica* and *A. vaccaria.* In the Northwest Atlantic *Aplysia* is found off the coast of Florida and Texas, where the most common species is *A. brasiliana;* occasionally *A. dactylomela* is also encountered here.

Of the four species commonly available off the American coasts—*A. californica, A. vaccaria, A. brasiliana,* and *A. dactylomela*—only *A. (Neoaplysia) californica* has been studied extensively, largely because Dr. Rimmon C. Fay of the Pacific Biomarine Laboratories, Inc. has made it available to scientists throughout the world. *A. californica, A. brasiliana,* and *A. juliana* can now be successfully cultured in the laboratory (Chapter 7).

The Shore Zones and Their *Aplysia* Populations

Because *Aplysia* feed on seaweed (Chapter 8) they are restricted to two types of shore environments: (1) the littoral (intertidal) and (2) the sublittoral zone (Figure 2-8). These zones make up two of the three shore environments described by Stephensen and Stephensen (1949; 1972). These authors have argued that despite great variability these three zonal patterns are almost universal. They are:

Supralittoral (maritime) zone. This zone is above the extreme high-water level of spring tides and is technically dry land, but it becomes wet during high tide from the splash of breaking waves.

Littoral (intertidal) zone. This zone lies between the extreme high- and extreme low-water levels of spring tides. The upper portion, the supralittoral fringe or upper shore, is usually dry except for days when high tide exceeds average range.[2] The middle portion, the midlittoral region or middle shore,

---

[2]The supralittoral fringe technically includes a portion on either side of the extreme high water level of spring tides and thus consists of the lower portion of the supralittoral zone as well as the upper margin of the littoral zone (Stephensen and Stephensen, 1972).

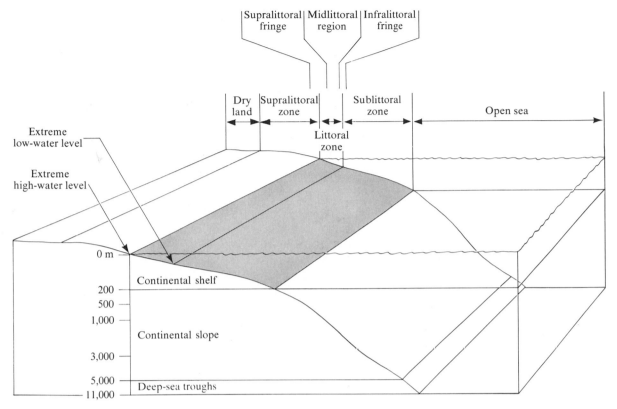

FIGURE 2-8.

The distribution of *Aplysia* (shown in gray area) within the three shore zones. The shore is divided into dry land and three tidal zones: supralittoral, littoral (intertidal), and sublittoral. *Aplysia* is found in only the latter two. [Based on data from Newell, 1970.]

is submerged and uncovered daily.[3] The lower portion, the sublittoral fringe or lower shore, is permanently covered except for the days when low tide exceeds average range.

[3]Lewis (1964) has suggested that the midlittoral region of Stephensen and Stephensen be divided into an upper region (which he calls the "littoral fringe") and a lower region (the "eulittoral region"). In addition, he suggests that the supralittoral zone be called the "maritime zone" because its indicators are primarily terrestrial and not marine organisms. Moreoever, Lewis has argued persuasively that biological indicators (in particular, characteristic organisms) are more satisfactory than tidal levels for defining the littoral zone.

Sublittoral (*infralittoral*) *zone.* This zone is below the littoral region and is never exposed to air. It extends to a depth of approximately 60–120m. At this depth the slope of the sea bottom rapidly becomes steeper and descends down the continental slope to the depth of the ocean floor.

The three zones exist because the boundary of land and sea creates humidity gradients and sharp differences between exposed (emersed) and submerged (immersed) areas. In addition, the tide and waves produce further, more erratic changes. Depending on the specific interaction of land and sea, the substratum of the littoral zones consists of rock, sand, or mud. In turn, the substratum determines the type of plants and animals that can survive there.

Aplysia that live in the littoral zone are often exposed to the air for several hours a day and are buffeted about by incoming and outgoing tides. Because of constant changes in the littoral environmental condition, animals living in this zone must withstand alternate exposure to air and immersion in the sea (Figure 2-9) and wide and rapid changes in temperature, humidity, salinity, pressure, mechanical stimulation, and wind exposure. In contrast to littoral animals, sublittoral animals lead a sheltered existence and are exposed to much less extreme fluctuations (Yonge, 1949a; Hardy, 1959; Kupfermann and Carew, 1974).

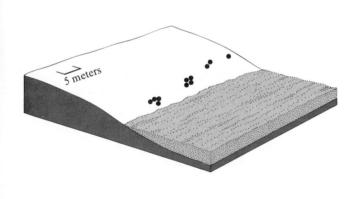

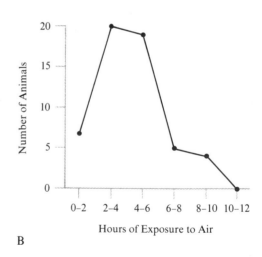

A   B

FIGURE 2-9.

Exposure to air of *A. californica* in the littoral zone. [After Kupfermann and Carew, 1974.]

A. Approximate distribution of *Aplysia* found exposed during a low-low tide at a tide-pool region several hundred meters north of the Scripps Institute, La Jolla, California. Animals tend to aggregate.

B. Frequency distribution of duration of exposure to air for a population of animals observed in tide-pool regions during low-low tides over a 20-day period.

Relationship of Littoral and Sublittoral Populations

Although the occurrence of populations of the same *Aplysia* species in two quite different shore environments has been known for some time, the relationship between such populations was poorly understood until recently. Garstang (1890) and Eales (1921) reported that *A. punctata* migrated in its life cycle from the sublittoral to the littoral zone. According to Eales, spawn are deposited in the littoral zone, where they hatch and release the planktonic veligers. The veligers are thought to then move offshore (presumably by dispersion) to metamorphose in the sublittoral zone, perhaps because the triggering substance for metamorphosis is found only in the sublittoral zone (see Chapter 7). Mature *Aplysia* gradually complete the cycle by returning to shore in the littoral zone (for a more recent discussion of this view, see Kay, 1964).

The necessity of this migration for the life cycle of the species has been questioned by Miller (1960, 1962) and by Carefoot (1967a). During the breeding season (April to August) Miller (1960) regularly found spawn masses of *A. punctata* in the sublittoral zone of the Irish Sea off the Isle of Man, down to a depth of 60 feet. Only rarely did he find *Aplysia* in the littoral zone. The finding of eggs in deep water suggests that shoreward (littoral) migration does not occur off the Isle of Man and that it is not an essential step in the life cycle of the species.

Carefoot has also found no shoreward migration (1967b) for *A. punctata* in the Trearddur Bay area of the Irish Sea (off the coast of Anglesey in northwest Wales), where the littoral zone, which is exposed only in low spring tides, is separated from the offshore sublittoral zone by a 1-km expanse of open sand. Carefoot found two independent and different colonies of animals in the sublittoral and littoral regions. There was no indication that larvae dispersed or migrated offshore to settle, nor was there evidence that mature *Aplysia* migrated toward the shore to spawn. The spawn were deposited and hatched in the sublittoral zone.

Although the data on the littoral colony are less complete, the great distance separating the two zones and the paucity of littoral animals make it unlikely that the littoral population results from a migration of sublittoral animals. More likely, the littoral animals developed locally or represent accidental displacements of sublittoral animals by wave action (Carefoot, 1967b). As a result of these studies, there seems no need to postulate spawning migration for *A. punctata.* The triggering substance for metamorphosis in *A. californica,* the seaweed *Laurencia pacifica* (Kriegstein, Castellucci, and Kandel, 1974), is present in both the littoral and sublittoral zones (Kupfermann and Carew, 1974), and therefore migration for purposes of metamorphosis seems unnecces-

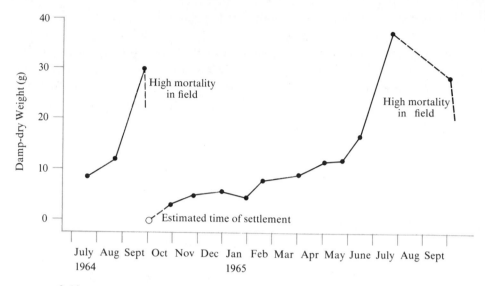

FIGURE 2-10.

The life cycle of *A. punctata* in the Trearddur Bay area off the northwest coast of Wales. The data plot the growth of sublittoral *Aplysia punctata* from July 1964 to October 1965. Each point represents the mean weight of about 115 animals, except for the September–October 1965 value, which represents 21 animals. [After Carefoot, 1967b.]

sary in this species as well.[4] However, studies on other *Aplysia* species are needed before a more general conclusion on this question can be reached.

The distribution of animal sizes found by both Miller (1960) and Carefoot (1967b) suggests an annual life cycle for *A. punctata* that begins with the appearance of new small animals in the autumn, with a peak in October (Figure 2-10). From February to September the number of small animals declined,

---

[4]Despite the great amount of neurobiological work done on *A. californica*, there are no systematic studies on the relation of sublittoral and littoral animals in this species. Winkler and Dawson (1963) and Beeman (1968) described *A. californica* as existing along the entire California coast. Beeman has found *A. californica* in the sublittoral zone down to about 100 feet. Littoral animals are found in four types of habitats: (1) on the open rocky coast of southern California, where they feed mainly on *Plocamium;* (2) in estuaries (protected rocky coves) and mud flats of northern California, where they often feed on the sea (eel) grass *Zostera;* (3) in kelp beds; and (4) in bays. In some northern California estuaries, such as the former Elkhorn Slough or the Bolsa Chica Slough, *Aplysia* existed in particularly great abundance and reached enormous size (MacGinitie, 1935). The water in these estuaries tended to be more variable in temperature but more constant in level than the coastal waters, and the green alga *Enteromorpha* was the dominant plant growth. Winkler suggests that quiet waters may have been beneficial for the survival of the larval stages. Kupfermann and Carew (1974) found *A. californica* in three southern California environments: (1) rocky littoral regions (Solana Beach); (2) sublittoral channels (Mission Bay Channel); and (3) sublittoral protected bays (Punta Bunda, Mexico). In contrast to the wide distribution of *A. californica, A. vaccaria* is found primarily along the rocky coasts and in kelp beds.

presumably representing the gradual maturation of the population. Large animals were found through July and August, when they disappeared, probably dying after they had spawned. Both casual observations in the field (MacGinitie and MacGinitie, 1968) and more detailed ones in the laboratory (Kriegstein *et al.*, 1974) support the hypothesis that *A. californica* also is an annual with maximum growth (in the field) in the late summer months, followed by death soon after spawning. Annual life cycles seem to be characteristic of other opisthobranchs and have been well documented in the nudibranchs *Adalaria, Archidoris, Goniodoris, Onchidoris,* and *Tritonia* (Thompson, 1958, 1962).

SUMMARY AND PERSPECTIVE

Aplysia were probably known to the ancients and have been studied in detail by naturalists since the Renaissance. A member of the subclass Opisthobranchia, the order Anaspidea and the family Aplysiidae, the genus *Aplysia* consists of five subgenera and 35 known species. These generally inhabit warm coastlands throughout the world. Approximately two-thirds of the known species occupy subtropical or tropical waters; only one species, *A. punctata,* has been found within the Arctic Circle, and three species are found throughout the world: *A. dactylomela, A. juliana,* and *A. parvula.*

    Aplysia feed on seaweed and are restricted therefore to two types of shore environments: the littoral (intertidal) zone (midlittoral region and sublittoral fringe), and the sublittoral zone. The littoral zone lies between the extreme high- and low-water levels of spring tides. *Aplysia* live in the middle and lower portion. The middle portion is submerged and uncovered daily. The lower portion is permanently covered except for the days when low tide exceeds average range. The sublittoral zone is below the littoral region and is never exposed to air. It extends gradually to an approximate depth of 200–400 feet. Sublittoral and littoral populations were once thought to represent different phases of the life cycle of *Aplysia*. Recent work on *A. punctata* suggests, however, that these are two separate populations of animals, each having an annual life cycle.

SELECTED READING

Eales, N. B. 1921. *Aplysia. Liverpool Mar. Biol. Comm., Mem. 24, Proc. Trans. Liverpool Biol. Soc., 35:*183–266. A small classic. This is the review that contemporary workers invariably consult when they begin work on *Aplysia*.

Eales, N. B. 1960. Revision of the world species of *Aplysia* (Gastropoda, Opisthobranchia). *Bull. Br. Mus. (Nat. Hist.) Zool., 5:*276–404. The best (and most complete) discussion of the genus *Aplysia*.

CHAPTER 3

Aplysia Among the Molluscs I
Sensory Capabilities and Effector Physiology

Behavior can be regarded as the coordinated activities of the somatic and visceral motor systems. To understand the neural control of an animal's behavior one must understand the structure of the sensory and effector systems of that animal. A comparative approach to behavior therefore requires an understanding of homologous sensory and effector structures in closely related species. In this chapter I shall review the physiology of the organ systems of *Aplysia* in comparison with those of other molluscs. I shall emphasize, in particular, the studies of sensory, somatic motor, and visceral motor physiology that are essential for later discussions of behavior.

BODY PLAN

In common with other bilaterally symmetrical animals, molluscs have an anteroposterior (head–foot) growth axis. In addition, they have a dorsoventral (visceral mass–mantle) growth axis, which is relatively independent of the anteroposterior growth axis and which leads to biradial enlargement through the addition of new shell around the edge of the mantle. The varieties of molluscan form are determined by the balance between the two growth axes or by the relative dominance of one over the other. For example, molluscs are generally bilaterally symmetrical, but changes in the position and growth of the visceral mass–mantle axis of gastropods lead to the alteration or even the loss of some symmetrical features.

The typical *Aplysia* body is moderately symmetrical and rounded, often appearing plump or flabby. Individuals of the various *Aplysia* species are large, particularly those belonging to the two Californian species, *A. californica* and *A. vaccaria.* Thus, *A. californica* measure up to 1 m in length and weigh up to 7 kg (MacGinitie, 1934; 1935). Some *A. vaccaria*—probably the largest gastropods in the world—weigh as much as 13 kg (Limbaugh, in Winkler and Dawson, 1963). Even the individuals of smaller species, such as *A. willcoxi, A. punctata, A. dactylomela,* and *A. juliana,* reach 100 to 400 g (Carefoot, 1970).

The molluscan body is covered by skin consisting of a single layer of ciliary epidermis containing many mucous and other gland cells. In addition to being a protective barrier, the skin is also a major sensory sheet for tactile and chemical senses. The body wall consists mainly of smooth muscle fibers, usually arranged not in clearly organized layers but in complex masses with no apparent orientation. As a result, the body is capable of extensive changes in shape.

In *Aplysia* the epithelial glands that are present in the skin secrete a great deal of mucus, which causes the animal to glisten and feel slimy to the touch. Skin color varies among species from red to olive green, brown, or black. In addition to species differences, the color depends in part on the particular seaweed eaten by the animals.

Aplysia has the characteristic features of the molluscan body plan: head–foot, mantle cavity, and visceral mass (Figure 3-1). I shall begin the discussion of the physiology of the organ systems by considering the head. I will then discuss all the sense organs together before reviewing the foot, the mantle cavity, and the organs of the visceral mass.

THE HEAD

Many molluscs, particularly gastropods, have a distinct and mobile head. The head contains a mouth and a muscular pharynx or buccal mass that houses the food-gathering organ, the *radula.* The head also carries relatively complex sense organs, such as the tentacles and eyes, and typically houses most of the ganglia of the central nervous system.

The head of *Aplysia* is well defined anteriorly but posteriorly it gradually merges into a constricted neck. Both head and neck are contractile and highly mobile and wave to-and-fro when the animal is searching for food. As the animal grazes on seaweed, the head is usually extended and held downward.

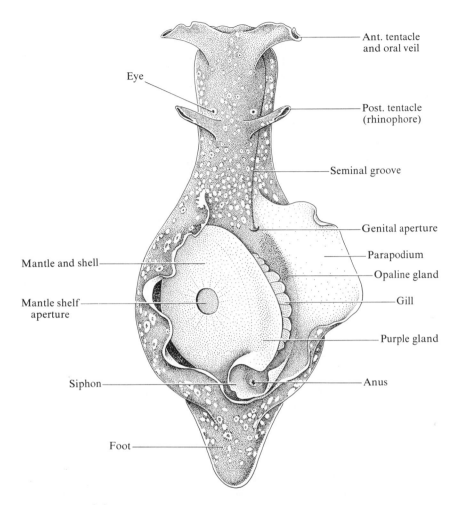

FIGURE 3-1.
Dorsal view of *Aplysia* (*Pruvotaplysia*) *punctata* showing main external structures.
[After Guiart. 1901.]

When disturbed, the animal can retract the head and the anterior portion of the foot by means of retractor muscles located on either side of the neck (Winkler, 1957).

On the front of the head are a pair of short, broad anterior tentacles that unite on the ventral surface to form an oral or labial veil over the mouth region (Figure 3-1). The anterior tentacles are sensitive to touch and chemical stimuli (Jahan-Parwar, 1972a; Preston and Lee, 1973) and are typically thrust out when the animal searches for food. Extracts or pieces of seaweed placed on the tentacles will arouse a hungry animal and produce searching and mouthing movements (Jordan, 1917; Frings and Frings, 1965; Jahan-Parwar, 1972a; Preston and Lee, 1973; Kupfermann, 1974a). At the base of the right anterior tentacle is the aperture from which the penis may protrude. This aperture is connected by means of the seminal groove to the common genital aperture (Figure 3-1).

On the back of the head are a pair of posterior tentacles, or rhinophores (nose bearers) resembling hare's ears. In some species, e.g., *A. californica,* the rhinophores are olfactory (chemosensory) in function and therefore true to their name. Touching the tip of the rhinophores with seaweed or with a stream of seaweed extract produces orienting responses to food in *A. californica* (Preston and Lee, 1973; Kupfermann, 1974a), and removing the rhinophores leads to loss of distance chemical reception (Audesirk, 1975).

THE SENSE ORGANS

The most evolved molluscs, the cephalopods, are highly visual and have a well developed eye and visual brain (Chapter 4, p. 119). Although *Aplysia* and other opisthobranchs have eyes, their visual sense is poorly developed. In nature they orient primarily by means of touch, chemosensation, and equilibrium reception.

The Eyes

A pair of small (300 to 600 μm) bluish-black eyes lies lateral and anterior to the rhinophores. They can withdraw into folds of skin when touched (Jacklet, 1969a). The eye is a closed vesicle composed of an internal spheroidal lens partially surrounded by a retina that is about 1 mm^2 in area (Figure 3-2). It contains 4,000 receptor neurons, each approximately 15 μm in diameter, and about 1,000 secondary neurons, each less than 20 μm (Jacklet, 1969a, 1973a, 1976; Hughes, 1970; Jacklet and Geronimo, 1971; Jacklet, Alvarez, and Bernstein, 1972).

There appear to be two types of receptor cells in the eye: a microvillous type and a ciliary type (Hughes, 1970). Both contain pigment granules in their distal segments and small (50–70 nm) clear vesicles in the remainder of the cell. The distal segments of the two types of receptor cells, the microvilli and the cilia, are thought to be the light-sensitive components; they interdigitate to form the

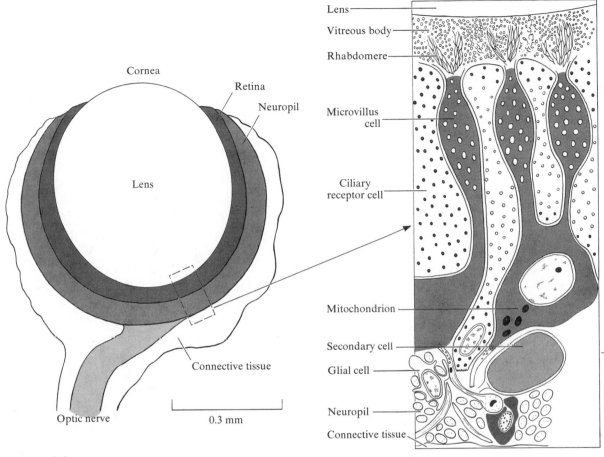

FIGURE 3-2.

The structure of the eye of *Aplysia*. The entire eye is enclosed by a connective tissue capsule. The enlargement at right shows two types of receptor cells in the outer layer of the retina: microvillous and ciliary cells. Both types of receptor cells have pigmented distal segments that project to the central lens. The inner layer of the retina consists of secondary cells whose fibers go into the optic nerve, fiber tracts, and neuropil (not shown in the enlargement). The optic nerve connects the retina with the cerebral ganglion. [After Jacklet and Geronimo, 1971; Jacklet, 1969b.]

rhabdomere (Jacklet, Alvarez, and Bernstein, 1972). The processes of the receptor cells, the secondary neurons, and the efferent neurons (see below) make up the retinal neuropil, where synaptic contacts are apparently made. The processes of the secondary neurons contain large (90–120 nm) dense-core granules thought to contain catecholamines (Luborsky-Moore, 1975; Luborsky-Moore and Jacklet, 1976). The axons of the secondary neurons (as well as perhaps some axons of receptor cells) emerge from the neuropil and enter a thin optic nerve (Jacklet and Geronimo, 1971; Jacklet, 1973a, 1976), which connects each eye to the cerebral ganglion on the same side (see Figure 4-12, p. 125). The optic nerve contains several thousand axons, some of which are efferent (Eskin, 1971).

According to Jacklet (1969a, 1973a) the cells of the eye show three types of response to light: graded depolarization, graded hyperpolarization, and action potentials. Jacklet attributes the graded responses to two types of receptor cells and the action potentials to the secondary neurons.[1] The synchronous action potentials of the secondary neurons are thought to give rise to the compound action potentials recorded in the optic nerve. Jacklet (1969a) suggests that this synchrony is achieved by means of electrical synapses between secondary neurons. Illuminating the eye gives rise to an "on" response consisting of a train of synchronized compound action potentials (Figure 3-3).[2] However, before one can accurately attribute the various types of electrical activity to the several types of retinal cells, better marking experiments will be necessary (for a beginning in this direction, see Jacklet, 1976).

The eye is under efferent (or central) control of the cerebral ganglion (Eskin, 1971). Eskin found that the frequency of firing and even the shape of the compound spike recorded in the optic nerve varies depending on whether the eye remains attached to the cerebral ganglia. In the attached eye the pattern of activity in the optic nerves is irregular, whereas in the detached eye it is regular (Figure 3-4A). The small action potentials are relatively synchronous in the two optic nerves of the animal, suggesting that they have a common source (Figure 3-4B). Small action potentials are absent in the isolated eye and can be abolished in the attached eye by cooling the cerebral ganglia, which blocks their neural activity, implying that the small spike represents efferent activity from the cerebral ganglia. Electrical stimulation of nerves from the cerebral ganglia inhibits the efferent nerve activity (Figure 3-4C). Eskin suggests that the efferent control allows nonvisual sensory input (e.g., mechanical stimulation of the head) to modulate the activity of both eyes as well as permitting one eye to modulate the activity of the other.

---

[1]Jacklet described only one morphological type of receptor cell and one type of non-neural support cell, in contrast to the two receptors described by Hughes (1970). The receptor described by Jacklet is apparently similar to the microvillous type described by Hughes. Jacklet recently noted (personal communication) that what he first called support cells may actually be ciliated receptors.

[2]Based on electro-retinogram data, Rayport and Wald (personal communication) determined that the *Aplysia* eye in two European species, *A. juliana* and *A. fasciata,* has a maximum spectral sensitivity at about 490 nm. There is evidence for a single receptor pigment. Comparison with other molluscs, such as squids and scallops (Wald and Seldin, 1968), suggests that the pigment is retinal.

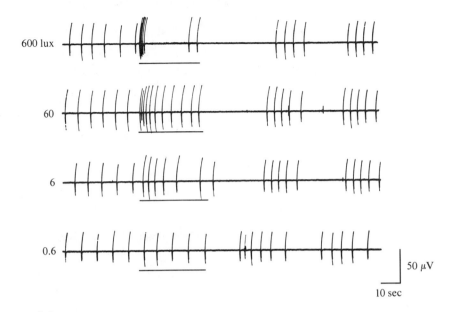

FIGURE 3-3.
Spontaneous and evoked compound optic nerve activity in *Aplysia*. Optic nerve potentials in response to illumination of the eye at four intensities of white light (the horizontal black bars indicate illumination). The top trace is about 600 lux and the succeeding traces are successive log units of lesser intensity. The eye is spontaneously active in the dark. This activity is enhanced by light and returns to normal after cessation of illumination. The potentials were recorded with a suction electrode on the optic nerve. Spike amplitude is 50 μV, negative upward. [From Jacklet, 1969b.]

A key finding in the study of the *Aplysia* eye has been the discovery of a circadian rhythm in the activity of the large compound spike (Jacklet, 1969b; for review see Lickey, Block, Hudson, and Smith, 1976). After an animal has been entrained on a 12-hour light, 12-hour dark cycle, the eye—whether in the intact animal or isolated *in vitro* with the optic nerve—shows a 24-hour circadian rhythm in frequency of firing of the compound spike (Jacklet, 1969b, 1971; Eskin, 1971). The rhythm can run free in the isolated eye for several days, even in total darkness, with the peak of activity occurring one hour after the projected dawn or onset of light of the entrainment light-dark cycle (Figure 3-5; Jacklet, 1969b; Eskin, 1971; Lickey *et al.*, 1976). The peak is less sharp in the eyes of animals previously kept in constant light (Jacklet, 1969b).

Eskin (1971) has found that this ocular clock can be entrained by the photoreceptors of the eye. He removed the eyes from animals exposed for five days

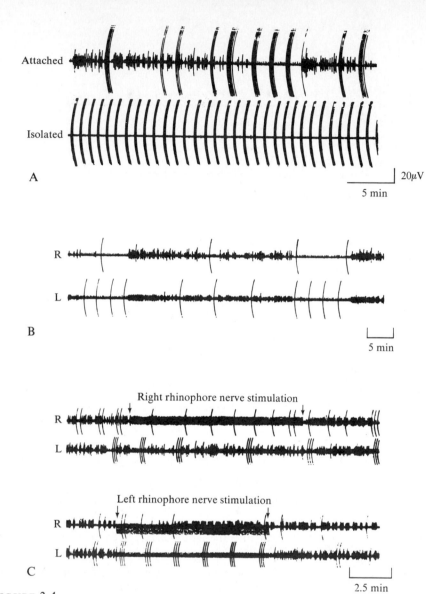

FIGURE 3-4.
Efferent activity in *Aplysia* optic nerve. [From Eskin, 1971.]

A. Simultaneous comparison of optic nerve activity in an attached eye and one that has been isolated from the cerebral ganglion. In the attached eye the large compound action potentials in the optic nerve are highly irregular. In the isolated eye the firing is regular. In the attached eye the optic nerve also shows small action potentials presumed to reflect efferent activity. These small action potentials are absent in the optic nerve record of the isolated eye.

B. Correlation of small spike (efferent) activity in the optic nerve of the attached right (R) and left (L) eyes. The small spike activity tends to be synchronous in the two eyes.

C. Inhibition of efferent activity in the optic nerve of the right (R) and left (L) eye following electrical stimulation of right or left rhinophore nerves.

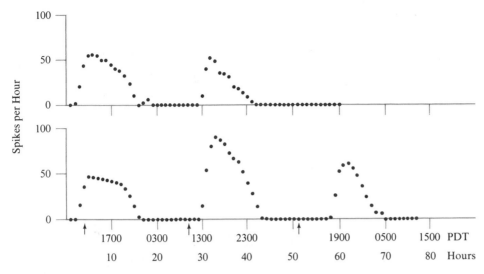

FIGURE 3-5.
Circadian rhythm of the compound optic nerve potential in the isolated eye of *Aplysia* carried at 15°C. The graph of impulse activation is from a pair of eyes from a single animal that had been on a schedule of 12 hours light, 12 hours dark for eight days. The eyes were tested in a culture medium. The black arrows on the time axis indicate the projected dark–light transition time for the eyes. [After Jacklet, 1969a.]

to a cycle of 12-hours light and 12-hours darkness and placed them in organ culture. Once in culture, he exposed one eye to the identical light-dark cycle it had experienced in the animal. The other eye he exposed to a cycle that was phase-advanced by 13 hours. By this means the experimental eye was entrained *in vitro* so that after five cycles it was 12 hours in advance of the control eye, a separation it retained stably.

Eskin (1971) also compared the speed of reentrainment in the intact animal and isolated eye. After entrainment of the intact animal on a 12-hour light, 12-hour dark cycle, the eye required only one exposure for regularization (entrainment) to a new and different 12-hour light–dark regime, which consisted of a 13-hour advance of the light–dark cycle relative to control. But the isolated eye required at least four or five cycle exposures for reentrainment to a similar phase advance. Moreover, the phase of the rhythm differed following *in vitro* and *in vivo* entrainment. Whereas the peak of the activity of the eye entrained *in vivo* occurred one hour after dawn, the peak of activity in the eye entrained *in vitro* occurred eight hours after dawn, suggesting that the circadian rhythm of the eye entrained *in vitro* has a longer free-running rhythm (Eskin, 1971). The cause of these differences is not known.[3]

Local illumination of photoreceptors of a single eye entrains the clock in that eye only; the illuminated ocular receptors cannot entrain the clock of the opposite eye (Lickey *et al.,* 1976). However, the circadian clock in the eye of *Aplysia* can be entrained by extraocular photoreceptors (Block, Hudson, and Lickey, 1974). These apparently influence the eye by way of the efferent fibers in the optic nerve. Thus, although the eye responds only weakly to red light, and responds well to white light, the circadian oscillator in the eye can be entrained by both red and white light. Entrainment of the eye by white light does not require that the optic nerve be intact, indicating that efferent activity is not essential for that entrainment. By contrast, entrainment of the eye by red light requires the intact optic nerve.

The existence of a circadian rhythm of electrical activity in the eye is of particular interest because *Aplysia* exhibits other circadian rhythms, for example locomotion and feeding. How these rhythms are interrelated will be considered in Chapter 8, p. 284.

The Tentacles: Chemosensation

Preston and Lee (1973) examined the animal's body surface using fine streams of seaweed extract to determine which sites contained the chemoreceptors that trigger the orienting responses to food (head waving). They triggered head waving from the anterior tentacles, the tips of the posterior tentacles (rhinophores), and, to a lesser degree, from the dorsal head region (Figure 3-6). All other areas, including the osphradium (see below) gave no responses. When seaweed was applied to the rhinophores and anterior tentacles, the first reaction was withdrawal, followed by arching of the neck to bring the head toward

---

[3]Jacklet and Geronimo (1971) found that surgically reducing the population of receptor and secondary neurons to 20 percent of normal, leaving only some cells near the base of the optic nerve, shortened the circadian period to as little as one hour. Jacklet and Geronimo proposed that the circadian rhythm does not result from a single master oscillator but from the interaction of a large population of secondary neurons. This population is thought to consist of noncircadian oscillators that achieve a circadian rhythm by means of the cooperative action of a large number of cells coupled by means of electrotonic connections. This idea has been questioned by Sener (1972; see also Strumwasser, 1974), who found that small pieces of the eye (50 μm $\times$ 50 μm) continue to show a circadian rhythm. Sener and Strumwasser therefore argue that the circadian rhythm could be driven by less than 100 cells and may even reside in a single cell.

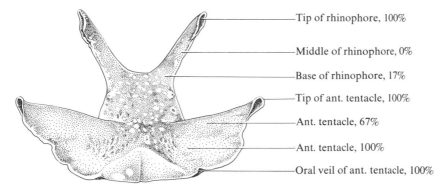

Tip of rhinophore, 100%

Middle of rhinophore, 0%

Base of rhinophore, 17%

Tip of ant. tentacle, 100%

Ant. tentacle, 67%

Ant. tentacle, 100%

Oral veil of ant. tentacle, 100%

FIGURE 3-6.
Chemoreceptor regions in *Aplysia* for food chemical stimulation. The percentages are the proportions of test animals showing positive responses to food or chemical stimulation in the areas shown. [After Preston and Lee, 1973.]

the site of the stimulation (Preston and Lee, 1973; Kupfermann, 1974a). The chemosensitivity of the anterior tentacles and the tip of the rhinophores makes the head-waving behavior of *Aplysia* an efficient means of food localization by allowing the animal to sense the direction of the chemical gradient it is exposed to.

Jahan-Parwar (1972a, b) has suggested that the active ingredients in seaweed to which *Aplysia californica* respond are the free amino acids aspartate and glutamate. These amino acids are effective in eliciting mouth-opening responses in concentrations of about 10^{-6} M. The lowest-threshold behavioral responses are produced by stimulating the inner surface of the anterior tentacles (the tentacular groove). By recording from dissected filaments of the nerve (from the cerebral ganglia) that supplies the tentacular groove, Jahan-Parwar (1972b) found that the individual axons (presumably connected to chemoreceptors) respond to low concentrations of aspartate, glutamate, or seaweed. These fibers only responded to other chemical stimuli in higher concentration, suggesting that the receptors may be fairly specific in their sensitivity to food attractants. But, paradoxically, the cerebral nerves innervating the mouth and lip areas, the most sensitive areas for eliciting the biting responses, did not appear to contain chemoresponsive fibers. These areas are presumably innervated by small fibers that are difficult to record from. Neurons in two symmetrical cell clusters in the caudal quadrants of the cerebral ganglia respond to appropriate stimulation of the chemoreceptors in the anterior tentacular groove.

The Osphradium: Osmoreception

The osphradium is a sensory organ found in all shelled marine molluscs. It lies in the mantle cavity in the direct path of the inhalant respiratory current and is thought to sense the chemical and physical properties of the seawater.

In *Aplysia* the osphradium is a small, 1–3 mm, yellow-brown epithelial patch located at the border between the floor (the dorsal body wall) and the roof of the mantle cavity immediately anterior to the attachment of the gill. The osphradium abuts the osphradial ganglion, a cell cluster thought by Merton (1920) to contain primary sensory neurons (Figure 3-7A). The osphradial ganglion lies at the end of the osphradial branch of the branchial nerve, one of the major peripheral nerves of the abdominal ganglion (see Figure 4-17B, p. 137). In addition to receiving innervation from the osphradial ganglion, the osphradium is perhaps also innervated directly by the abdominal ganglion through branches of the branchial nerve. Stinnakre and Tauc (1969) have studied osphradial function in *A. californica* using electrophysiological techniques (Figure 3-7B). They applied chemical solutions of differing osmotic strength to the osphradium while recording from cell R15 in the abdominal ganglion, and found that its characteristic firing pattern, which consists of recurrent bursts of action potentials, was inhibited by hypoosmotic solutions of seawater (Figure 3-8). A five percent reduction in osmolarity was sufficient to produce an effect. By contrast, large increases in osmolarity were ineffective; however, removal of hyperosmotic solutions and return to isoosmotic conditions caused inhibition. These findings have recently been extended by Kupfermann and Weiss (1974, 1976), who found that cell R15 releases a substance (apparently a low molecular weight polypeptide) that controls salt or water balance. When normal animals are placed in 95 percent hypoosmotic seawater they gain little weight (less than 5 percent). But when animals are injected with an extract of a single R15 cell, they increase their body weight by up to 10 percent by taking up water.

FIGURE 3-7.
The osphradium of *Aplysia*.

A. Transverse section of the osphradium and the osphradial ganglion of *Aplysia punctata*. [After Merton, 1920.]

B. Diagram of the chamber used by Stinnakre and Tauc to study the response of the osphradium to osmotic stimuli. The chamber has two compartments, one each for the ganglion and osphradium. Also shown are the recording and polarizing systems (see Figure 3-8). Changes in the osmolarity of the bathing solution were restricted to osphradium chamber. [After Stinnakre and Tauc, 1969.]

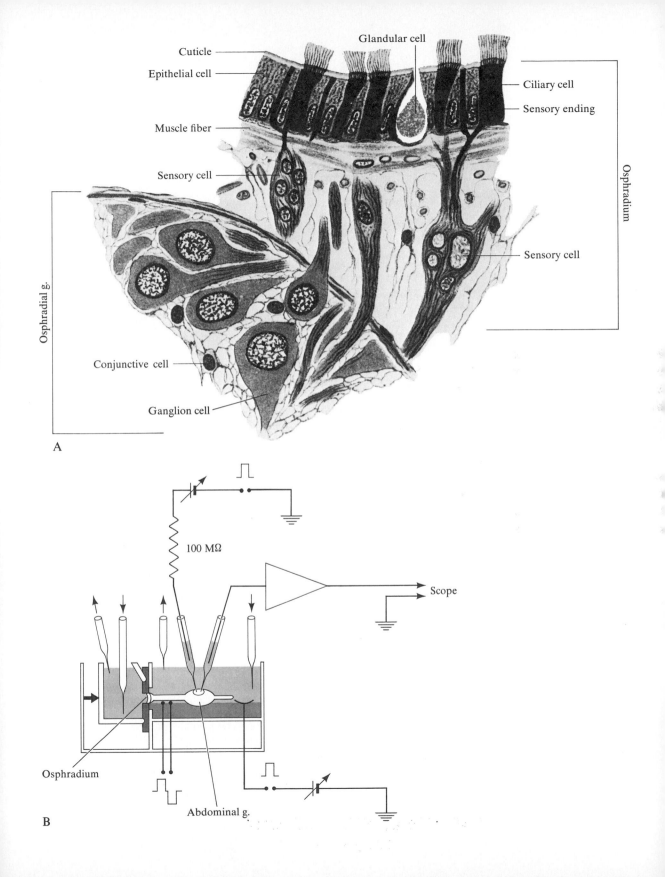

Cuticle

Epithelial cell

Glandular cell

Ciliary cell

Sensory ending

Muscle fiber

Sensory cell

Osphradium

Sensory cell

Osphradial g.

Conjunctive cell

Ganglion cell

A

100 MΩ

Scope

Osphradium

Abdominal g.

B

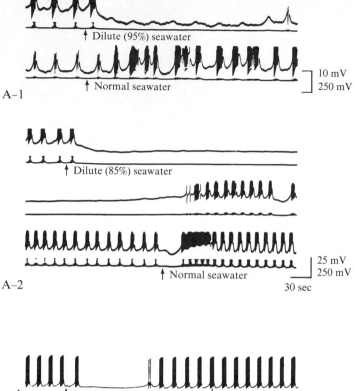

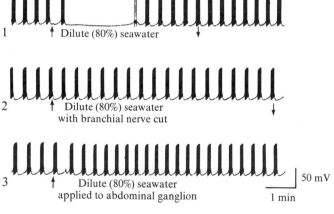

FIGURE 3-8.

Intracellularly recorded response of cell R15 in the abdominal ganglion of *Aplysia* to application of dilute seawater to the osphradium. [From Stinnakre and Tauc, 1969.]

A. **(1)** Application of slightly hypotonic (95%) seawater to the osphradium produces a slowing that is reversed when normal seawater is restored. **(2)** Application of 85% seawater causes a more profound inhibition. (Two different amplifications have been used; the top trace is high gain, the bottom trace low gain.) The burst activity comes back even before the return to normal seawater, presumably due to adaptation of the osphradial receptors. With return to normal seawater, the rebound sometimes has a short accelerating effect.

B. Application of slightly hypotonic seawater to the osphradium with and without branchial nerve connection. Diluted seawater (80%) was put in contact with the osphradium in the whole preparation **(1)** and after the branchial nerve had been cut **(2).** Diluted seawater (80%) was substituted for the normal seawater bathing the visceral ganglion **(3).**

Cell R15 thus seems to secrete a hormone that is analogous to the antidiuretic hormone of vertebrates; the hormone may act to transport salt across the skin into the body. In hypotonic seawater solutions the animal has a tendency to gain water, which upsets its internal osmolarity. Compensating for this effect, the external hypotonic solution acts on the osphradium to inhibit the firing of R15, thereby inhibiting hormone release and promoting the loss of salt (see Figure 3-18 and p. 87).

The finding of Stinnakre and Tauc that hypoosmotic solutions inhibit cell R15 have been confirmed by Jahan-Parwar, Smith, and von Baumgarten (1969), who also found that the cell was inhibited by seaweed extracts applied to the osphradium. Moreover, these extracts also excited other cells in the ganglion (R6, R8, R13, and R14) all of which belong to the white-cell cluster, a cell group believed on morphological grounds to be neurosecretory (Chapter 5; and see Frazier *et al.*, 1967). In addition to its osmosensitivity, the osphradium may also serve a chemosensory function in *A. californica,* as it does in some other gastropods, sensing food and other chemical stimuli (Copeland, 1918; Wölper, 1950; Brown and Noble, 1960; Bailey and Laverack, 1966). For example, the cephalaspid opisthobranch *Bulla* normally emerges from the sand when food is placed in the water; but it fails to emerge after the osphradial nerve has been cut. However, unlike *Bulla,* extracts of seaweed applied to the osphradium of *A. juliana* and *A. californica* do not produce head-orienting responses or radular movement (Frings and Frings, 1965; Preston and Lee, 1973). Thus, if the osphradium of *Aplysia* is involved in food sensing, its effects are subtle. Alternatively, seaweed extract could be serving as an osmotic rather than chemical stimulant.

The Statocysts: Equilibrium Reception

The statocysts, sense organs of spatial orientation, are important for locomotion and swimming. They are located in a connective tissue sheath at the rostral portion of each pedal ganglion (Coggeshall, 1969; Dijkgraaf and Hessels, 1969; Wolff, 1973a, b; Wiederhold, 1974; McKee and Wiederhold, 1974). Each statocyst is a hollow sphere whose wall consists of 13 ciliated nerve cell bodies (the so-called hair cells, which in *A. californica* are about 100 μm $\times$ 160 μm each) and many smaller supporting cells. Each nerve cell body contains about 700 cilia (Wiederhold, 1974). The sphere encloses a central space that contains statolymph and several hundred small stones, the *statoconia* (Figure 3-9A). The statocysts function as orientation indicators, much like the otoliths of mammals. The statoconia fall through the statolymph, deflecting the cilia of the receptor

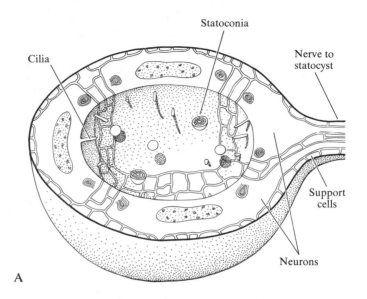

Cilia

Statoconia

Nerve to
statocyst

Support
cells

Neurons

A

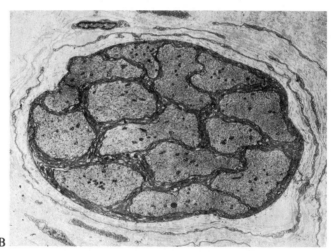

B

FIGURE 3-9.
The statocyst.

A. Horizontal section through a statocyst. Four large neurons are shown and the section passes through the nuclei of three. Small supporting cells fill the interstices between the large neurons. The proximal part of the nerve to the statocyst can be seen. Cilia jut from the neuronal perikarya into the lumen of the statocyst. Statoconia, two of which are horizontally sectioned, and statolymph fill the rest of the lumen. [After Coggeshall, 1969.]

B. A low-power electron micrograph of the statocyst nerve. The profiles of the 13 axons, one from each cell, are clearly seen. Interspersed between the axons are innumerable support cell processes. This particular nerve has dimensions of approximately 24 × 16 μm. (× 1,800) [From Coggeshall, 1969.]

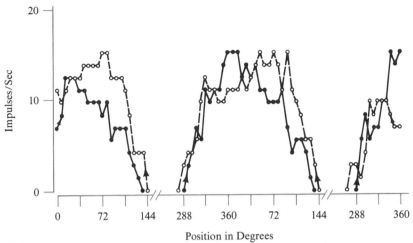

C–1

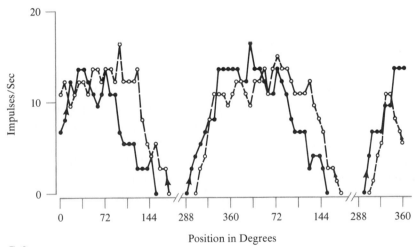

C–2

C. Response of a single statocyst receptor cell of *Aplysia limacina* to rotations about the longitudinal **(1)** and transverse axis **(2).** Two consecutive 360° rotations were performed in each direction with a rotation velocity of 10°/sec. Solid curves, to be read from left to right, represent rotations to the left **(1)** and forward **(2).** Dashed curves, to be read from right to left, represent rotations in the opposite directions. The sense cell was active only within the angular range indicated. First dot of each rotation series represents the "steady frequency" in normal position. [After Wolff, 1973a.]

cells forming the bottom third of the sphere, thereby stimulating the neurons of the statocyst and indicating the direction of gravity or acceleration. Wiederhold (1974) obtained intracellular recordings from single receptor neurons and found that rotating the statocyst in the cell (thus moving the cell upward, i.e., away from the statoconia) produces either no potential change or a slight hyperpolarization. Rotating the statocyst so that the receptor cell moves downward produces a large depolarizing receptor potential that triggers action potentials. The receptor potential appears to be due to a conductance increase to Na^+ ions (Gallin and Wiederhold, 1977; Wiederhold, 1977). The statoconia seem to move in a random fashion even when the statocyst is stationary. This movement appears to be initiated by the motion of the cilia of the receptor cells suggesting that these cilia are both sensory and motile (Wolff 1973a, b; Wiederhold, 1974).

Wolff (1973a) found that any given body position of *A. limacina* (= *fasciata*) causes five or six receptor cells of the statocyst to be activated. Any deviation about the horizontal axis changes the population of active cells. During a 360° rotation each cell becomes active only within a specific *angular range* (Figure 3-9C). A given cell responds with the same frequency of action potentials to any position in space that falls within its particular angular range. Differences between positions are therefore not signaled by the activity of a specific cell; rather, position is signaled by the spatial pattern of activity in the population of responding cells. The direction of movement by which a particular position is achieved is signaled by the time sequence of activity in the responding cells.

The sensory hair cells of the statocyst of the nudibranch *Hermissenda crassicornis* have also been studied with intracellular techniques (Alkon and Bak, 1973; Detwiler and Alkon, 1973). As in *Aplysia*, mechanical stimuli cause a graded depolarization. In some cells this depolarizing receptor potential is followed by a hyperpolarizing postsynaptic potential due to interaction between hair cells in the same statocyst. Cells that are 180° apart with respect to each other along the circumference of the statocyst invariably show reciprocal inhibition (Detwiler and Alkon, 1973; for sensory interaction of signals from the statocyst and the eye, see Chapter 9, p. 374).

In some molluscs the nervous system remaining following the removal of the statocyst is capable of compensating for the reflex balance and postural adjustments normally mediated by the statocyst (Wolff, 1973b). For example, removal of one statocyst from the prosobranch heteropod *Pterotrachea* disturbs swimming by reducing muscle tone on the operated side so that the animal loses its balance (Friedrich, 1932). Most of the symptoms, however, are eventually compensated for. After a while the animal regains the ability to right itself (righting reflex) although strong stimulation will lead to a recurrence of symptoms. These disorders and their compensation may involve neuronal sprouting and readjustments in central connections. *Aplysia fasciata* shows more restricted symptoms following statocyst removal; it loses the righting reflex on the affected side (Dijkgraaf and Hessels, 1969). It would be interesting to see if these defects occur in other *Aplysia* species and to what degree compensation occurs. Are the compensations more dramatic in species that swim as well as crawl, such as *A. brasiliana* and *A. dactylomela*? If so, the compensations could be studied with cellular techniques and might provide an interesting example of differences in neuronal compensation in different *Aplysia* species.

As we will see later (Chapter 7) the statocysts of *Aplysia* develop in an interesting way. The statocysts are the earliest neural structure to develop, each appearing as a six-celled organ when the embryo reaches the 300-cell stage (Saunders and Poole, 1910). The number rapidly increases to 13, presumably by only one division of most cells. Thereafter, the number of nerve cells remains constant throughout larval development and adult life (Coggeshall, 1969; Dijkgraaf and Hessels, 1969). Thus, the neurons of the statocyst are the first population in *Aplysia* that has been successfully followed throughout the life cycle and shown to be invariant in number. Even the size of each nerve cell body remains constant (Coggeshall, 1969); only the volume of the axoplasm increases as the statocyst nerve to the cerebral ganglion increases in length.

Mechanoreception

Aplysia, like most molluscs, is well equipped with mechanoreceptors. In fact, mechanosensation may well be the best-developed sensory capacity in *Aplysia*. *Aplysia* is responsive to touch pressure and noxious stimuli on all parts of the body, particularly its appendages: the anterior tentacles, rhinophores, siphon, parapodia. Mechanical stimuli initiate brisk withdrawal responses.

The best-studied group of mechanosensory neurons are those that innervate the siphon of *A. californica* (Byrne *et al.*, 1974). This compact group consists of about 24 neurons, located together on the ventral surface of the left abdominal hemiganglion. The cells have receptive fields in the siphon skin that are organized in an overlapping fashion that characterizes the organization of invertebrate sensory systems. Some sensory cells have a receptive field that covers a small part of the siphon; other cells have a receptive field that overlaps these and covers a large area, and finally, one cell has a receptive field that covers the whole siphon (for further discussion see *Cellular Basis of Behavior*).

The mechanoreceptors that innervate the mantle shelf and purple gland of *A. californica* are also centrally located and their pattern of innervation is similar to that of the siphon mechanoreceptors (Byrne *et al.,* 1974). The cell bodies of the mechanoreceptor neurons of the clam *Spisula* and the opisthobranch *Tritonia* are also located within the central nervous system (Prior, 1972a; Getting, 1975).

THE FOOT AND PARAPODIA:
THE SOMATIC MOTOR APPARATUS

The foot of molluscs is used both as a holdfast, fixing the body of the animal to the substratum, and as a locomotor structure, propelling the body over the ground (Figure 3-10; Lissman, 1945a, b; Morton, 1964; Gray, 1968). To act as a holdfast, the foot relies upon the adhesiveness of the mucus it secretes as well as upon suction (Lissman, 1945a; Morton, 1964).

The molluscan foot has considerable expansive powers. It is perfused with hemolymph, which acts as a fluid skeleton that can penetrate into the small spaces or sinuses among the muscles of the sole (Morton, 1964; Wells, 1968). When a gastropod moves over a surface, most of the ventral surface of the foot is fixed to the substratum while other regions of the body move (Gray, 1968). To propel the body forward, waves of muscular contractions pass along the length of the foot, separated by relaxation.

The gastropod foot consists of: (1) thin sheets of transverse and longitudinal muscle fibers, which form the sole and provide rhythmic propulsive waves; and (2) the muscles of the side of the body and of the shell. In the pulmonate *Limax* the foot consists of three longitudinal muscle bands. The two lateral bands make contact with the substratum while the medial band produces pedal waves (Prior and Gelperin, 1974). In *Agriolimax* (Jones, 1973) there are three layers; the most dorsal layer is compact and longitudinal, and from this longitudinal layer run two kinds of oblique fibers. The *posterior oblique fibers* run from the

FIGURE 3-10.
Relation of foot musculature to shell.

A. Types of insertion of the head–foot muscle in the mantle region of various groups of molluscs. [After Morton, 1963.]

B. *Aplysia punctata.* Semidiagrammatic dorsal representation of the columellar and other longitudinal muscles and dorsoventral musculature of the head–foot region, which has been opened dorsally and drawn as though pinned flat. All the viscera have been removed. [From Brace, 1977c.]

Ancestral Mollusc

Bivalve

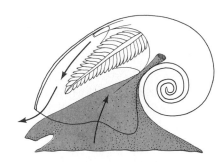

Gastropod

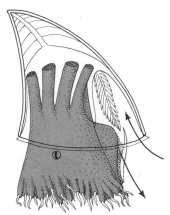

Cephalopod

A

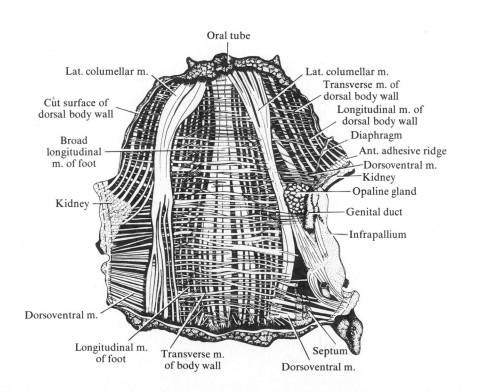

Oral tube

Lat. columellar m.

Lat. columellar m.
Transverse m. of
dorsal body wall

Cut surface of
dorsal body wall

Longitudinal m. of
dorsal body wall

Diaphragm

Broad
longitudinal
m. of foot

Ant. adhesive ridge

Dorsoventral m.

Kidney

Opaline gland

Kidney

Genital duct

Infrapallium

Dorsoventral m.

Longitudinal m.
of foot

Transverse m.
of body wall

Septum

Dorsoventral m.

longitudinal band forward and downward to the pedal epithelium, exerting a posteriorly directed force on the epithelium. The *anterior oblique fibers* run backward and downward and exert an anteriorly directed force on the epithelium. Several waves propagate consecutively along the sole at any time. The anterior oblique muscles shorten during the passage of a wave and these fibers are thought to be responsible for producing the pedal waves. They contract and pull the pedal epithelium upward and forward. They are then thought to relax as the wave of pedal contraction moves forward, which now allows the sole to be pushed down (by hemocoelomic pressure) after having been moved forward. Whereas the anterior oblique fibers contract over a short region of a single wave and pull the sole forward relative to the body, the posterior oblique fibers contract over a comparatively large region between waves and pull the body forward relative to the sole. Thus, the anatomical arrangement and the sequence of contraction explain the propagation of the pedal wave.

In animals that have shells the muscles of the foot are attached to the shell by a series of muscles called muscles of the shell (*columellar muscles*). In most molluscs the shell muscles are usually paired, but in most gastropods only a single large columellar muscle attaches to the shell (Figure 3-10A). The foot is highly modified in some species of molluscs. The most dramatic adaptive changes are found in cephalopods, in which the foot is subdivided into tentacles or arms that surround the mouth.

In *Aplysia* the foot is long and not clearly delineated from the body. Its anterior end, or propodium, extends to the mouth while its posterior end projects beyond the visceral mass as a short tail, or *metapodium*. The tail is adhesive and in some species (especially *A. juliana* and other members of the subgenus *Aplysia*) it has developed a specialized suction disc at the posterior end of the animal. As a result, *Aplysia* can cling to a substratum while swinging its head and body freely so as to make undulating movements in search of food (Jordan, 1917; Frings and Frings, 1965; Kupfermann, 1974a).

The muscles of the foot are composed of a dorsal longitudinal and a ventral transverse layer (Eales, 1921; Winkler, 1957). The dorsal longitudinal muscle layer is thick and runs the length of the foot. Near the lateral margin, the muscles from the lateral walls pass between these longitudinal strands to the ventral layer. The transverse ventral layer contains short vertical muscles that are thought to produce the local suction. A broad longitudinal muscle runs laterally on either side along the entire length of the animal from the head (oral tube) to the tail (Brace, 1977a, c). Some transverse fibers diverge from this muscle and insert in the shell. According to Eales, the fibers (on each side) make up the retractor muscle of the mantle. Brace argues that the lateral tracts derive from what were originally the columellar muscles (Figure 3-10B). The locomotor patterns of *Aplysia* and other opisthobranchs are considered further in Chapter 8.

The foot and body wall extend laterally into two fleshy expansions, the *parapodia* (see Figure 3-1). In several species, including *A. brasiliana, A. dactylomela, A. fasciata,* and *A. juliana,* the parapodia are well developed and are used for forward swimming. Animals that swim will cling and crawl on a firm substratum, but when the substratum is removed they will swim. (We will consider details of swimming patterns in Chapter 8.)

The outstretched parapodia are large. In an animal 22 cm long and 19 cm wide from parapodial edge to parapodial edge, the surface area of a single parapodium is 115.35 cm^2 (Neu, 1932). Since the parapodia have a large surface area and the animal is only slightly heavier than seawater, minimal downpressure on the water is necessary to stay afloat. The parapodia are controlled by the pedal ganglia (Hening, Carew and Kandel, 1976). Swimming has been postulated to be mediated by a chain of reflexes (see Chapters 4 and 8, and Fröhlich, 1910a, b).

A hemal meshwork similar to that found within the foot also serves as a hydraulic skeleton in the movement of some appendages (Yonge, 1949b; Morris, 1950; Russell-Hunter, 1968). In many species the tentacles and the siphon can be rapidly withdrawn by muscular contraction, but they are only slowly extended hydraulically as hemolymph is shifted from one part of the body to another. Thus, molluscan movements often use retractor muscles, sometimes with only remote antagonists; the transmission function for the action of the antagonists is mediated by the hydraulic skeleton.

THE MANTLE CAVITY: THE RESPIRATORY SPACE

Primitive molluscs differ from their turbellarian–nemertine ancestors primarily in their thickened ventral body wall, greatly increased body size, and body form (Chapter 1). These changes in molluscs are correlated with two advances: the development of gills for respiratory exchanges and a more effective circulatory system. According to Fretter and Graham (1962) the primitive respiratory (mantle) cavity was a shallow groove between the foot and the mantle skin that housed the gill and into which the kidneys and anus discharged. With further evolution, the cavity enlarged so that it also served as a protective retreat into which the large foot and the head could be withdrawn (Yonge, 1947; Morton and Yonge, 1964).

The mantle cavity of modern molluscs typically contains also the anus and the openings of the kidneys and hypobranchial gland (and their possible homolog, the ink gland), all of which discharge into the exhalant current (Figure 3-11A). The osphradia, usually a paired sense organ, lie in the inhalant current and are believed to sample the physical and chemical properties of the water (Yonge, 1947).

The mantle cavity of *Aplysia* is a crescent-shaped space that is open anteriorly and dorsolaterally where the parapodial edges are not fused, and is partially closed posteriorly by the fusion of the parapodia (Figure 3-1). It contains a single osphradium, the genital aperture, the discharge pore for the kidney, the anus (see below), the purple gland and the opening of the duct of the opaline gland (Figure 3-11B). The genital aperture lies at the anterior end of the mantle cavity. A seminal groove (Figure 3-1) arises from it and runs forward to the penis, which is at the base of the right anterior tentacle. The kidney pore is situated on the underside of the overhanging roof of the mantle cavity, or suprapallium, just behind the posterior attachment point of the gill (Figure 3-11B).

In *Aplysia* the mantle cavity is surrounded by the vital organs of the visceral mass: the heart, kidney, gut, and ovotestis, with the gill projecting into the mantle cavity. Before metamorphosis these organs are fully protected by a shell (Chapter 7). But after metamorphosis the body rapidly outgrows the shell, and the mantle margins fold around the edge of the shell. The right lateral edge of the mantle margin remains free and covers the gill while the left margin connects to the foot, so that the mantle of the adult *Aplysia* forms a projecting shelf (the *mantle shelf*) that is fixed on the left but free on the right side (Figure 3-11B).

The mantle shelf contains a residual shell, which is largely hidden from view by the overgrowth of the mantle; the only portion of the shell exposed is a small central area known as the aperture. In *A. californica* the mantle usually covers the shell (there is no aperture), making the shell an internal structure. Although the shell is important for protection of the larval and postlarval animals (see Figure 7-1, p. 219), it probably serves only a minor protective function in the adult. In closely related anaspid genera, *Petralifera* and *Notarchus*, the shell present in the larva disappears completely in the adult (Figure 2-5, p. 36).

In the postmetamorphic animal a variety of defensive contractile movements of the head, neck, siphon, mantle shelf, gill, and tail are designed to protect this vital region. The parapodia provide further protection for the visceral organs by projecting dorsally to cover the mantle and visceral hump (Figures 3-1 and 3-11B). In addition, the parapodia serve a respiratory and cleansing function by aiding in the circulation of water through the mantle cavity. The parapodia are not joined anteriorly; their anterior margins provide an orifice that acts as an

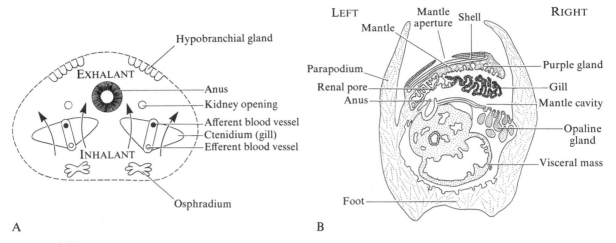

FIGURE 3-11.
The molluscan mantle cavity.

A. A posterior view of the typical molluscan mantle (pallial) complex showing how water circulates through the mantle cavity. The flow is ventral (inhalant) to dorsal (exhalant). [After Russell-Hunter, 1968.]

B. Transverse section through the mantle cavity of *Aplysia punctata* (posterior view), illustrating mantle aperture, shell, kidney, gill (collapsed), mantle, parapodia, purple gland, opaline gland, anus, visceral mass, and foot. [After Guiart, 1901.]

inhalant siphon, which allows fresh seawater to enter and flow through the mantle cavity, cleansing the cavity and supplying oxygen to the gill. This water is expelled from the mantle cavity via two posterior ports: (1) the pseudo-siphon, an exhalant funnel formed by the flaring of the parapodial lobes just before they join the foot, and (2) the true siphon, a thick, fleshy extension of the mantle shelf (Figure 3-1). At the base of the true siphon lies the anus, which is controlled by a muscular sphincter (Figure 3-1).

Most *Aplysia* species contain a purple (Blochmann's) gland on the underside of the free edge of the mantle shelf, where it occupies a clearly demarcated, pitted, purplish-brown area (Figures 3-1 and 3-11B). The purple gland is inner-vated by the abdominal ganglion; in *A. californica* it is controlled by three identified cells that are electrically coupled to one another (cells L14A, B, and C; see Carew and Kandel, 1977a; and Chapter 9 in *Cellular Basis of Behavior*). The gland gives off a dark purplish secretion that could serve as a defensive screen for animals in small tide pools. The purple gland may be the homolog of the hypobranchial gland of more primitive molluscs (see Figure 1-1, p. 7; and see Morton, 1958a). The duct of the opaline (Bohadsch) gland ends in the floor of the mantle cavity; this gland discharges a white secretion of unknown function, when the animal receives a noxious stimulus.

The Gill: The Respiratory Organ

In primitive molluscs the respiratory organ takes the form of a pair of gills (*ctenidia*), each with a central axis containing a dorsal afferent vein and a ventral efferent vein. On both sides of the gill axis are alternating rows of elongated triangular filaments or plates (Figure 3-12A). In all modern molluscs the mantle cavity is placed so that the inhalant current of seawater comes into contact first with the side of the filament containing the efferent vein. From there, oxygenated blood flows back to the heart for systemic distribution. The dorsal surface contains the afferent vein that carries deoxygenated blood (Figure 3-12B). Blood flow through gill filaments is therefore from dorsal to ventral; water flow is from ventral to dorsal. The cilia that cover the lateral surfaces of the filaments are arranged so that their beating produces a countercurrent of water through the mantle cavity and through the gill filaments, opposite to the flow of blood.[4] This countercurrent of oxygenated water and deoxygenated blood produces an efficient mechanism of oxygenation of blood (Yonge, 1947).

The gill of *Aplysia* is normally covered by the mantle shelf, and, as in other molluscs, it is perfused with blood by means of two veins that the mantle cavity lining has taken with it and that connect the gill with the circulation of the body. The afferent branchial vein attaches to the posterior lateral portion of the mantle cavity (the posterior insertion) and runs along the inner surface of the gill. The efferent branchial vein attaches to the anterior portion of the mantle cavity and runs along the outer side (Figure 3-13F). By contraction and relaxation of one or the other vein, the gill can be rotated on its axis or extended beyond the edge of the mantle shelf (*Cellular Basis of Behavior*, p. 379). The gill of *Aplysia* is an extension of the lining of the mantle cavity and is folded like a fan. In *A. californica* the gill is made up of a series of about 12 *pinnules* (folds) arranged side by side (Carew et al., 1974). Each pinnule forms a small chamber connecting the afferent vein to the efferent vein (Figure 3-13C). Because each pinnule is formed by the sinusoidal folding of a single sheet of tissue, it is continuous across the midline of the gill with an adjacent pinnule on the opposite side. This can be visualized by representing the afferent and efferent veins by a pair of parallel tubes, as in Figure 3-13.

In advanced molluscs muscular contractions often replace ciliary respiratory mechanisms. In the clam *Yoldia* and the lamellibranch *Malletia* the gill attaches to the roof of the mantle cavity by means of a muscular membrane that

---

[4]The frontal and abfrontal cilia that fringe the ventral and dorsal surfaces of the filaments act as cleaners and aid in the filter-feeding by which bivalves and mesogastropod prosobranchs sort food particles of different sizes for delivery to the mouth.

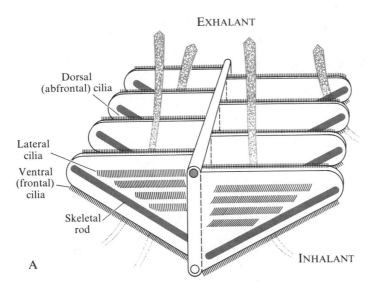

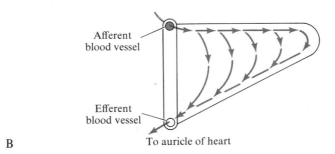

FIGURE 3-12.
The molluscan gill. [After Russell-Hunter, 1968.]

A. Stereogram showing the water current from ventral (inhalant) to dorsal (exhalant) between adjacent gill plates. The water current is created by the lateral cilia and is counter to the direction of the blood flow. The dorsal (abfrontal) and ventral (frontal) cilia are for cleaning the gill.

B. Blood flow through a single gill plate illustrated in part A to show the counter-current flow.

relaxes as each pulse of blood enters the gill (moving the gill downward) and contracts as the blood leaves the gill (moving the gill upward), thereby helping ventilate the gill filaments (Drew, 1899; Yonge, 1947). Some molluscs, e.g., the anaspid opisthobranch *Notarchus punctata,* use pumping movements for

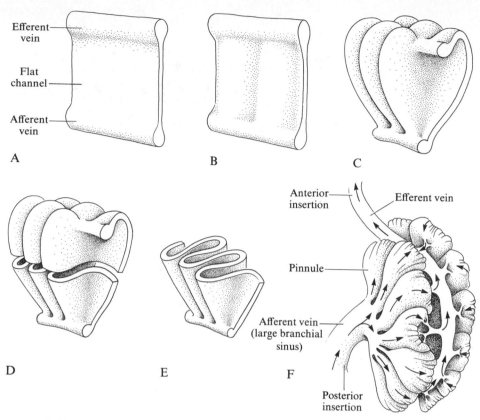

FIGURE 3-13.
The gill of *Aplysia*. [After Carew *et al.*, 1974.]

A–C. Series of perspective drawings showing progressive folding of the type that might account for the adult gill structure. The efferent and afferent veins communicate lengthwise through a flat, large-surface channel. This channel assumes pleated folds perpendicular to the longitudinal axis of the veins because the efferent vein is longer than the afferent. As the elaboration of the folds is carried further, each fold becomes shaped like a wedge and begins to overgrow the efferent vein, tending to obscure it from view.

D–E. Horizontal cutaway (E) of the unfolded gill at the level indicated in D illustrates that despite the infolding, a single cleft interconnects the veins. In the mature gill the wedge-shaped pinnules interconnect the veins and partly cover the efferent vein by their expansion at that end.

F. Circulation of blood through the gill. Deoxygenated blood is carried from the hemocoel via the afferent vein, and oxygenated blood is returned to the heart via the efferent vein.

locomotion as well as swimming, thereby adjusting the rate of gas exchange over the gill surface to the degree of activity (Bauer, 1929; Purchon, 1968). Respiratory pumping mechanisms are found also in *Aplysia* (Chapter 8). These advanced mechanisms for the control of respiration permit more complex behavior in these animals by increasing total respiratory capacity and rates of oxygenation.

Muscle Groups of the Gill

Most gill movements are due to contractions of the muscles of the two major veins (Carew *et al.*, 1974; Kupfermann *et al.*, 1974). The efferent vein of the gill has three thick layers of muscle: (1) longitudinal; (2) inner-circular; and (3) outer-circular. The longitudinal muscle bundles range up to 500–600 μm in thickness and are the most prominent. They lie intercalated between an almost continuous layer of outer-circular muscle (approximately 90 μm in thickness) and intermittent bands of inner-circular muscle (Figure 3-14). Ex-

FIGURE 3-14.
Muscles of the gill of *Aplysia*. The major muscle groups of the afferent and efferent veins and pinnules of the gill. [After Carew *et al.*, 1974.]

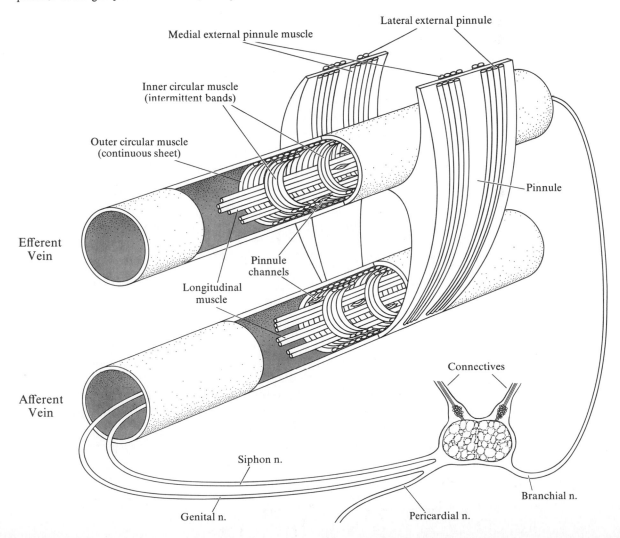

ternal to the outer-circular muscle is a layer of spongy connective tissue and a ciliated columnar epithelium similar to and continuous with that in the pinnules.

The afferent vein is similar in structure. It has prominent longitudinal muscle fiber bundles and, associated with them, longitudinal nerves. Small circular muscles are found internal and external to the longitudinal muscles, but the circular muscles of the afferent vein are much smaller than the corresponding ones in the efferent vein. The external connective tissue and epithelium are much the same as in the efferent vein, but the lining of the lumen of the afferent vein differs from that in the efferent vein. The lining consists of several layers of cuboidal cells, especially thick on the surface adjacent to the openings into the pinnules, whereas the efferent vein has no cuboidal cell layers. From the afferent vein a thin muscle strip, the lateral external pinnule muscle, extends along the whole lateral surface of the pinnules up to their top (Figure 3-14). From the efferent vein the medial external pinnule muscles extend the short distance along the medial surface between the efferent vein and the top of the pinnules as well as down some distance into the inner folds of the pinnules. Close to either vein, the pinnules contain a considerable amount of muscle and connective tissue; the epithelial lining of the pinnules is cuboidal. Further out toward the crest of the pinnule the muscle and connective tissue disappear, so that the muscular extensions from each vein are not continuous with each other. Both veins are innervated by neurons that lie in the abdominal ganglion (Carew *et al.,* 1974). Additional neurons (presumably motor cells) lie within the sheaths of the axons innervating the gill (Peretz, 1970; Carew *et al.,* 1974).

THE VISCERAL MASS: INTERNAL ORGAN SYSTEMS

Cardiorespiratory System

CIRCULATION

As in vertebrates, the principal functions of the circulatory system in molluscs (Figure 3-15) are: (1) distribution of blood and thereby nutrients to tissue; (2) maintenance of adequate concentrations of critical hormones and gases; and (3) removal of waste products of metabolism. In molluscs the blood also serves a skeletal function in locomotion and posture.

Like other molluscs (except cephalopods), *Aplysia* has an open circulatory system (Figure 3-16A; Eales, 1921; Winkler, 1957; Wright, 1960; for an early description see Milne-Edwards, 1849). The arterial system is closed and

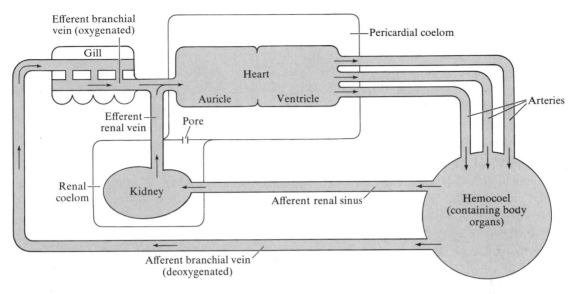

FIGURE 3-15.
The molluscan circulatory system. Auricles, gills, and kidney are usually paired. The circulatory system is open and consists of arteries leading to open tissue spaces, such as the hemocoel, which contains the major organs of the body and which is drained by collecting sinuses. [After Villee *et al.,* 1963.]

consists of arteries leading from the heart to the body tissues, but the venous system is open and consists of sinuses, lacunae, and veins. There are no capillaries; the arteries discharge into large lacunae within the tissues. Because the lacunae are devoid of epithelium, the blood comes in direct contact with the tissues. From the tissues the blood travels back to the body cavity surrounding the gut and the other organs. The body cavity is not a true coelomic cavity (a liquid-filled cavity in the mesoderm lined with endothelium). Rather, it is a *hemocoel,* a huge sinus, an enlarged venous space containing blood.[5] The body cavity, also called the *common ventral abdominal sinus,* thus serves as a principal reservoir for the blood, which is squeezed in and out of the spongy muscular walls.

From the hemocoel blood is collected by venous sinuses, wide spaces that convey the blood through the kidney and the gill on the way back to the heart.

[5]The hemocoel is a pseudocoel, a remnant of the primary body cavity, the *blastocoel.* (During formation of the gut the blastocoel is reduced, but some component of the space between the endoderm and the ectoderm remains.) There is no endothelial lining to the walls of the pseudocoel. The coelom is a secondary body cavity, a new space in the mesoderm lined by endothelium (called mesothelium).

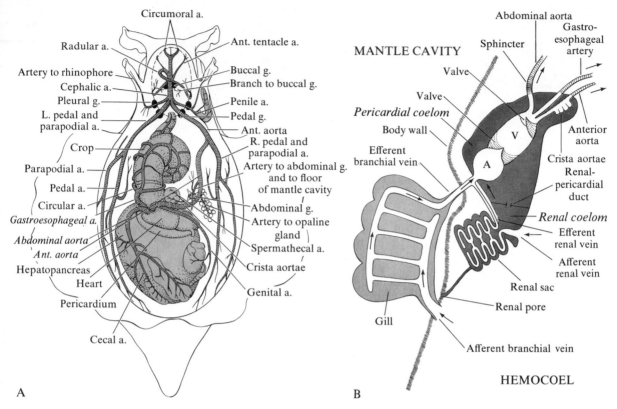

FIGURE 3-16.
The circulation of *Aplysia*.

A. Heart and arterial system of *Aplysia californica* based on perfusion with vinyl dyes. [After Rubinson, unpublished observations.]

B. Diagram of the flow of circulation through the gill, kidney, and heart in *Aplysia*. [From Kandel, 1976; after Mayeri *et al.*, 1974.]

One of the collecting sinuses is the afferent branchial vein (also called the large branchial sinus), the main afferent gill vessel. It results from the union of five smaller branchial sinuses and lies immediately beneath the floor of the mantle cavity posterior to the gill. The afferent branchial vein enters the gill on its inner concave side and extends the entire length of its surface (Figure 3-16B). This vein has large gaping orifices from which deoxygenated blood passes to the gill pinnules. Blood is then collected and delivered into the auricle of the heart by the efferent branchial vein, which lies along the outer portion of the gill. Like the afferent vein, the efferent vein communicates with the pinnules by wide openings. The afferent and the efferent veins are neurally controlled by various combinations of neurons (see *Cellular Basis of Behavior,* p. 354).

Not all the blood from the common ventral abdominal sinus enters the gill for aeration. A short afferent renal vein brings a small quantity of blood from the left portion of the hemocoel directly to the kidney (Figure 3-16B). The renal sinuses discharge into the kidney, from which the blood is collected by the efferent renal sinus, which opens into the base of the efferent branchial vein as the latter enters the auricle. *Aplysia* thus has a renal portal system that enables some venous blood to be cleared of nitrogenous and other wastes by the kidney before it returns to the heart. The blood entering the heart is therefore mixed. The greater part comes from the gill and is oxygenated; a smaller part comes from the kidney purified of wastes but not oxygenated (see p. 86 for a discussion of renal filtration).

THE HEART

The heart is a pump that maintains a pulsatile flow of blood through the circulatory system to the tissue. It is located within a pericardial cavity (Figure 3-16B), which together with the renal cavity (with which it communicates) makes up the total coelomic cavity of molluscs. The heart usually consists of one or two auricles (depending upon the number of gills) and a single ventricle (for detailed reviews see von Skramlik, 1941; Krijgsman and Divaris, 1955; Hill and Welsh, 1966).

The molluscan heart, like the vertebrate heart, is *myogenic;* it beats spontaneously in the absence of neural regulation. The nervous system influences heart activity through the visceral or abdominal ganglion, thus modulating activity in concert with behavior. Acetylcholine usually mediates inhibition and the biogenic monoamine serotonin usually mediates excitation (Welsh, 1956; Hill and Welsh, 1966; Liebeswar *et al.,* 1975). This provides an interesting parallel with the vertebrate heart, which is inhibited by acetylcholine and excited by another biogenic monoamine, norepinephrine.

The heart of *Aplysia* has two chambers, an auricle and a ventricle, and lies on the left side of the animal near the gill (Figures 3-15 and 3-16B). As indicated above, the auricle receives blood from the gill through the efferent gill vein and from the kidney through the renal vein. The auricle is larger than the ventricle and has thinner muscular walls. Two semilunar valves prevent backflow from ventricle to auricle. The ventricle gives rise to the aorta and a third semilunar valve prevents the reflux of blood back into the ventricle. The auricular–ventricular valve appears to be under neural control; the aortic valve is not. Two of the three aortae are also under neural control (Mayeri *et al.,* 1974).

The heart leads to the periphery by means of three major branches of the aorta (Figures 3-15 and 3-16). The *anterior aorta* supplies the buccal mass, the anterior region of the gut, the mantle and its glands, the foot and parapodia, the head, the tentacles, the radula, the nervous system, and the reproductive organs other than the ovotestis. Between this aortic branch and the ventricle lie sacculations, or flaps, of unknown function called the *crista aortae,* one lying ventral and one dorsal. The *gastroesophageal artery* supplies the stomach and the esophagus. The *abdominal* or *posterior aorta* supplies the digestive gland and the ovotestis (Eales, 1921).

PRESSURE GRADIENTS AND CARDIAC FILLING

The maximum pressure that can be exerted by contraction of the ventricle of the heart of *Aplysia* is about 40 mm of water (Wright, 1960). By contrast, the pressure in the efferent branchial vein is 15 mm of water when the gill is relaxed and 25 mm of water when the gill is contracted (Wright, 1960). Pressure in the hemocoel is about 20 mm of water when the animal is relaxed and 30 mm when it is contracted. Pericardial pressure is 20 mm of water, equal to that of the hemocoel.

All parts of the auricle appear to contract simultaneously and to drive blood into the ventricle. The mechanism that prevents reflux into veins is not known, but Krijgsman and Divaris (1955) postulate that there is a sphincter at the apex of the auricle that closes during contraction. Well-developed atrioventricular valves prevent reflux during ventricular contraction. Wright (1960) suggests that the cardiac pacemaker region exists at the ventricular node. By recording from the heart. Wright found that a spike-like prepotential, localized to the region of the suspected node, preceded the muscle action potential.

Molluscs do not have external structures that facilitate the expansion of the heart. Refilling may be due to the change in pericardial pressure that is responsible for the considerable increase in size of the auricle and ventricle when they fill. When the wall of the pericardium is opened, the diastolic dilation stops (see Krijgsman and Divaris, 1955); although systolic contractions may continue for a while, the pattern of normal contraction is altered. Because venous pressure is too low to account for diastolic filling, and because filling seems so dependent upon an intact pericardium, Ramsay (1952) and Krijgsman and Divaris (1955) have explained arterial filling hemodynamically by assuming a rigid pericardial membrane and therefore a constant pericardial volume. According to this *constant volume hypothesis* the pericardium, which is filled with fluid, is closed except for its narrow outlet at the renopericardial duct (Figure 3-16B). When the ventricle contracts, it expels blood and reduces the volume of fluid in the space. This causes a decrease in the hydrostatic pressure

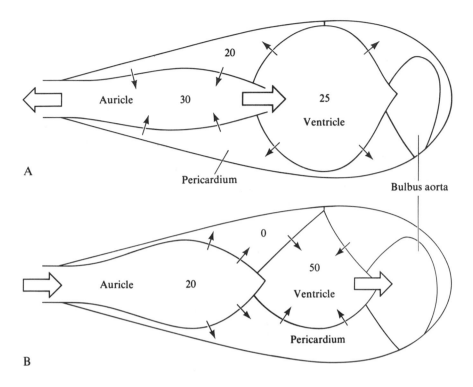

FIGURE 3-17.
The constant pericardial volume hypothesis. Diagrammatic transverse section of the heart of *Patella* showing the pressures present in the different parts of the heart and the pericardium. **A.** The condition at ventricular diastole. **B.** The condition at ventricular systole. The large arrows represent flow of blood; the small arrows represent the forces acting on the auricle and ventricle. The numbers indicate water pressure in mm. [After Jones, 1970.]

in the pericardial space that cannot be compensated for by an increase in the volume of the pericardial fluid because the renopericardial pore is too small; nor can the change in hydrostatic pressure collapse the wall of the rigid pericardial membrane. As a result, the decrease in hydrostatic pressure is compensated for by an expansion of the auricle.

In an attempt to test this hypothesis, Jones (1970) measured the pressure within the heart and pericardium of *Patella vulgata*. He found a pressure gradient during systole of 30 mm of water from the branchial vein to the aorta. During auricular systole the auricular pressure was 30 mm, whereas the ventricular diastolic pressure was 25 mm; during ventricular systole the ventricular pressure was 50 mm, whereas the atrial diastolic pressure was 20 mm (Figure 3-17). Thus, as in the vertebrate heart, the molluscan heart circulates blood by generating a pressure gradient from the arterial to the venous end of the vascular system. Jones (1970) found that the pressure in the pericardium is lower than in the heart: 20 mm of water during ventricular diastole and 0 mm during ventricular systole (Figure 3-17). This reduction in pressure in the pericardial space during ventricular systole is consistent with the constant volume hypothesis.

An independent and more direct test of this hypothesis has been provided by Civil and Thompson (1972) using an artificial pericardium in *Helix pomatia*. They found that isolated hearts, free of the pericardium, would beat well when mounted in a perfusion apparatus with a simulated venous pressure of 8 cm of saline. The heart did not translocate fluid, however, and stopping the perfusion led to stoppage of the heart beat. If the isolated heart was mounted in an artificial pericardium (an appropriately stoppered glass tube with inflow and outflow tubes connected to the two ends of the heart), the heart translocated fluid and its rate of contraction was independent of venous pressure. Opening the artificial pericardium in an attempt to simulate opening of the renopericardial canal reduced diastolic filling and led to arrest of heart rate. These results support the hypothesis of a constant pericardial volume.

REGULATION OF CARDIAC OUTPUT

Cardiac output, the volume of blood expelled into the circulation per unit time, is dependent on two factors: stroke volume (which in turn is dependent on filling pressure and arterial pressure) and heart rate. As arterial pressure increases, cardiac output decreases. Maximal stroke volume in *Aplysia* is less than 0.6 cc (for details see Wright, 1960). Each of these variables is controlled by myogenic as well as neural factors.

The arteriovenous difference in oxygen content of the blood is 2.5 volume percent (cc of oxygen per 100 cc of blood). A 600-g animal uses 0.1 cc of oxygen per minute (Wright, 1960). Given an average stroke volume of 0.3 cc, a heart rate of 13 beats per minute will supply sufficient oxygen to an animal in the steady state. At rest, the heart rate in an adult *Aplysia* is between 12 and 35 beats per minute (Bottazzi and Enriques, 1900; Straub, 1904b; Feinstein *et al.*, 1977; Dieringer, Koester, and Weiss, 1978). In juveniles heart rate is much higher, reaching 100 to 120 beats per minute (Kriegstein, 1977b).

Heart rate can be altered by intrinsic regulation as well as neuronally by reflex and central commands (see Koester *et al.*, 1974; Feinstein *et al.*, 1977; Dieringer, Koester, and Weiss, 1978; and *Cellular Basis of Behavior*, Chapter 10). For example, Schoenlein (1894) and Bottazzi and Enriques (1900) described neural excitation of the heart after stimulation of the central end of a cut branchial nerve. Dieringer and others (1978) found an acceleration after a variety of stimuli: noxious stimulation of the parapodia or the siphon, feeding, spontaneous activity, e.g., locomotion, increases in temperature, and hypoxia. Increases in heart rate can also be induced by exposure of the animal to air (Feinstein *et al.,* 1977) and changes in carbon dioxide levels (Straub, 1901). An increase in carbon dioxide tension produces a large increase in the rate of the isolated heart, leading to sustained contractions followed by a prolonged slowing (Straub, 1901). Many of these changes seem to be due to intrinsic regulation; the responses are unaltered by cutting the pericardial nerve, which carries the axons of the motor cells that lie in the abdominal ganglion (Dieringer *et al.,* 1978). However, certain effects, such as the increase associated with food arousal, are largely mediated by the abdominal ganglion (Dieringer *et al.,* 1978).

BLOOD AND RESPIRATORY PIGMENTS

The blood (or hemolymph) of *A. californica* is a faintly purple liquid. Mostly water, it contains approximately two percent protein, consisting primarily of four large molecular species: hemocyanin, acetylcholinesterase, erythrocruorin, and a hemagglutinin (Giller and Schwartz, 1971b; Bevelaqua *et al.,* 1975). Hemocyanin is a multimeric copper–protein complex of high molecular weight. The copper is bound to the protein and makes up 0.26 percent of its dry weight; because of the copper, hemocyanin combines reversibly with oxygen and functions as a respiratory pigment (Ghiretti, 1966). The protein is apparently lacking in some European species of *Aplysia* (Ghiretti, 1966). Erythrocruorin is a large iron-containing respiratory protein having a molecular weight of two to three million (Bevelaqua *et al.,* 1975). Despite the presence of these pigments, the oxygen dissolved directly in the blood is alone probably sufficient to oxygenate the tissues of *Aplysia* (Winkler, 1955).

Little is known about energy sources for *Aplysia,* but the main circulating reducing sugar in the hemolymph is trehalose (Blankenship, unpubl.).

Urinary System: Water Balance

URINE FORMATION

The urinary system regulates the constancy of the internal environment by governing the water content of the body, regulating the electrolyte and acid-base balance of the body fluids, and by conserving valuable nutrients and excreting wastes. The molluscan kidney is a filtration kidney, as is that of vertebrates. However, the basic units of the renal system differ radically from the vertebrate nephron, which contains within a single structure the sites for filtration, reabsorption, and secretion. In molluscs the sites for these functions are separated (for review of excretion in molluscs see Potts, 1967; Riegel, 1972).

Filtration into the pericardium occurs across the walls of the auricle, the walls of the ventricle (through a layer of cells that resemble the podocytes of the vertebrate glomerulus), or across the walls of the crista aortae (Martin and Harrison, 1966; Hyman, 1967; Andrews and Little, 1971). Reabsorption and secretion occur in the kidney, a folded epithelial membrane (or coelomic duct) that forms an invagination of the coelomic cavity and drains the pericardium to the exterior (Goodrich, 1946; Martin and Harrison, 1966; Potts, 1967; Tiffany, 1974). The kidney is well supplied with blood flowing between the body and the gill. In some land pulmonates there is also a ureter of ectodermal origin. The walls of the kidney are deeply folded and resemble the epithelium of the vertebrate proximal tubules (in which salt, water, and glucose are reabsorbed and nitrogenous compounds are secreted). Like the vertebrate tubules, most of the epithelial cells of the molluscan kidney are cuboidal, contain granules or concretions, and have a brush border. The basal membrane, thought to be involved in active reabsorption, is usually folded and there are mitochondria between the folds (Potts, 1967).

These simple filtration and glandular structures carry out the three major processes of urine formation in ways that resemble the function of vertebrate kidneys (Martin and Harrison, 1966; Kirschner, 1967). The filtration rate (approximately 1–10 percent of body weight per hour) is comparable to that of vertebrates. The ultrafiltrate is expelled into the pericardium and then flows via the renopericardial duct into the kidney. Here the tubular fluid is modified by selective reabsorption of some electrolytes and organic substances, such as glucose, and by secretion. Thus, some test substances (phenol red, paraminohippuric acid, and creatine) are secreted into the filtrate before the urine is expelled, and there is a close similarity between these renal transport systems and those in vertebrates. Like vertebrate kidneys, the molluscan kidney filters the nonmetabolized test sugar insulin and actively reabsorbs glucose. This reabsorption can be inhibited with phlorizin (Martin and Harrison, 1966; Kirschner, 1967; Riegel, 1972).

In marine molluscs (octopods, *Sepia*) there is little or no reabsorption of NaCl, but in freshwater species (*Anodonta, Viviparus*) all the major ions are reabsorbed (Riegel, 1972). Sulfate and hydrogen are excreted in the kidneys of *Octopus* and other marine species in which the urine is more acid than the blood (Vorwohl, 1961). The kidney of the land pulmonate snails *Helix* and

Achatina can reabsorb water; about 25–50 percent of the perfusate is reabsorbed (Vorwohl, 1961).

Blood pressure, the colloid osmotic pressure of the blood, and intrarenal hydrostatic pressure affect formation of urine in molluscs primarily by controlling filtration rate, much as they do in vertebrates. For example, in the prosobranch gastropod *Viviparus* pericardial fluid formation and urine flow vary directly with arterial blood pressure (Little, 1965), and in the giant pulmonate snail *Achatina* retrograde pressure of 12 cm of water stops urine flow by decreasing filtration (Martin, Stewart and Harrison, 1965).

One important distinction between the molluscan and the vertebrate kidney is the ability of the molluscan kidney cells to form and excrete solid concretions. This is best developed in land pulmonates, in which most nitrogenous waste is excreted in solid form, but the opisthobranchs also can excrete large granules (Potts, 1967). However, molluscs resemble vertebrates in some aspects of nitrogen excretion. For example, aquatic species of molluscs concentrate ammonia in the urine in proportion to the concentration of hydrogen ions and some land forms excrete uric acid. Some aquatic species may also eliminate ammonia through the gill or the digestive gland. In addition, as an apparent compensation for their limited renal tissue, molluscs, particularly land pulmonates, have developed a relatively high tolerance for ammonia (2–5 mM compared to 0.1 mM in cold-blooded aquatic vertebrates). Thus, blood ammonia in some molluscs is 30 times higher than in fish and 1,000 times higher than in mammals (Meglitsch, 1972).

The urinary system of *Aplysia* consists of a single triangular yellow-colored kidney, asymmetrically placed to the left of the visceral hump. The kidney contains the renal sac, a section of the coelom that communicates with the mantle cavity through the *renal pore,* a muscular aperture at the base of the gill through which the renal excretion is discharged to the outside of the animal (Figure 3-16B).

The kidney is a relatively simple, thin-walled sac with walls projecting into the lumen of the cavity as fairly regular parallel lamellae with numerous folds. The folds are covered with epithelium and contain blood spaces (Hyman, 1967). Blood enters the kidney from a dorsally placed afferent renal vein (sinus) that collects from the left anterior part of the ventral abdominal sinus (Figure 3-16B). The afferent renal sinus branches repeatedly over the surface of the renal sac. The blood vessels then rejoin to form a large sinus, the efferent renal sinus, which leaves the kidney on the right side and discharges into the efferent branchial sinus where the latter enters the auricle. The kidney is innervated by the branchial and pericardial nerves (Eales, 1921; Goodrich, 1946).

In *Aplysia,* as in other molluscs, formation of urine presumably begins with the production of a blood ultrafiltrate by its perfusion across the wall of the atrium into the pericardial space (Figure 3-16B). The pericardium itself is glandular and some reabsorption may occur across its wall (Martin and Harrison, 1966). The fluid is drained from the pericardial coelom through the renopericardial duct into the lumen of the kidney (Figure 3-16B). The pumping of the ultrafiltrate from one compartment of the coelom to another may be accomplished by the muscular contraction of the pericardium, which can be regulated by the discharge of the pericardial motor cells located within the abdominal ganglion (Koester and Kandel, 1977). In turn, the secretion of the kidney is excreted into the mantle cavity through the urinary pore. Cells L7 and LC_K contract the kidney, which may aid excretion (Koester and Kandel, 1977; Mayeri, unpubl. observ.).

OSMOTIC REGULATION

Van Weel (1957) found that when *A. juliana* was exposed to 95 percent seawater, a modest hypoosmotic stress consistent with that found in its natural environment, the animals increased in weight to a new steady state, but the increase was somewhat less than 5 percent (Figure 3-18A).[6] This was accompanied by an increase in oxygen consumption, which was at first steep and then declined to 10 percent above control. In the steady state the salinity of the blood in 95 percent seawater was slightly higher than the surrounding medium, indicating that *Aplysia* is capable of osmotic regulation and of maintaining an osmotic gradient, by renal or gill function. That gradient is achieved by expending energy, as is suggested by the increased oxygen consumption. Returning the animal to 100 percent seawater produced a rapid and large initial drop in weight—far greater than would be expected from an osmoadjustor—followed by a gradual return (over six to seven hours) to its original weight. The weight

---

[6]On the basis of experiments using rather radical changes in the osmolarity of seawater, Bethe (1926, 1930, 1934) originally proposed that *Aplysia* was an osmoregulator that reached new levels of osmotic equilibrium passively, without expenditure of energy. Bethe found that if an *Aplysia* was kept in dilute (80 percent) seawater, a rapid increase in weight occurred as a result of an osmotically determined influx of water. This was followed by a decrease in weight to a new equilibrium, six to eight hours later, after some salt had diffused out of the body. If an *Aplysia* was transferred to an isotonic sugar solution, it lost weight by diffusion of salt out of the body so that the blood became hypotonic and water left the body. When the animal was returned to normal seawater it gained weight, sometimes even above its initial level. This gain presumably occurred because in the isotonic sugar solution sugar had diffused into the blood through the skin, making the animal at least temporarily hypertonic to seawater. Upon return to normal seawater the sugar diffused out slowly, causing the animal to take up water (Bethe, 1930). Van Weel (1957) repeated Bethe's experiments and found that these large changes in osmolarity produced irreparable damage to the animal. This observation casts doubt on Bethe's experiments.

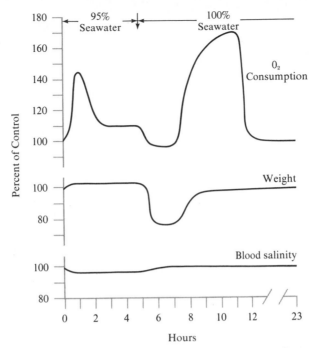

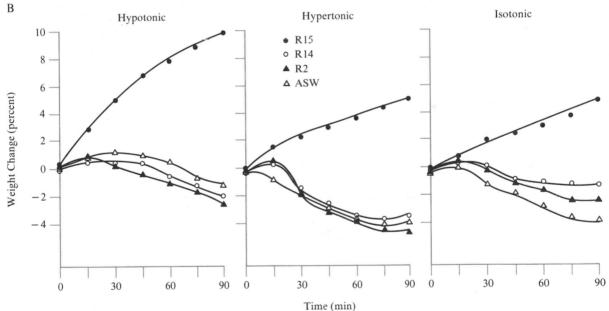

FIGURE 3-18.
Osmotic responses of *Aplysia* to changes in the osmolarity of the surrounding seawater.

A. Changes in weight, consumption of oxygen, and salinity of blood caused by 95 percent seawater and subsequent immersion in 100 percent seawater. [After van Weel, 1957.]

B. Osmotic effects of hormone released by cell R15. Weight change as a function of time after injection of homogenates of cells R15, R14, R2, and artificial seawater (ASW). Animals were maintained in 5 percent hypotonic seawater, 5 percent hypertonic seawater, or isotonic seawater. Each curve represents the mean of five animals. [After Kupfermann and Weiss, 1976.]

loss was associated with a sharp drop in oxygen consumption, which rose sharply, however, when the animal began to gain weight again (Figure 3-18A). The weight gain that gradually follows upon the return to normal seawater cannot be explained by osmotic pressure because at this point the blood and seawater are isoosmotic. As is evident from the increase in oxygen consumption, water seems to be actively taken up and retained, perhaps through the intestinal tract and the kidney.[7]

As described earlier (p. 58), Kupfermann and Weiss (1976) found that osmotic regulation involves a hormone released from cell R15 that may function in a manner analogous to the antidiuretic hormone released by the supraoptic and paraventricular nuclei of the vertebrate hypothalamus (Figure 3-18B). The release of the R15 hormone is inhibited by hypotonic seawater. The inhibition is mediated by an osmosensitive organ, the osphradium (Stinnakre and Tauc, 1969; and see p. 58). Thus, whereas the osmolarity control system of mammals usually functions as a servocontrol feedback system and the hormone is commonly released in response to changes in the *internal* environment, the osmolarity control mechanisms in *Aplysia* seem to function as a reflex and release of the R15 hormone is shut off by changes in the *external* environment.

Digestive System

The feeding habits of molluscs are extraordinarily varied. Almost every type of feeding pattern is encountered: carnivorous, microphageous and macrophageous herbivorous, and ciliary feeding. Ancestral molluscs are thought to have been herbivores that grazed on plants (Chapter 1), and in general the monoplacophorans and polyplacophorans retain this type of feeding pattern. The bivalves feed on filtered particles, primarily small algae, diatoms, and flagellates, whereas the scaphopods and the cephalopods are exclusively carnivorous. Gastropods are both herbivorous and carnivorous. For example, most opisthobranchs are carnivorous, grazing on sessile animals or devouring larger, more

---

[7]Taurine, which is present in high concentration in molluscan tissues, particularly in marine forms, may also play a role in osmoregulation. In the axons of squid and cuttlefish aspartic acid and taurine are the amino acids found in highest concentration (Lewis, 1952). Taurine is completely absorbed by *Aplysia* (Carefoot, 1967c). Lewis suggested that these amino acids may have a dual function: to provide the anions required to balance the internal cations in cells and to supply extra solutes for intracellular osmoregulation that oppose water movement between cells and body fluids (also see Florkin, 1966).

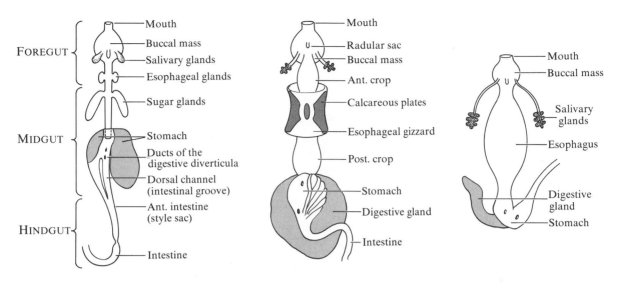

FIGURE 3-19.
The digestive system of various molluscs. [After Owen, 1966b.]

In primitive molluscs, such as the aplacophorans and polyplacophorans, the digestive system consists of a mouth, a buccal mass containing a radula, a tubular esophagus, and an intestine opening posteriorly as an anus. Three glands feed into the system. The salivary glands open into the buccal cavity, the esophageal glands open into the esophagus, and the digestive diverticula (midgut glands) open into the stomach.

In gastropods, such as opisthobranchs and pulmonates, the alimentary canal is similar. The salivary glands open into the buccal cavity and there are no esophageal glands. In many opisthobranchs and pulmonates the esophagus is dilated to form a crop in which food is stored and digested. The crop is often divided into three portions: an anterior crop, a muscular (esophageal) gizzard (with calcareous plates), and a posterior crop. In opisthobranchs and pulmonates there is also a simplification of the gastric region due to increased extracellular digestion.

active forms by esophageal suction; however, some of the more primitive forms are herbivorous. The digestive tract of molluscs can be divided into a foregut, a midgut, and a hindgut (Figure 3-19).[8] Here we will consider first the generalized gastropod digestive system and then that of *Aplysia*.

The gastropod *foregut* is involved in food capture and ingestion; it consists of a buccal-cavity mass—comprising the radula, odontophore, and buccal mass—

[8]The digestive system of molluscs, and particularly of opisthobranchs, has been extensively studied. See Howells (1942), Graham (1949), Forrest (1953), Gascoigne (1956), Fretter and Graham (1962), Hurst (1965), Bidder (1966), Owen (1966a), Rose (1971), Rudman (1971, 1972), Fretter and Peake (1975), and Thompson (1976).

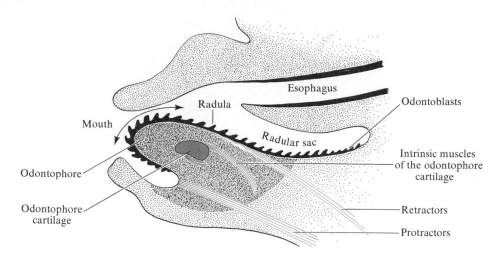

FIGURE 3-20.
The gastropod foregut. Longitudinal section of radula with radular sac. [After Runham, 1963.]

and the esophagus (Figure 3-20). The muscular odontophore is a tongue-like structure supported by cartilage and covered dorsally by the radula, a chitinous tooth-bearing surface. The alternating contractions of the antagonist buccal muscles, which attach to the radula, produce rhythmic movements of the chitinous ribbon. In this way, the radula scrapes food particles and also serves as a conveyor belt to transport the food down the gut. Among opisthobranchs the foregut works in quite different ways to capture prey. In certain pteropods the prey is grasped by various cephalic weapons: hooks, spines, tentacles, and teeth. In some sacoglossans the teeth are dagger-like in action and split open the cell walls of the algae they feed upon, while a pulsatile buccal pump sucks out the cell contents. In other sacoglossans esophageal suction plays a major role in feeding; these animals have a strong band of circular muscles extending from the mouth to various levels of the oral tube wall (Starmühlner, 1956). Esophageal suction is also well developed in the carnivorous cephalaspid opisthobranch *Navanax* (Chapter 8) but plays only a minor role in the feeding of other gastropods. Finally, some mesogastropod prosobranchs use the ciliary mechanism of their enlarged gill to develop feeding currents for suspension feeding (Owen, 1966a). In most molluscs food is moved by cilia or by specific radular swallowing movements from the buccal cavity into the esophagus, where mucus is secreted into the food by paired salivary glands.

The resulting bolus is moved into the *midgut,* first into the esophageal crop and then into the sorting region of the stomach. Smaller pieces are digested here, while larger ones may be crushed in a muscular mill, the triturating stomach. Following reduction, the particles are again sorted. Some are sent to the digestive region and are digested extracellularly by enzymes secreted by the digestive glands and intracellularly by phagocytosis. Some particles are rejected and guided to the intestine. The stomach has a prominent diverticulum, the digestive or midgut gland (Figure 3-19), which originally was paired. This gland secretes extracellular digestive enzymes (amylase, cellulose, protease, and lipases) and is thought to be in part analogous to the vertebrate pancreas (Meglitsch, 1972). The digestive diverticulum of molluscs may also serve as an organ of absorption, phagocytosis, food storage, and excretion (Owen, 1966b).

The *hindgut* or intestine is often long. Although it may participate in solute reabsorption, its main role is in water reabsorption and in the formation of compact feces.

The anatomy and physiology of the digestive system of *Aplysia punctata* and *A. californica* (Figure 3-21) have been studied by Mazzarelli (1893), MacFarland (1909), Eales (1921), Howells (1942), Susswein (1975), and Susswein and Kupfermann (1974). The foregut consists of a buccal mass, which contains the feeding organ, the odontophore, two salivary glands discharging over it, and an esophagus. The midgut consists of a large crop, two (anterior and posterior) gizzards or triturating stomachs, and a true stomach, which receives the ducts of the digestive gland and a cecum. The hindgut is made up of an intestine and a rectum.

FOREGUT

The *buccal mass* is a muscular organ that is partially thrust out during feeding to grasp seaweed. At its anterior end lies the mouth, a narrow slit-like opening that leads into the buccal cavity. The odontophore projects into the posterior end of the buccal cavity. The esophagus arises posteriodorsally where the ducts of the salivary glands enter the buccal mass (Figures 3-21 and 3-22).

Food enters the mouth and is moved over the top of the *odontophore,* which extends dorsally as a V-shaped fold, the two odontophore halves. The odontophore halves carry the *radula,* a broad tooth-bearing ribbon (Figure 3-20), which consists of a membrane in which numerous chitinous teeth are arranged in close and regularly spaced transverse rows. The central tooth (rhochidion) of each row is spine-shaped. The lateral teeth on each side of the central tooth

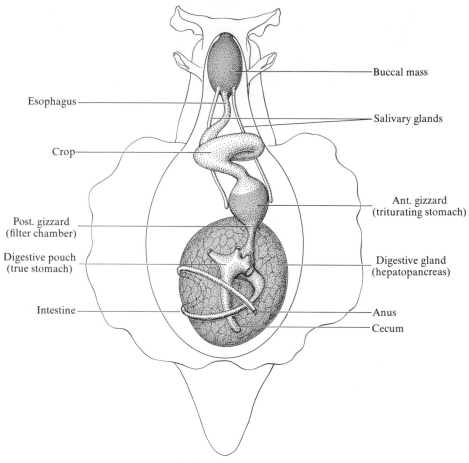

Esophagus

Crop

Post. gizzard
(filter chamber)

Digestive pouch
(true stomach)

Intestine

Buccal mass

Salivary glands

Ant. gizzard
(triturating stomach)

Digestive gland
(hepatopancreas)

Anus

Cecum

FIGURE 3-21.
Dorsal view of the digestive system of *Aplysia.*

are alike and those on each side are mirror images of each other.[9] As the mouth opens to grasp seaweed, the odontophore tips forward until the V-fold points outward. The halves open and close like a pair of jaws to grasp the seaweed (Figure 8-23, p. 305). The odontophore then returns to its original flat shape, and its retraction pulls the seaweed with it. Depending on the size of the ani-

---

[9]The number of teeth in a row is expressed as a formula that has been used as a molluscan taxonomic characteristic. The formula lists the number of each kind of tooth from right to left on each side of the central tooth, the counts being made at the wide point in the radula. Thus, the formula for an adult radula of *A. vaccaria* is 80-1-80, indicating that there are 80 laterals on each side of the central tooth (Winkler, 1957). The formula for *A. californica* is 50-1-50 (Winkler, 1957) and that for *A. punctata* 15-1-15 (Eales, 1921).

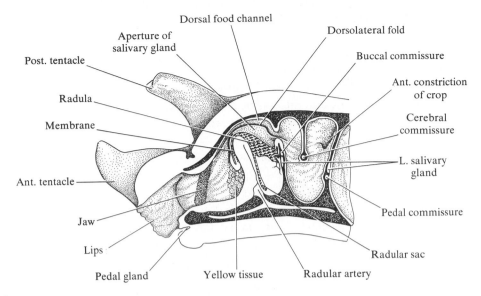

Dorsal food channel

Aperture of
salivary gland

Post. tentacle

Radula

Membrane

Ant. tentacle

Jaw

Lips

Pedal gland

Yellow tissue

Radular artery

Dorsolateral fold

Buccal commissure

Ant. constriction
of crop

Cerebral
commissure

L. salivary
gland

Pedal commissure

Radular sac

A

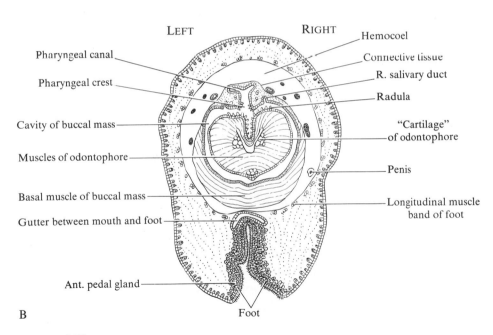

LEFT   RIGHT

Pharyngeal canal

Pharyngeal crest

Cavity of buccal mass

Muscles of odontophore

Basal muscle of buccal mass

Gutter between mouth and foot

Ant. pedal gland

Foot

Hemocoel

Connective tissue

R. salivary duct

Radula

"Cartilage"
of odontophore

Penis

Longitudinal muscle
band of foot

B

FIGURE 3-22.
Mouth and buccal mass of *Aplysia*.

A. Sagittal section through the head region of *Aplysia*. The cut surfaces have been left blank. The body cavity is shown in black. [After Howells, 1942.]

B. Longitudinal section through the head of *Aplysia* showing the interior of the buccal mass. The pharyngeal crests, arising from the walls of the pharyngeal canal, are not shown. [After Eales, 1921.]

mal, a backward thrust of the odontophore pulls in 1 mm or more of seaweed (Howells, 1942). In *A. punctata* the mouth is open during these odontophore movements, but after 1–2 cm of seaweed have been taken into the buccal cavity, the jaws close and the tension produced by the pull of the radular teeth breaks it. These movements are quite rapid and permit the animal to take large pieces of seaweed into its crop.

Feeding in *Aplysia* also involves swallowing, a protraction–retraction sequence distinct from biting in which the odontophore does not protract fully. Touching seaweed to the lips elicits a biting response followed by a series of swallowing movements that pull the food in. The swallowing movement seems to be initiated by receptors in the inner surface of the mouth or buccal cavity (Kupfermann, 1974a). Seaweed then passes into the upper part of the buccal cavity, from which it is conveyed to the esophagus partly by cilia and partly by muscular action. The *Salivary glands* discharge into the buccal cavity (Figures 3-21 and 3-22) and secrete amylase, cellulase, and proteases over the surface of the odontophore (Howells, 1942; Stone and Morton, 1958; Koningsor, McLean, and Hunsaker, 1972).

The esophagus is short, thin, and highly muscular. It extends from the posterior part of the buccal mass, through the circumesophageal complex of nerves and ganglia, and ends in a large distensible, thin-walled crop in which food is stored. The esophagus also contains a sphincter that prevents reflux of food from the gut back to the buccal mass. This sphincter is innervated by the esophageal nerve; cutting the nerve eliminates sphincter contraction (Susswein, 1975).

The buccal mass has two groups of extrinsic muscles, the protractors and the retractors, which move the entire organ, and six intrinsic muscles, some of which operate the odontophore (Eales, 1921; Howells, 1942). The *protractors* insert at various levels in the body wall surrounding the mouth and in the buccal tube and pull the odontophore forward; the *retractors* insert in the body wall posterior to the buccal mass (Hyman, 1967) and pull the odontophore backward (Figure 3-23A). The *intrinsic buccal muscles* are dark red, comparable in color to the red, tonic muscles of mammals. These muscles contain the respiratory pigment myoglobin, a globular monomer protein similar to mammalian myoglobin (Rossi-Fanelli and Antonini, 1957; Antonini and Brunori, 1971).[10] This pigment facilitates the diffusion of oxygen into the muscle (for

[10]Hemoglobin and myoglobin contain heme (iron-containing) proteins. They are able to undergo a reversible reaction with molecular oxygen as a result of their ferroporphyrin prosthetic group. The term "hemoglobin" is used when this respiratory pigment is found in the blood; the term "myoglobin" or "intracellular hemoglobin" is used when the respiratory pigment is found in other tissues, usually muscle or nerve. Hemoglobins and myoglobins differ in the number of units or monomers that make up the protein. Hemoglobins are made up of four units (tetrameric);

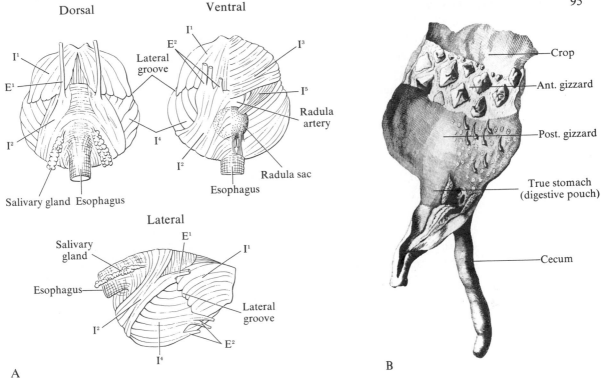

Dorsal

Ventral

I¹

E²

Lateral groove

I³

I¹

E¹

I¹

I⁵

Radula artery

I²

I⁴

I⁵

I²

Radula sac

Salivary gland Esophagus

Esophagus

Lateral

Salivary gland

E¹

I¹

Esophagus

Lateral groove

I²

E²

I⁴

A

Crop

Ant. gizzard

Post. gizzard

True stomach (digestive pouch)

Cecum

B

FIGURE 3-23.
Buccal mass and stomachs of *Aplysia*.

A. Muscles of the buccal mass. In the ventral view the superficial intrinsic muscles I¹ and I² have been cut away to show the underlying muscles (I³, I⁴, and I⁵). E¹ and E² are the extrinsic (protractor) muscles; I¹ to I⁵ are the intrinsic (retractor) muscles. [After Howells, 1942.]

B. Dissection of part of the digestive system, showing the anterior gizzard (triturating stomach), crop, posterior gizzard, and true stomach. [From Cuvier, 1817.]

greater detail see Read, 1966; Wittenberg, 1970; and below). In gastropods the myoglobin of the buccal muscle has a greater affinity for oxygen than does the hemocyanin of molluscan blood and can therefore extract oxygen trans-

---

myoglobins are made up of a single unit (monomeric). (For discussion see Wittenberg, 1970; Antonini and Brunori, 1971.)

The molecular weight of *Aplysia* myoglobin is 16,000 to 18,000. An *Aplysia* myoglobin molecule contains only one histidine residue to which the heme group is attached. This is of interest because all other known myoglobin residues contain, in addition to a proximal histidine to which the heme group is attached, one or more distal histidine residues that are thought to shield or buffer the proximal site. In combining with oxygen and carbon dioxide, *Aplysia* myoglobin again resembles mammalian myoglobin, but its dissociation constant for oxygen is seven times faster. Myoglobin accounts for five percent of the dry weight of the buccal muscle. (For discussion see Wittenberg *et al.,* 1965; Wittenberg, 1970; Antonini and Brunori, 1971.)

ported to the muscle by the circulating respiratory pigment (Manwell, 1960; Rossi-Fanelli and Antonini, 1957; Read, 1966). This might be useful to *Aplysia,* a herbivorous grazer whose buccal muscles are often active for long periods of time. Thus, the myoglobin of the buccal mass may serve as an oxygen bank.

MIDGUT

The esophagus forms a single structural and functional unit with the *crop* (Figure 3-21; Howells, 1942). The crop leads to the *anterior gizzard,* a triturative stomach. This is a short section of gut surrounded by a thick band of circular, red muscle fibers and lined on its inner surface with four or five rows each with 12–20 large pyramidal teeth that compress and grind the contents in the gut (Figure 3-23B). The muscular, thick-walled anterior gizzard leads to the thin-walled *posterior gizzard,* which contains some small, pointed teeth. The posterior gizzard narrows as the digestive tract enters the *midgut digestive gland* (or hepatopancreas) and ends in the *true stomach,* a small digestive pouch. The posterior part of the posterior gizzard is embedded in the digestive gland (Eales, 1921; Marcus, 1953).

Several ducts bring digestive juices into the stomach from various parts of the digestive gland. In addition, a long tube-like diverticulum secretes the mucous strings that tie the fecal pellets together in the digestive pouch. The blind end of the *cecum* is located at the posterior margin of the digestive gland.

Digestion begins when salivary secretions act on food in the buccal mass. Digestion continues in the crop where secretions from the digestive diverticula in the stomach are regurgitated by contraction of the stomach walls. The digestive gland secretes the main digestive enzymes for extracellular digestion. But it also carries on intracellular digestion and is the main site of absorption. Moreover, it may also participate in excretion. Its digestive cells are thought to be capable of secretion, digestion, absorption, and excretion (Hyman, 1967).

HINDGUT

Ciliary currents convey pieces of seaweed directly from the gizzard to the intestine. The fluid of the gut, however, is drawn into the stomach through a filter of cilia. The partly digested food emerges from the cecum as a compact rod and is carried across the posterior wall of the stomach into the intestine, which loops one and a half times around the digestive gland in which it is embedded (Figure 3-21). Finally, the fecal mass is consolidated and cemented by the secretion of the intestinal glands. The whole mass is propelled toward the *anus* by the combined action of cilia and muscles. A fecal string is ejected by means of the muscular activity of the anal walls. The contraction of the anal sphincter separates this fecal string into pellets (Howells, 1942).

Reproductive System

Most molluscan species are sexually dimorphic, but pulmonates and opisthobranchs are usually hermaphroditic (Figure 3-24). In sexually dimorphic molluscan species the males and females usually have a similar appearance, although the female is somewhat larger than the male. The eggs are sometimes cast directly into the sea for fertilization, but in opisthobranchs fertilization is internal and eggs are sheltered in the female and then deposited in gelatinous masses.

In hermaphroditic molluscs the ova and sperm are generally produced in the same follicle of the ovotestis (gonad). The ova pass through the common hermaphroditic duct (gonoduct) to the seminal receptacle (receptaculum seminis), a pouch where sperm from another individual (allosperm) is stored. Following fertilization, the ovum is coated in the albumen and mucous glands. It then continues along the female portion of the hermaphroditic duct to a thick-walled muscular vagina that discharges into the mantle cavity (Figure 3-24; Fretter and Graham, 1964; for a detailed comparative discussion of opisthobranchs see Ghiselin, 1965).

In *Aplysia californica* the ovotestis, called the *hermaphroditic gland,* is an orange-yellow (sometimes greenish) gland embedded in the posterior dorsal surface of the brownish midgut gland (Eales, 1921; Thompson and Bebbington, 1969; Coggeshall, 1970). The ovotestis is made up of a large number of follicles, each of which opens into a division of the *small hermaphroditic duct* (Figure 3-25). The connective tissue and muscle cells of the small hermaphroditic duct contain many small neurons (Coggeshall, 1972). The follicles are bordered by Ansel's membrane, which consists of the basal lamina, or basement membrane, of the follicular epithelium and some associated muscle cells.

The fate of newly developed autosperm (the animal's own sperm) and its path out of the male component of the reproductive tract have been studied by Beeman (1970) in the closely related aplysiid *Phyllaplysia taylori* using radio-labeled thymidine. Spermatogenesis takes about 10 days. The autosperm formed in the ovotestis is collected and transported to the *ampulla* (or seminal vesicle), the main part of the small hermaphroditic duct (Figure 3-26). Here

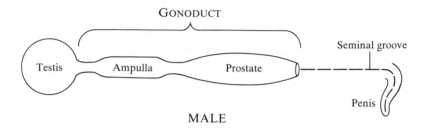

MALE

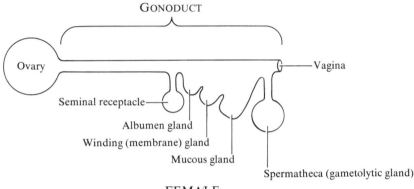

FEMALE

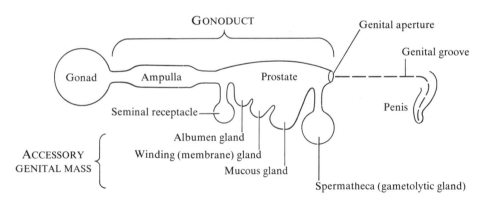

HERMAPHRODITIC

FIGURE 3-24.
Main features of the opisthobranch reproductive system. Relationship between the male and female systems and the hermaphroditic system thought to be characteristic of the opisthobranch ancestor. The hermaphroditic system is also fairly characteristic of molluscs in general. [After Ghiselin, 1965.]

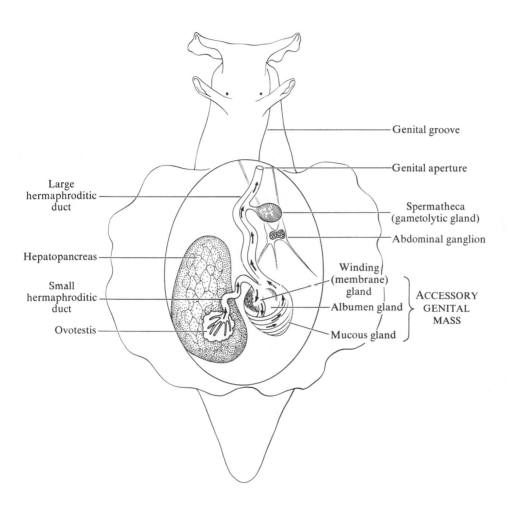

Genital groove

Genital aperture

Large
hermaphroditic
duct

Spermatheca
(gametolytic gland)

Abdominal ganglion

Hepatopancreas

Winding
(membrane)
gland

ACCESSORY
GENITAL
MASS

Small
hermaphroditic
duct

Albumen gland

Ovotestis

Mucous gland

FIGURE 3-25.
The reproductive system of *Aplysia californica*.

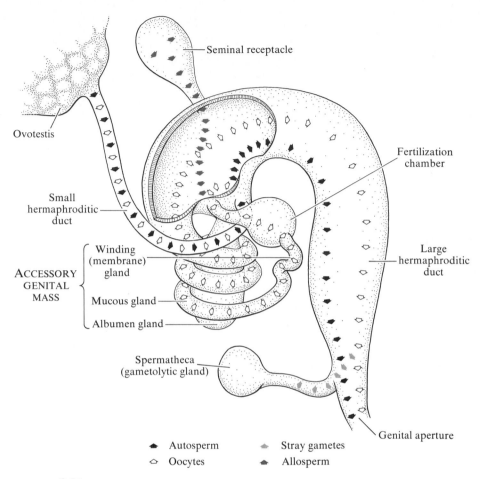

FIGURE 3-26.
The flow of gametes in the reproductive system of *Aplysia californica*. Arrows show the course of the outgoing gametes (autosperm and oocytes) during copulation and egg-laying (which may occur simultaneously) as well as the course of allosperm and stray gametes. [After Blankenship, unpublished.]

mature autosperm is stored prior to copulation. The small hermaphroditic duct is attached to the surface of the mucous and albumen gland complex and then loops around it to form the *internal autospermal groove* (spermatic duct) of the *large hermaphroditic duct*. Unlabeled sperm in the seminal vesicle are totally replaced within 30 days by thymidine-labeled sperm. Since 10 days are required for spermatogenesis, animals allowed to copulate retain sperm in the ampulla less than 20 days (Figure 3-27). During copulation peristaltic waves

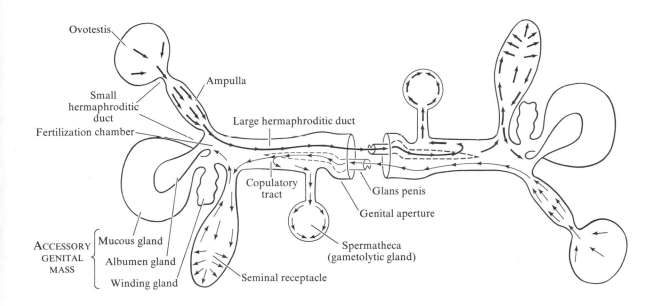

FIGURE 3-27.
The reproductive systems of a radio-labeled *Phyllaplysia taylori* (left) and an unlabeled animal (right) during copulation. The dark arrows indicate the course of the labeled sperm, the light arrows the unlabeled sperm. Not drawn to scale. [After Beeman, 1970.]

of contraction force sperm from the seminal vesicle into the internal auto-spermal groove (Figure 3-28). The internal autospermal groove leads past the *spermatheca*, or gametolytic gland, to the *external autospermal groove.* At the junction with the spermatheca, the prostate gland adds its secretions to the sperm, which then travels to the anterolateral body wall of the penis.

In *A. californica* the penis is usually retracted into an inverted U by muscles that attach to its sheath and lie at its base. The muscles working together accomplish the erection and extension of the penis during copulation (Figure 3-28). Allosperm injected into the copulatory tract of the large hermaphroditic duct of the female will move into the *seminal receptacle* (spermatocyst) near the end of copulation. Here the allosperm are stored until egg-laying, when the muscular contraction of the seminal receptacle ejects the allosperm and brings them into contact with the oocytes.

Allosperm in the seminal receptacle are mobile, whereas autosperm in the small hermaphroditic duct are not. It is thought that only sperm that are mobile

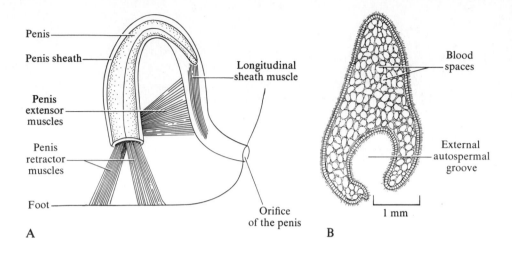

Penis

Penis sheath

**Penis
extensor
muscles**

Penis
retractor
muscles

Foot

Longitudinal
sheath muscle

Orifice
of the penis

A

Blood
spaces

External
autospermal
groove

1 mm

B

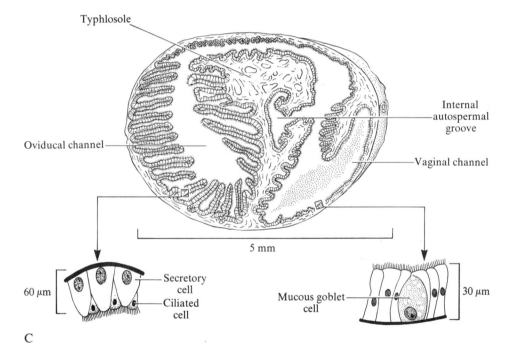

Typhlosole

Oviducal channel

Internal
autospermal
groove

Vaginal channel

5 mm

60 μm

Secretory
cell

Ciliated
cell

Mucous goblet
cell

30 μm

C

FIGURE 3-28.
Copulatory apparatus of *Aplysia*.

A. The muscles controlling the extrusion of the penis of *A. californica* during copulation. [After Winkler, 1957.]

B. Section through the penis of *Aplysia fasciata*, near the tip, showing the external autospermal groove and the central area, rich in blood-spaces.

C. Section through the reproductive organs of *Aplysia fasciata*, showing the division of the hermaphroditic duct into three channels: the vaginal and oviductal channels and the internal autospermal groove. In this species the internal autospermal groove is nonglandular. Insets show the endothelium in different parts of the duct. [After Thompson and Bebbington, 1969.]

can fertilize, and this may explain the lack of self-fertilization in *Aplysia,* even though ova and autosperm meet in the small hermaphroditic duct (Coggeshall, 1972). Sperm activation seems to occur only after copulation (Thompson and Bebbington, 1969). Within five hours of copulation some of the allosperm have buried their heads in the lining of the seminal receptacle. Excess allosperm now move to the spermatheca where they are destroyed (Figures 3-26 and 3-27).

Mature oocytes are generally released from the ovotestis during the night (Thompson and Bebbington, 1969), probably as a result of contraction of the extrinsic muscles of the follicle (Thompson and Bebbington, 1969; Coggeshall, 1970). The ova proceed at a speed of 0.1 mm/sec through collecting tubules to the seminal vesicle, where they undergo their first meiotic division. Division stops after the breakdown of the germinal vesicle membrane. The ova are propelled further (at a rate of 0.2 mm/sec) by ciliary movement through the mass of nonmobile autosperm in the seminal vesicle to the small hermaphroditic duct. From here they are conveyed (at a rate of 1 mm/sec) to the *albumen gland,* which pours a viscous secretion over them (Eales, 1921; Thompson and Bebbington, 1969; Coggeshall, 1972). In the albumen gland the ova encounter great quantities of active allosperm that have been extruded from the nearby seminal receptacle. Peristaltic movements of the albumen gland force the ova and sperm into the *fertilization chamber;* a sphincter muscle at the entrance presumably acts as a valve that allows only one or a few ova to enter at any time (Thompson and Bebbington, 1969). From the fertilization chamber the mixture of gametes (ova and allosperm) enter the *winding* (membrane) *gland.* In *A. californica* the winding gland consists of 5–10 blind, closely packed coiled tubules that project from the fertilization chamber. Coggeshall (1972) considers them glandular extensions or diverticulae of the fertilization chamber. The ova swirl into and out of these tubes and back into the fertilization chamber, where fertilization begins. They then leave the winding gland to enter the *white mucous gland,* two spiral tubes (each divided into an ascending and descending channel) that encircle the *albumen gland* (Coggeshall, 1972). As the fertilized eggs enter the mucous gland, they are placed into egg capsules. In *A. californica* the number of eggs placed in a capsule varies as a function of age (P. Kandel and T. Capo, personal communication). Small young animals have only 3–10 ova per capsule; larger animals have up to 100 per capsule.

As the egg capsules progress in the mucous gland, a long egg strand, or *cordon,* is formed of many capsules. The substance of the cordon becomes thicker and harder by the addition of large amounts of jelly as the cordon progresses through the gland. The strand enters the ascending channel of the first spiral of the mucous gland, passes to the apex of the gland, and returns to the base through the descending channel. It then enters the ascending channel of the second spiral, ascends to the apex once again, and returns through the descending channel of the second spiral. Eventually it enters the oviductal channel of the large hermaphroditic duct, which runs parallel to and is effectively separated from the internal autospermal groove (Figure 3-28C). In the large hermaphroditic duct an adhesive secretion is added that facilitates the attachment of the egg strip to the substrate. The egg strand is released from the *genital aperture* and moves along the external genital groove to be deposited on the substrate by the animal's weaving head movements. Spawn usually emerge at approximately 0.5 cm per minute.

The distal portion of the large hermaphroditic duct contains, in addition to the oviductal channel and internal autospermal groove, a third, vaginal channel which receives the penis of another animal during copulation (Figure 3-28C). This vaginal channel leads to the seminal receptacle, a blind diverticulum of the large hermaphroditic duct, where the allosperm are stored (Figure 3-26).

The ovotestis of *A. punctata* contains a variety of steroid hormones: pregnenolone, progesterone, 17α-hydroxyprogesterone, testosterone, and oestradiol (Lupo di Prisco, Dessi' Fulgheri, and Tomasucci, 1973). The function of these hormones is unknown.

INNERVATION OF EFFECTOR STRUCTURES

The effector organs of molluscs involve a variety of muscle types. As we have seen in *Aplysia,* the muscle types range from those of the buccal mass, a red muscle made up of smooth, obliquely striated, muscle fibers (Rosenbluth, 1972 and unpubl. observ.), to rhythmically active muscle fibers of the heart, the various muscles of the gill veins, the muscle sphincters of the aorta, the longitudinal and transverse muscles of the foot, as well as the completely non-oriented muscles of the siphon and body wall.

Unlike the well-studied insects and crustacea, neuromuscular physiology in most molluscs has not been extensively explored. The best-studied cases are the byssus retractor muscle of the clam *Mytilus* (Twarog, 1967), the buccal muscle of the pulmonate *Helisoma* (Kater, 1974), and the studies of *Aplysia:* on the gill (Carew *et al.,* 1974), heart (Mayeri *et al.,* 1974), aortic sphincter (Mayeri *et al.,* 1974), and buccal mass (Cohen *et al.,* 1974, 1978; Orkand and Orkand, 1975).

From the studies of *Aplysia* a number of findings have emerged. First, many muscles in *Aplysia* consist of muscle fibers that are electrically coupled to one another. Second, the muscle fibers are typically (in the buccal mass, gill, and aortic sphincter) *polyneuronally innervated,* i.e., different motor neurons, using different transmitters, synapse on each fiber. Third, there is peripheral inhibition (in the gill, heart, and buccal mass). Fourth, many organs are innervated by both central motor neurons, whose cell bodies are located in central ganglia, and by peripheral motor neurons, whose cell bodies lie close to the organ innervated. The relation of the central to the peripheral innervation in *Aplysia* will be further considered in Chapter 6.

SUMMARY AND PERSPECTIVE

Aplysia has the characteristic molluscan morphotype: a head, foot, mantle, and visceral mass. The head contains an anterior and posterior pair of tentacles. Both pairs are chemosensory; the anterior pair participates directly in feeding. The head also contains two small eyes; these are not well developed but they have a circadian rhythm of their electrical activity. The remaining sensory apparatus of *Aplysia* consists of an osmoreceptor organ (the osphradium), statocysts, and mechanoreceptors. The osphradium is involved in sensing the osmolarity of the seawater bathing the mantle cavity. Body position and acceleration are sensed by the symmetrical statocyst organs. The entire body surface, the buccal mass, and presumably the anterior gut contain mechanoreceptor structures that are sensitive to tactile as well as noxious stimuli.

The foot is long and extends posteriorly from the mouth, along the whole length of the animal and projects posteriorly as a short tail, or metapodium. The mantle region consists of the mantle cavity and its organs, the gill, osphradium, and siphon, and the openings of the opaline gland, the alimentary, urinary, and reproductive tracts. On the edge of the mantle shelf is the purple gland, which secretes a dark ink when the animal is disturbed. The opaline gland releases a white secretion of unknown function. The shell is reduced and vestigial.

The visceral mass contains most of the gastrointestinal tract, the heart, and the urinary and reproductive systems. The digestive system consists of (1) a buccal mass (which carries an odontophore); (2) two salivary glands that discharge into the buccal mass; (3) an esophagus that dilates into a large crop; (4) an anterior gizzard or triturating stomach and a posterior gizzard; (5) a true stomach that receives the ducts of the digestive gland; (6) a cecum; (7) an intestine; and (8) a rectum.

The cardiovascular system of *Aplysia* consists of a heart, a closed arterial system, and an open venous system. The heart has two chambers—an auricle and a ventricle. It is myogenic and its rate and strength of contraction are under neural control.

The urinary system consists of a single asymmetrically placed kidney. Filtration presumably occurs across the walls of the auricle into the pericardial cavity. *Aplysia* has some capability for osmotic regulation. An identified neuroendocrine cell, R15, releases a hormone that may be analogous to the antidiuretic hormone released by the supraoptic and paraventricular nuclei of the vertebrate hypothalamus. This hormone participates in regulating water reabsorption. The osphradium mediates a hypoosmotic reflex leading to an inhibition of the secretion of the water-retaining hormone released by R15.

Aplysia is hermaphroditic; any individual can serve as male or female for any other, but an animal cannot copulate with itself.

Substantial progress has recently been made in understanding the nature of several sensory receptors that transmit information from the animal's environment to its central nervous system, especially the eyes, statocyst, osphradium, and some mechanoreceptors. What is needed in the immediate future is a better understanding of the physiology of *Aplysia's* organ systems. In some areas, such as regulation of circulation and respiration, the developing neurophysiological insights are simply not being matched by adequate functional understanding. Here knowledge of physiological operation is required to further understand the function of neural regulation. Moreover, systematic studies of the comparative anatomy and physiology of the various *Aplysia* species would be a valuable background for subsequent studies of comparative behavior among *Aplysia* species and their anaspid relatives.

SELECTED READING

Wilbur, K. M., and C. M. Yonge, eds. 1964; 1966. *Physiology of Mollusca,* 2 vols. New York: Academic Press. A good reference work.

Thompson, T. E. 1976. *Biology of Opisthobranch Molluscs,* Vol. I. London: Ray Society. A discussion of modern systematics. Contains a good review of molluscan reproduction.

CHAPTER 4

Aplysia Among the Molluscs II
The Nervous System

Throughout the animal kingdom the more advanced phyla tend to have greater learning capabilities than more primitive phyla, as well as larger and more concentrated nervous systems. Are the increased learning capabilities related to the size of the brain (and the greater number of nerve cells), to the concentration of ganglionic masses (and different types of interconnectivity), or to both? This fascinating question can probably best be approached by comparative studies within a phylum. The evolutionary development of the brain within a phylum parallels to a remarkable degree the evolutionary development across phyla.

This is clearly evident in a large phylum, such as Mollusca. Molluscs range from small, simple wormlike forms to large, highly complex ones. The most primitive molluscs lack a ganglionated central nervous system and have restricted behavioral capabilities, whereas the most advanced molluscs have a highly concentrated central nervous system and remarkable learning capabilities. Similar but less dramatic evolutionary developments in the morphology of the nervous system occur *within* each molluscan class, within each subclass, and even within many orders. The existence of this evolutionary progression among closely related genera provides an opportunity for experimentally analyzing the adaptive significance of such a progression and for studying the processes by which it occurs.

An understanding of the gross plan of the nervous system of a species is also essential for a neurobiological analysis of behavior. One needs to know the pathways by which sensory information enters the nervous system and the connections used for conveying information between ganglia. One must also establish which ganglia control the various behavioral responses and the pathways by which motor commands to effector organs leave the central nervous system. Finally, in molluscs, one must separate the contributions to behavior of the central and peripheral nervous systems.

In this chapter I shall first review the major trends in the comparative anatomy of the molluscan central nervous system. Next, I shall focus on the central nervous system of *Aplysia californica*. Finally, I shall describe some interspecific variation in the plan of the *Aplysia* nervous system and some interfamilial variation in the plan of the anaspid nervous system. The progression within the order Anaspidea, and to a somewhat lesser degree even within the genus *Aplysia*, shows the trends that characterize the evolution of the nervous system across phylogeny: *concentration*, the tendency toward ganglionic fusion; and *cephalization*, the anterior migration of what in more primitive forms are posterior ganglionic masses.

THE MOLLUSCAN CENTRAL NERVOUS SYSTEMS

The nervous system of molluscs ranges greatly in complexity and number of central neurons (Figure 4-1).[1] The most primitive molluscs, the monoplacophorans, have simple cordlike nervous systems comparable in structure to the ladder-like nerve cords of flatworms (Figure 4-2). The most advanced molluscs, the cephalopods, have complex brains comparable in size and capability to those of lower vertebrates.

In the primitive molluscs—the monoplacophorans, polyplacophorans, and aplacophorans—the nervous system consists of two pairs of longitudinal cords, one lateral and one ventral, cross-connected by numerous commissures (Figure 4-2). The nervous system of the polyplacophorans is particularly primitive and lacks central ganglia. By contrast, in the monoplacophorans (*Neopilina*) and aplacophorans the cerebral commissure is thickened and becomes the locus for cerebral ganglia. Except for the cerebrals, however, the nerve cells are not generally clustered together into ganglia, but are distributed throughout the nerve cords as they are in the polyplacophorans. The polyplacophorans, apla-

---

[1]For an excellent treatment of the molluscan nervous system see Hoffmann (1939) and Bullock and Horridge (1965).

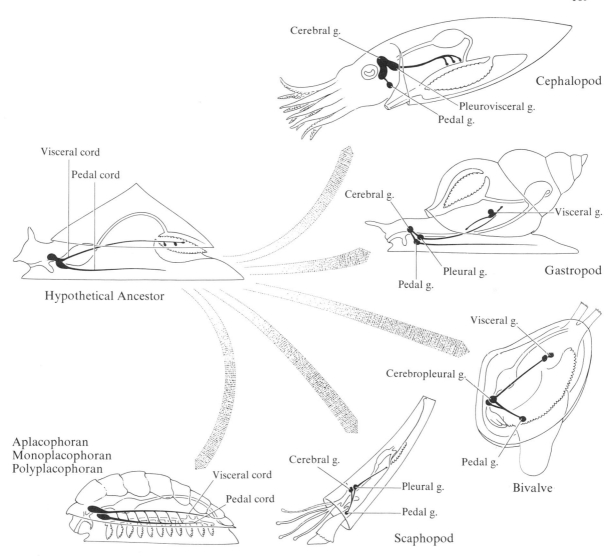

FIGURE 4-1.
The nervous systems of the seven classes of molluscs and a hypothetical ancestor in outline. There is a gradual tendency in these nervous systems toward (1) ganglion formation; (2) ganglionic fusion; and (3) anterior migration of the ganglia toward the head. [After Naef, 1911.]

cophorans, and monoplacophorans also lack complex sense organs; some species lack even statocysts.

The bivalves (*Pelecypoda*), a sedentary class of filter feeders, also lack complex sense organs; they even lack a head (Figure 4-3). Their nervous system is bilaterally symmetrical, consisting of three relatively small paired ganglia—

110

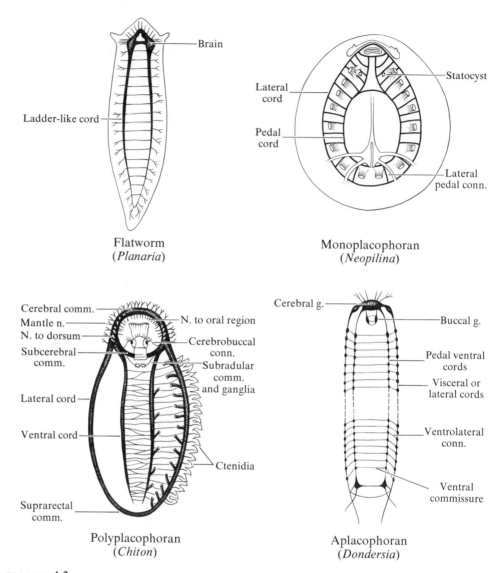

Flatworm
(*Planaria*)

Monoplacophoran
(*Neopilina*)

Polyplacophoran
(*Chiton*)

Aplacophoran
(*Dondersia*)

FIGURE 4-2.

Comparison of the nervous systems of primitive molluscs with that of flatworms. The nervous system of the monoplacophorans and the polyplacophorans has regular and complete commissures and very few central ganglia. The aplacophoran nervous system also has regular and complete commissures. In addition, there are clear ganglia, of which the cerebral and buccal are particularly prominent. [After Buchsbaum, 1965; and Bullock and Horridge, 1965, based on Wirén, 1892; and van Lummel, 1930.]

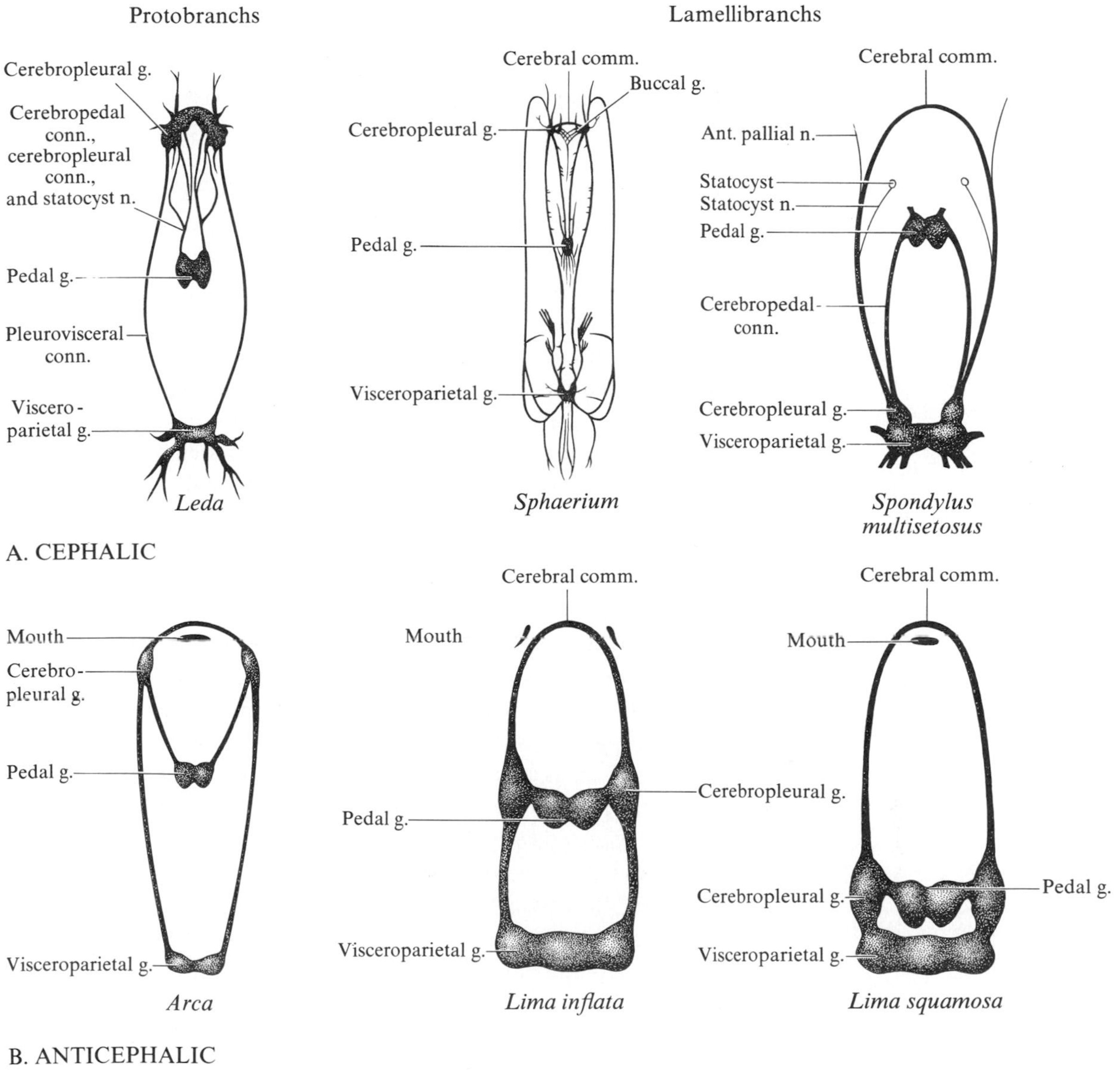

Protobranchs | Lamellibranchs

A. CEPHALIC

Leda — Cerebropleural g.; Cerebropedal conn., cerebropleural conn., and statocyst n.; Pedal g.; Pleurovisceral conn.; Viscero-parietal g.

Sphaerium — Cerebral comm.; Buccal g.; Cerebropleural g.; Pedal g.; Visceroparietal g.

Spondylus multisetosus — Cerebral comm.; Ant. pallial n.; Statocyst; Statocyst n.; Pedal g.; Cerebropedal conn.; Cerebropleural g.; Visceroparietal g.

B. ANTICEPHALIC

Arca — Mouth; Cerebro-pleural g.; Pedal g.; Visceroparietal g.

Lima inflata — Cerebral comm.; Mouth; Cerebropleural g.; Pedal g.; Visceroparietal g.

Lima squamosa — Cerebral comm.; Mouth; Cerebropleural g.; Pedal g.; Visceroparietal g.

FIGURE 4-3.
The bivalve nervous system.

A. Cephalic ganglionic concentration in bivalves. A comparison of the protobranch *Leda* [after Stempell, 1898] and the lamellibranchs *Sphaerium* (*Musculium*) [after Willigen, 1920] and *Spondylus multisetosus* [after Bullock and Horridge, 1965].

B. Anticephalic ganglionic concentration in bivalves. A comparison of the protobranch *Arca* and the lamellibranchs *Lima inflata* and *Lima squamosa*. Note the progressive migration of the head ganglia caudally toward the large visceroparietal ganglia. [After Bullock and Horridge, 1965.]

the cerebrals, pleurals (generally fused) and pedals—and the relatively large paired visceroparietals. In bivalves one begins to see the concentration and fusion of the three paired ganglia typical of all higher orders of molluscs. The main head ganglia tend to move away from the cerebrals and to fuse with the large viscerals.

The gastropods show the greatest diversity in nervous structure among the molluscan classes. Primitive prosobranch gastropods, such as *Haliotis*, resemble primitive monoplacophorans in having two long nerve cords running anteroposteriorly: a visceral cord comparable to the lateral cord and a pedal cord comparable to the ventral cord (Figure 4-4). However, primitive gastropods have fewer commissures than do monoplacophorans, and the cells are almost entirely clustered in paired ganglia; the connectives are largely free of nerve cell bodies.

Some of the most distinctive differences in the nervous systems of gastropods arise from torsion, detorsion, and ganglionic fusion. In torsion the visceral cords are crossed in a figure eight (also called *streptoneury* or *chiastoneury*) causing the supraintestinal (or supraesophageal) ganglion to be on the left and the subintestinal (or subesophageal) ganglion to be on the right (Figure 4-4; see also Figure 1-6, p. 19). With detorsion or shortening of the connectives (also called *euthyneury* or *orthoneury*), bilateral symmetry is partially regained.[2] The connectives become uncrossed and the subintestinal ganglion lies on the left, close to the left pleural, whereas the supraintestinal ganglion is on the right, near the right pleural. Actual detorsion is encountered only in opisthobranchs. Pulmonates do not detort; they regain symmetry primarily by a shortening of the connectives and ganglionic fusion (Figure 4-6). Since torsion is believed to have occurred early in gastropod evolution, detorsion and shortening of connectives with ganglionic fusion are thought to have occurred later and to be secondary to torsion.

Of the three subclasses of gastropods, the prosobranchs are the most primitive; the opisthobranchs and pulmonates probably originate from them (Fretter and Graham, 1962; Brace, 1977a, c). Crossed connectives (streptoneury) are largely limited to prosobranchs; they occur in almost all prosobranchs (Figure 4-4) but only in a few primitive opisthobranchs and pulmonates (Figures 4-5 and 4-6).

Although some prosobranch subclasses are more primitive than others, each subclass contains species that vary greatly in complexity from primitive to fairly evolved. Moreover, all subclasses share common mechanisms for achiev-

[2]Several aspects of the nervous system as well as most of the visceral mass remain nonsymmetrical bilaterally.

Archeogastropod

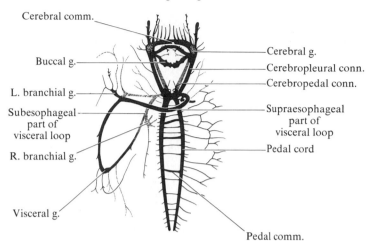

Haliotis tuberculata

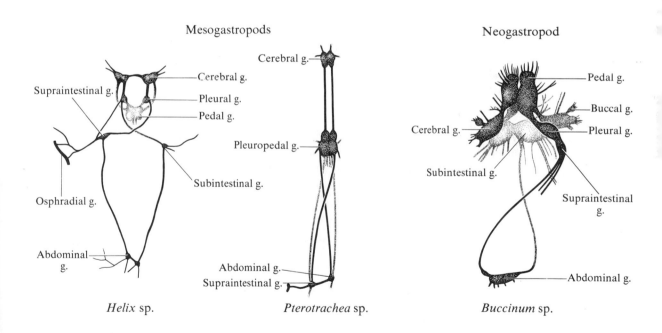

FIGURE 4-4.
The torted nervous systems of representative prosobranchs. A comparison of the archeogastropod *Haliotis tuberculata* [after Crofts, 1929], the mesogastropods *Helix* sp. and *Pterotrachea* sp. [after Spengel, 1881], and the neogastropod *Buccinum* sp. [after Bullock and Horridge, 1965].

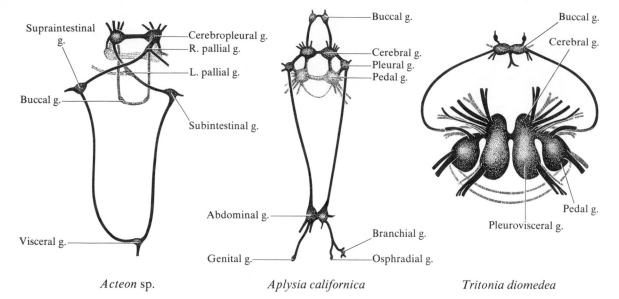

FIGURE 4-5.

The opisthobranch nervous system. The nervous system of the primitive cephalaspidean *Acteon* sp. shows torsion characteristic of prosobranchs [after Bouvier, 1891]. The nervous system of the more advanced anaspid *Aplysia californica* is already detorted [after Spengel, 1881]. The still more advanced nudibranch *Tritonia diomedea* shows the high degree of ganglionic concentration characteristic of higher opisthobranchs [after Willows, 1973].

ing advanced levels of organization. Most striking is the independent tendency, within each subclass, toward a more concentrated nervous system. This is usually paralleled by the cephalization of the circumesophageal ring of ganglia resulting from an anterior migration of the posterior (parietal, pedal, and pleural) ganglia and their ultimate fusion with the cerebrals, and by a concomitant concentration of the ganglia of the visceral loop. The greatest concentration and ganglionic fusion among the prosobranchs occurs in the neogastropods (Figure 4-4). As we shall see later, extensive ganglionic concentration also occurs in the nudibranchs and the stylommatophoran pulmonates. Although cephalization and concentration might appear to be required for more complex or rapid neural coordination, the highest degree of cephalization and concentration in prosobranchs is found in small forms, suggesting that these processes represent, in part, an adaptation to size. But there seems to be little doubt that when migration occurs it is toward the scene of action. Thus, when ganglionic migration and concentration occur in bivalves, where the head is poorly de-

Basommatophorans

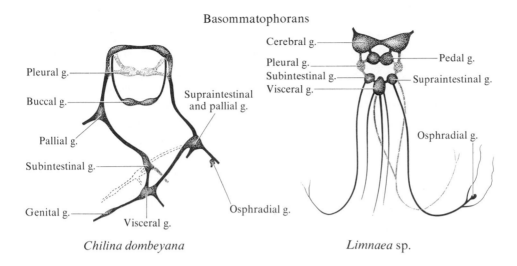

Pleural g.

Buccal g.

Pallial g.

Subintestinal g.

Genital g.

Visceral g.

Supraintestinal and pallial g.

Osphradial g.

Chilina dombeyana

Cerebral g.

Pleural g.

Subintestinal g.

Visceral g.

Pedal g.

Supraintestinal g.

Osphradial g.

Limnaea sp.

Stylommatophoran

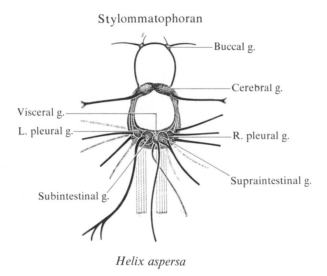

Buccal g.

Cerebral g.

Visceral g.

L. pleural g.

R. pleural g.

Supraintestinal g.

Subintestinal g.

Helix aspersa

FIGURE 4-6.
The pulmonate nervous system. The nervous system of the primitive basommatophoran *Chilina dombeyana* shows intermittent torsion and detorsion and thus resembles the torsion of prosobranchs (see Figure 4-4). The dotted line is the position *in situ,* seen from above; torsion (chiastoneury) depends on the position of the soft body parts [after Haeckel, 1913]. In the more advanced stylommatophoran pulmonates, e.g., *Helix aspersa,* the ganglia are characteristically concentrated around the esophagus *Limnaea* [after Spengel, 1881; *Helix* after Bullock and Horridge, 1965].

veloped or altogether lacking, it is anticephalic, away from the cerebrals and toward the massive visceral ganglia (Figure 4-3).

Among opisthobranchs, crossed connectives are present only in the most primitive order, Cephalaspidea, e.g., *Acteon* (Figure 4-5), which therefore resemble the mesogastropod prosobranchs (Figure 4-4). Some primitive anaspidean genera e.g., *Akera,* are also torted. But the more advanced anaspidean genera, e.g., *Aplysia,* and all other orders of opisthobranchs are partially or fully detorted.

In the more advanced opisthobranchs the nervous system is nearly symmetrical. The only remnant of torsion is the unequal size of some peripheral nerves that results from the unequal development of peripheral structures. In addition, there is a tendency for the abdominal (or pleurovisceral) ganglion to lie away from the midline. Also, the pallials, a pair of ganglia absent in prosobranchs, are found in the visceral loop of opisthobranchs between the pleurals and the intestinals (Figure 4-5).[3]

Pulmonates have the most consistently concentrated gastropod nervous system (Figure 4-6). Typically, it consists of a thick circumesophageal ring of paired cerebral, buccal, pedal, pleural, and a single visceral ganglion that represents a fusion of the subintestinal (left parietal) and visceral ganglia. There is usually no distinct supraintestinal (right parietal) ganglion and no crossing of connectives. One exception is the genus *Chilina,* which has the clearest vestiges of its prosobranch ancestry. Its connectives can cross in a figure-eight and then uncross with normal changes in the position of the viscera (Figure 4-6).

In addition to cephalization, the history of gastropod phylogeny is characterized by progressive changes in the position of the abdominal ganglia and, in higher forms, their ultimate fusion. Torsion twists the pleurovisceral connectives and brings the two intestinal ganglia forward so that the ganglion on one side is in line with the pallial ganglion of the other side. But torsion does not alter the rest of the nervous system (Figure 1-6, p. 19). Therefore, as might be expected, all attempts at regaining symmetry—whether by detorsion, by shortening of connectives, by a zygosis (a process postulated for pulmonates that is thought to lead to the development of a new connection between the pleural ganglion and the subintestinal ganglion of the same side), or by ganglionic fusion—primarily affect the several ganglia of the visceral loop. As a result, different species with similar head ganglia vary considerably in the location

---

[3]The pallial ganglia are additional ganglionic masses found only in primitive opisthobranchs and a few pulmonates. These often fuse with other ganglia. For example, in *Akera* the right pallial has fused with the supraintestinal and the left is thought to have fused with the left pleural (Brace, 1977b).

Order Sepiida

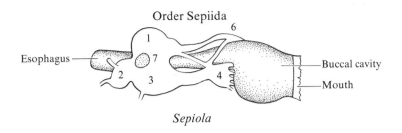

Sepiola

Order Teuthidida

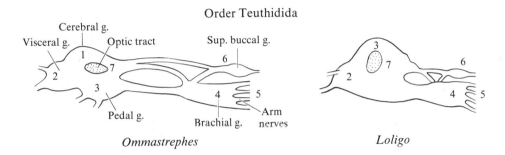

Ommastrephes   *Loligo*

Order Octopoda

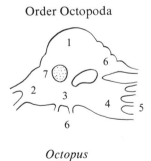

Octopus

FIGURE 4-7.
The cephalopod central nervous system showing progressive degrees of condensation.
[After Pelseneer, 1888.]

and degree of torsion of their abdominal ganglia. One example, which we will consider later, is found in cephalaspidean and anaspidean opisthobranchs. These animals have fused the several ganglia of the abdominal complex in various ways. The tendency toward fusion is evident among all opisthobranchs, and progressive concentration occurs both by shortening of the connectives and fusion of the various ganglia of the visceral loop. The visceral ganglion is usually the last to fuse with others. A parallel example is found in the different patterns of condensation of the abdominal ganglionic complex in pulmonates.

Gastropod phylogeny, as revealed by the variations in the ganglia of the abdominal complex, is useful to the evolutionary biologist interested in the relationship between the nervous system and behavior as a function of different environmental pressures. For example, it would be interesting to know what behavioral advantages derive from various patterns of fusion. How are the basic circuits mediating a given behavior (for example, defensive withdrawal of the siphon and gill, egg-laying, and inking) modified by torsion, detorsion, or ganglionic fusion? Such studies could take advantage of the rich evolutionary parallelism of molluscs by relating the evolution of neuronal function to the evolution of behavior in a closely related group of animals that differ subtly yet significantly in nervous organization. In particular, these studies might clarify what advantages (if any) are offered by ganglionic concentration and fusion, two processes that are so characteristic of the evolution of the nervous system of vertebrates.

The molluscan brain achieves its greatest development in cephalopods (Figures 4-7 and 4-8). Here the connectives are greatly shortened and the ganglia extremely concentrated. These brains approach the complexity of the brain of lower vertebrates in number of nerve cells (10^8), in concentration of ganglionic mass, and in perceptual and learning capability (see Chapter 9; and Young, 1971). Nonetheless, the cephalopod brain has a distinctly molluscan plan. As in advanced gastropods, a group of ganglia is concentrated around the esophagus (Figure 4-7). These ganglia are homologous to the standard cerebral, pedal, pleural, buccal, and visceral. Generally, the ganglia have fused into a complex brain. The supraesophageal region contains the cerebral ganglia and their subdivisions, e.g., the optic and olfactory lobes, and the superior and inferior buccal ganglia. The subesophageal region consists of paired pleural, pedal, and visceral ganglia and their subdivisions, e.g., the branchial and infundibular parts of the pedal ganglion. In addition, new central ganglia (olfactory and peduncular) and peripheral ganglia (stellate and axial cords of the arms) have been added (Figure 4-8). Finally, certain sensory receptors—especially the eyes, olfactory organs, and statocysts—have become highly developed.

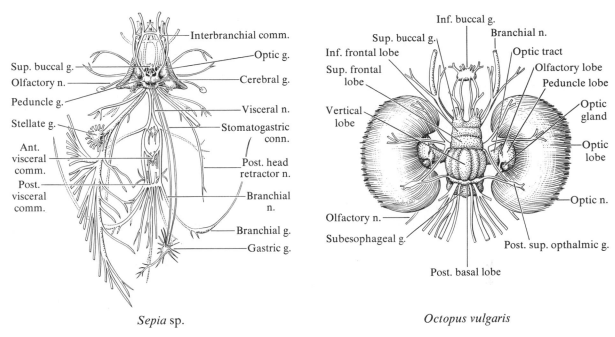

Sepia sp.

Octopus vulgaris

FIGURE 4-8.
The cephalopod nervous system. Dorsal view of *Sepia* sp.; right stellate ganglion and fin nerves have been omitted; stomatogastric system in black [after Bullock and Horridge, 1965]. Dorsal view of *Octopus vulgaris* [after Young, 1971].

For example, *Sepia* and *Loligo* have eyes that are comparable in complexity to those of vertebrates and are capable of forming images. The similarity of the cephalopod and vertebrate eyes is all the more interesting because they derive from somewhat different embryonic sites; the cephalopod eye derives directly from surface ectoderm whereas the vertebrate eye arises indirectly from the surface ectoderm by means of the embryonic neural tube.

APLYSIA NERVOUS SYSTEM

The nervous system of *Aplysia* first attracted neural scientists because of the large size of its nerve cell bodies (Mazzarelli, 1893; Bergh, 1902; Arvanitaki and Cardot, 1941).[4] But *Aplysia* and other gastropods also offer additional

[4]The best description of the *Aplysia* nervous system, as well as that of other opisthobranchs, is by H. Hoffmann (1939). A more restricted treatment in English is available in MacFarland (1909) and in Eales (1921).

experimental advantages. The ganglia are widely spaced, so that interganglionic commissures and connectives are easy to sever. This facilitates the analysis of the behavioral functions mediated by the individual ganglia. The nervous system is readily accessible upon opening the animal up and parts of it can be exposed without undue damage to other organs. The nervous system is avascular (Chapter 5) and can be readily maintained outside the animal in artificial culture media for several days (Strumwasser and Bahr, 1966). Finally, the ganglia are made up of a rind of cell somata and a central core of neuropil so that there is a very clear segregation of the synaptic region and the cell bodies, which are free of synapses (Chapter 5).

Aplysia is also useful for comparative studies because of the animal's interesting transitional position in the phylogeny of opisthobranchs. *Aplysia* is intermediate between the asymmetrical and torted form characteristic of the more primitive prosobranchs and the symmetrical detorted form characteristic of the more advanced opisthobranchs. The torted nervous system is still found among some of the primitive opisthobranchs, such as the cephalaspids *Acteon, Scaphander,* and *Philine,* and anaspid *Akera* (Figure 4-9). The visceral cords (connectives) of these forms are long, and twisted into a figure eight. In the pulmonates (Figure 4-6) the cords are uncrossed and often shortened, so that the ganglia of the visceral cords all come to lie around the esophagus close to the head ganglia. In the intermediate form *Aplysia,* the visceral connectives are untwisted but they are still long and the abdominal ganglion lies some distance behind the head ganglia.

As a result of torsion and detorsion the side of the body on which the various ganglia of the abdominal complex come to lie varies in different opisthobranchs. In the idealized primitive opisthobranchs (Figure 4-10), the subintestinal ganglion is on the left, the supraintestinal on the right, and the visceral on the midline. In torted opisthobranchs, such as *Acteon* (Figure 4-10), the subintestinal and the visceral come to lie on the right and the supraintestinal on the left. In partially detorted opisthobranchs, such as *Aplysia,* the subintestinal is thought to return to the left and the supraintestinal to the right. The symmetrical pallial ganglia do not change sides as a result of the torsional and detorsional changes.

In *Aplysia,* as in most molluscs, somatic and visceral motor function are controlled by different parts of the central nervous system as first pointed out by Bottazzi and Enriques (1900). The somatic functions are controlled largely by the head ganglia, whereas the visceral functions are controlled primarily by the abdominal ganglion. I will therefore consider these two parts of the central nervous system separately.

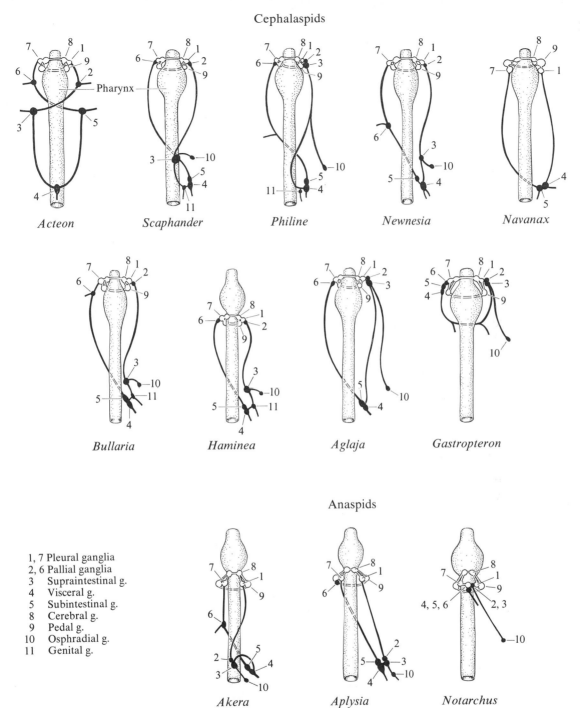

FIGURE 4-9.
The central ganglia of cephalaspids and anaspids showing combinations of fusion of the various head and abdominal ganglia. [After Guiart, 1901; Hoffmann, 1939.]

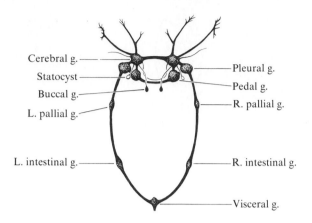

A. Idealized Primitive Opisthobranch

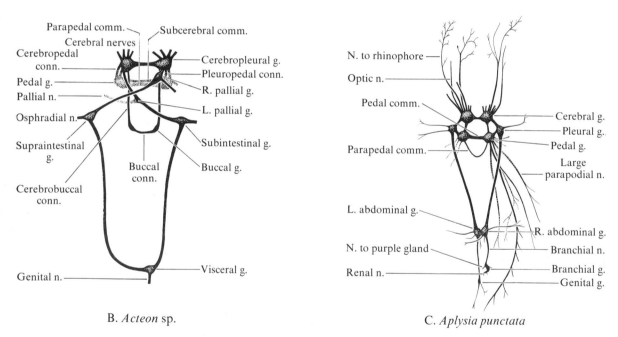

B. *Acteon* sp.   C. *Aplysia punctata*

FIGURE 4-10.

Comparison of an idealized primitive opisthobranch with *Acteon* and *Aplysia*.

A. Idealized nervous system of a primitive untorted opisthobranch. [After Guiart, 1901.]

B. Nervous system of the primitive cephalaspid opisthobranch *Acteon* sp. The nervous system is torted so that the connections are crossed and the subintestinal and visceral ganglia lie on the right side and the supraintestinal ganglion on the left. [After Bouvier, 1891.]

C. Nervous system of the anaspid *A. punctata,* one of the most primitive *Aplysia* species. The nervous system is detorted and the connectives uncrossed. [After Eales, 1921.]

Head Ganglia

The circumesophageal ring is made up of eight head ganglia; two cerebral ganglia (which are fused in most species), two buccal ganglia, two pedal ganglia, and two pleural ganglia. They are arranged in symmetrical pairs and are joined by connectives.[5]

CEREBRAL GANGLIA

The cerebral ganglia innervate a number of sensory structures: the anterior tentacles, mouth, eyes, and rhinophores (Figure 4-11). In addition, each cerebral ganglion sends a nerve to the ipsilateral statocyst, which lies on the surface of the pedal ganglion. The right cerebral ganglion also innervates the penis, although the major innervation comes from the right pedal. The cerebral ganglia are connected to the buccal and pedal ganglia by connectives, and to each other by a short commissure, which is not visible in most species because of the fusion of the two ganglia (Figure 4-12).

Removal of the cerebral ganglia in *A. limacina* (*-fasciata*) is reported to increase excitability and lead to incessant locomotor activity (Jordan, 1901). Animals so treated will beat their parapodia and swim back and forth in an aquarium for several weeks until they become exhausted and die. Unilateral extirpation, or cutting the cerebropedal connective on only one side, causes the animal to circle toward the intact side (Jordan, 1901). This may be due to deafferentation of the statocyst on the lesioned side.[6]

The cerebral ganglia also seem to control aspects of feeding. Jordan (1901, 1929) found that following removal of the cerebral ganglia animals no longer eat. Jahan-Parwar (1972a) found that individual fibers of the anterior tentacular nerve, which supplies the tentacular groove, are excited by seaweed extracts and by aspartate and glutamate, amino acids found in high concentration

---

[5]The innervation of the central ganglia in *A. depilans* was studied by Mazarelli (1893), in *A. dactylomela* and *A. cervina* by MacFarland (1909), and in *A. punctata* by Eales (1921). Hoffmann (1939) provided an excellent chart summarizing these several studies. Where differences exist, I have also indicated the innervation of *A. californica* based upon the experiences in our laboratory. The motor and reflex functions of peripheral nerves and ganglia are reviewed by Ten Cate (1928, 1931) and Bullock and Horridge (1965).

[6]There is also an increased tonus and reflex responsiveness on the lesioned side. If equal stimuli are applied to the two parapodia, the one on the lesioned side will contract more briskly. Both locomotion and increased reflex responsiveness disappear following removal of the pedal ganglia (Jordan, 1929). Topical application of strychnine to the cerebral ganglia causes hyperexcitability and loss of coordinated movements (Fröhlich, 1910b), but the mechanism of action of this effect is not known.

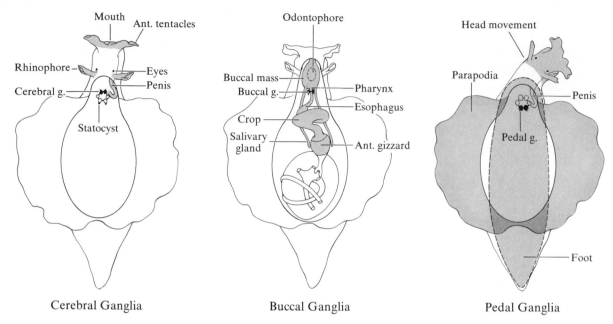

FIGURE 4-11.
The areas of innervation from the cerebral, buccal, and pedal ganglia of *Aplysia.*

in seaweed. Kupfermann (1974b) found that interruption of the cerebrobuccal connectives abolishes the biting phase of feeding. Since these consummatory biting movements are under the control of the buccal ganglia (Kuperfmann and Cohen, 1971), the cerebral ganglia may command or provide sensory input for odontophore protraction and retraction. As we will discuss below, two cerebral ganglion cells that are important in the control of feeding are the symmetrical metacerebral cells (Weiss, Cohen, and Kupfermann, 1975). These cells act as modulatory command cells and facilitate the central action of the feeding motor command cells and radula protractor motor neurons in the buccal ganglion. In addition, the metacerebral cells act on the buccal muscles directly to facilitate the action of the radula protractor motor cells.

The cerebral ganglia may also have a role in mediating the appetitive component of the food searching response since these responses are also eliminated by cutting the cerebropedal connectives (Kupfermann, 1974b). The cerebral and pedal ganglia control the penile movements of copulation, and stimulation of the right cerebral ganglion leads to peristaltic contraction of the penis sheath (Bottazzi and Enriques, 1900). In some species, e.g., *A. punctata,* the anterolateral surface of each cerebral ganglion includes a distinct subganglion, the optic

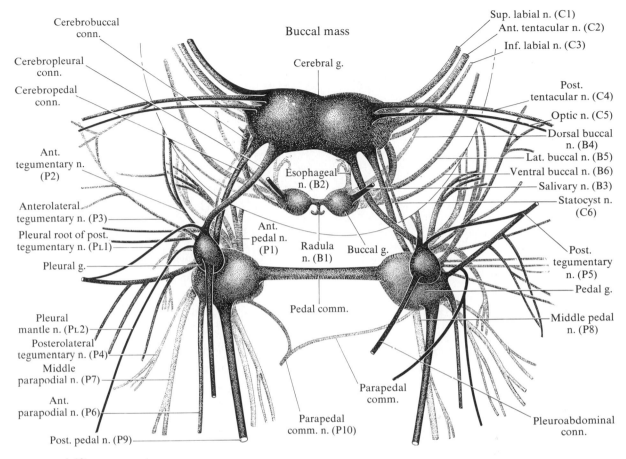

Cerebrobuccal conn.

Buccal mass

Sup. labial n. (C1)
Ant. tentacular n. (C2)
Inf. labial n. (C3)

Cerebropleural conn.

Cerebral g.

Cerebropedal conn.

Post. tentacular n. (C4)

Optic n. (C5)

Dorsal buccal n. (B4)
Lat. buccal n. (B5)
Ventral buccal n. (B6)
Salivary n. (B3)

Esophageal n. (B2)

Statocyst n. (C6)

Ant. tegumentary n. (P2)

Anterolateral tegumentary n. (P3)

Ant. pedal n. (P1)

Radula n. (B1)

Buccal g.

Post. tegumentary n. (P5)

Pleural root of post. tegumentary n. (PL1)

Pedal g.

Pleural g.

Pedal comm.

Middle pedal n. (P8)

Pleural mantle n. (PL2)

Posterolateral tegumentary n. (P4)

Middle parapodial n. (P7)

Ant. parapodial n. (P6)

Parapedal comm.

Post. pedal n. (P9)

Parapedal comm. n. (P10)

Pleuroabdominal conn.

FIGURE 4-12.

The cerebral, buccal, pedal, and pleural ganglia of *Aplysia*. [After Jahan-Parwar, unpublished observations.]

NB. In this and subsequent illustrations of the *Aplysia* nervous system I shall include in parentheses following the common names of nerves the identifying labels based on Hoffmann's nomenclature (1939). This system uses a capital letter to indicate the site of origin (C = cerebral ganglia; B = buccal; P = pedal; PL = pleural; etc.), followed by a number (P1, P2, P3, etc.) which indicates the area of innervation (the sequence of the numbers is rostral to caudal). Divisions of a major nerve are indicated by lower case letters, e.g., the three main branches of P3 are called P3a, P3b, and P3c.

ganglion, which is quite large considering the small size of the eye (Figure 4-13; Mazzarelli, 1893; Eales, 1921).

With the exception of the two large metacerebral cells at the rostral pole of the dorsal surface (which are about 250 μm in diameter), the cerebral ganglia are composed primarily of smaller cells (about 30–50 μm). Detailed electrophysiological and biochemical studies of the metacerebral cells in *Aplysia* have shown them to be homologous to those in pulmonates (see Chapter 10; Figure

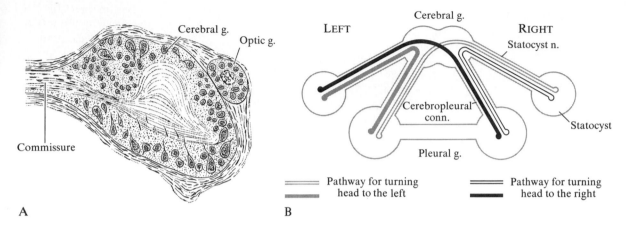

FIGURE 4-13.
The optic ganglion and statocyst of *A. punctata.* **A.** The optic ganglion [after Hoffmann, 1939]. **B.** Connections of the statocysts to the pedal ganglia.

10-3, p. 396; Kunze, 1918; Kandel and Tauc, 1965a; Sakharov, 1970; Weinrich *et al.,* 1972; Eisenstadt *et al.,* 1973; Weiss and Kupfermann, 1976). Except for Anderson's study (1967) of the sensory responses of some unidentified cells and the more recent studies of Jahan-Parwar (1976), little is known about the functional properties of other cells in these ganglia (for available studies see Anderson, 1967; Jahan-Parwar, 1976; Rossner, 1974; Jahan-Parwar and Fredman, unpubl. observ.).

In *Aplysia californica* six major nerves come off each cerebral ganglion (Figure 4-12)[7]:

The superior labial nerve (C1) arises in the anterior ventral portion of the ganglion and runs to the anterior part of the mouth and the skin of the anterior tentacles. A branch innervates the orifice of the buccal mass on the dorsal side of the buccal bulb.

The anterior tentacular nerve (C2) arises from the anterolateral side of the ganglion and innervates the anterior (or labial) tentacle. It bifurcates into a thick branch and a thin branch. The thick branch distributes to the musculature of the oral veil, while the thin branch goes directly to the base of the anterior tentacle.

[7]The discussion of the nerves of the major ganglia of *Aplysia* is primarily based on *A. californica.* In the discussion that follows I shall indicate in parentheses following the common names the identifying labels based on the nomenclature developed by Hoffmann (1939).

The inferior labial nerve (C3) runs to the posterior edge of the mouth. On the right side a branch goes to the anterior part of the penis and penis sheath.

The posterior tentacular nerve (C4) runs straight to the base of the posterior tentacles (rhinophores). A branch of this nerve anastomoses with the antero-lateral tegumentary nerve (P2) of the pedal ganglion.

The optic nerve (C5) is a slender nerve emerging from the dorsal surface of the ganglion and running parallel to the posterior tentacular nerve. It goes directly to the eye located at the base of the posterior tentacle.

The statocyst nerve (C6) is a very small nerve that extends from the cerebral ganglion, along the cerebropedal connective, to the statocysts.

BUCCAL GANGLIA

The buccal ganglia innervate the pharynx, salivary glands, esophagus, crop, and the gizzard, as well as the muscles of the buccal mass that control protraction and retraction of the odontophore (Figures 4-11, 4-14). The buccal ganglia are the smallest in volume of the major ganglia, but they contain many fairly large cells ranging in size from 100 to 200 μm in diameter (Gardner, 1971; Gardner and Kandel, 1972). The ganglia lie on the ventral side of the buccal mass, anterior to the cerebral ganglia, to which they are connected by means of the cerebrobuccal connectives (Figures 4-12 and 4-14). The two symmetrically placed buccal ganglia interconnect through the buccal commissure.

Individual identified motor cells in the buccal ganglia (Gardner, 1969, 1971) produce protraction and retraction of the odontophore (Cohen, Weiss and Kupfermann, 1978). The ganglia modulate the peristaltic rhythm of the esophagus (Bottazzi and Enriques, 1900). Electrical stimulation of the buccal ganglia interrupts the spontaneous rhythm of the esophagus and leads to rapid and powerful longitudinal contraction and transverse dilation of the esophagus, contraction of the crop, and, occasionally, intestinal movements.

The buccal ganglia give rise to six major nerves (Figure 4-14):

The radular nerve (B1) arises as a single nerve from the buccal commissure then splits into left and right nerves to innervate the radular sac from its ventral side.

The esophageal nerve (B2) sends some branches backward along the esophagus to innervate the esophageal sphincter and form an anastomotic network that accompanies the gut. It also sends branches to the buccal musculature and to the salivary gland.

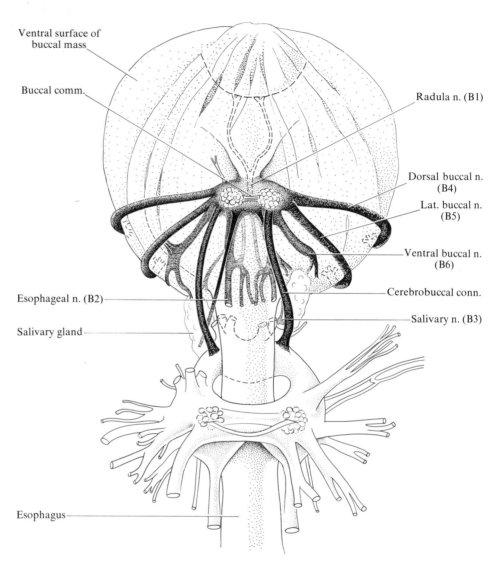

Ventral surface of
buccal mass

Buccal comm.

Radula n. (B1)

Dorsal buccal n.
(B4)

Lat. buccal n.
(B5)

Ventral buccal n.
(B6)

Esophageal n. (B2)

Cerebrobuccal conn.

Salivary n. (B3)

Salivary gland

Esophagus

FIGURE 4-14.
The buccal ganglia of *A. californica*. [After Strumwasser, unpublished.]

The salivary gland nerve (B3) innervates the salivary gland.

The third buccal (*ventral pharyngeal*) *nerve* (B4) supplies the basal muscles of the buccal mass, the deep muscles of the radula, the anterolateral muscles of the pharyngeal bulb, and the deep ventral muscles of the pharynx. One branch innervates a tendon-like structure that connects the buccal mass to the body wall.

The second buccal (*dorsal pharyngeal*) *nerve* (B5) supplies the lateral surface muscles of the buccal mass and also dips into the deep muscles of the buccal mass.

The first buccal nerve (B6) goes to the buccal musculature. Sometimes it also anastomoses with the esophageal nerve.

PEDAL GANGLIA

The pedal ganglia (Figure 4-12) are large and lie ventrolaterally. They supply the foot, parapodia, head, caudal part of the penis, penis sheath, and penis retractor muscles (Figure 4-11). Because the two ganglia are relatively far apart, the pedal and parapedal commissures that unite them are long, the parapedal commissure being the longer of the two. The anterior aorta passes between the two commissures. Each pedal ganglion lies close to the ipsilateral pleural ganglion and sends a very short (usually not clearly visible) connective to the ipsilateral pleural ganglion, as well as a stout connective to the ipsilateral cerebral ganglion.

The anterior portion of each pedal ganglion contains the statocyst, a sensory organ concerned with equilibrium (Chapter 3). The statocyst also receives a slender nerve from the cerebral ganglion on the same side. The pedal ganglia are not completely symmetrical; the right ganglion is larger and gives off more peripheral nerve branches than the left (Eales, 1921). These nerves go to asymmetrical structures, such as the penis, genital groove, and mantle.

The pedal ganglia control the foot and parapodia, as well as swimming and head movements. Removal of the pedal ganglia in *A. punctata* arrests swimming and locomotion (Jordan, 1910). Electrical stimulation of the cerebropedal connectives or of the pedal ganglia causes the parapodia to beat and the foot to undulate (Jordan, 1910; Fröhlich, 1910b). Some components of these actions can be produced in *A. californica* by stimulating individual cells (Hening *et al.*, 1976). After the pedal commisures are cut, the effects of stimulation of one pedal ganglion are limited to the stimulated side (Jordan, 1910).

The mechanism that produces rhythmic beating of the parapodia in the species that swim and the traveling contraction wave of the foot during locomotion is not known. Fröhlich (1910c) suggested that parapodial swimming movements involve a chain of reflexes controlled by the pedal ganglia. He found that in the intact animal the rostral part of the parapodium has the lowest threshold to mechanical stimulation and will respond first; this response in turn could initiate reflex actions in the adjacent part until the wave has moved across the whole parapodium.[8] However, these experiments cannot eliminate a central program and this question should be re-examined with modern techniques.

Ten Cate (1928) examined defensive withdrawal of the parapodia to tactile stimulation and found that the withdrawal did not seem to require feedback from peripheral structures.[9] Ten Cate (1928, 1931) therefore proposed that defensive parapodial movements involve a central-pattern generator, not the chained reflexes of the sort that may be involved in swimming.

In addition to controlling movement, the pedal (and perhaps the cerebral) ganglia may control muscle tone of the foot. Jordan (1929) found that muscle deprived of its connection with the pedal ganglia is tonically contracted, and suggested that the pedal ganglia have an inhibitory effect on muscle tone.

The pedal ganglia also seem to control head movement. Cutting the cerebro-pedal connectives eliminates the head-waving component (anterior foot movements) of the feeding response (Kupfermann, 1974b). The right pedal ganglion also innervates the retractor muscles of the penis and, in conjunction with the cerebrals, controls penile eversion. Stimulation of the pedal ganglia also leads to secretion by the mucus glands of the skin and foot (Bottazzi and Enriques, 1900).

---

[8]Fröhlich (1910c) cut each parapodium of *A. punctata* into thirds. One cut separated the areas innervated by the anterior parapodial nerve by severing the rostral one-third of the parapodia from the caudal two-thirds. The other cut separated the area of innervation of the posterior parapodial nerve by severing the caudal one-third from the rostral two-thirds. As long as all three parapodial strips were innervated, stimulation of the anterior strip elicited contraction of that strip as well as of the other two. When the connection between the middle strip and the pedal ganglion was severed, however, the middle strip no longer responded to stimulation of the anterior strip, and, unless the stimulus strength was markedly increased, neither did the posterior strip. After removal of the pedal ganglia, mechanical stimulation of part of the intact parapodia or electrical stimulation of an individual parapodial nerve produced only localized contraction.

[9]Using an experimental design similar to that of Fröhlich (see note 9), Ten Cate stimulated the central end of a parapodial nerve on one side and obtained reflex withdrawal of both parapodia. These bilateral responses persisted after denervation of the middle segment of the parapodium on the stimulated side, even though the denervated segment itself did not contract. Whether the movement was initiated rostrally by stimulating the anterior parapodial nerve, or caudally by stimulating the posterior parapodial nerve, the contraction invariably skipped the denervated middle segment and spread to involve the distal segment.

In all *Aplysia* species the pedal ganglia make up the largest neural aggregation of the central nervous system. The nerve cells are also fairly large, many ranging from 150 to 300 μm in diameter, but they have not been extensively studied with cellular techniques (for a beginning see Dorsett, 1968; Hughes, 1970; Weevers, 1971; Hening, Carew, and Kandel, 1976, 1977; Hening, unpublished observ.).

The pedal ganglia give off nine bilaterally paired nerves and one unpaired nerve which eminates from the parapedal connective. The description below derives from the studies of Jahan-Parwar (unpublished) in *A. californica*.

Three paired nerves emerge laterally from the anterior surface of each ganglion (Figure 4-12):

The anterior pedal nerve (P1) emerges most ventrally from the ganglion. It usually has four large branches. The most medial of these innervates the anterior part of the foot in the region of the mouth. The most lateral branch goes to the caudal part of the anterior foot, and the other two innervate the area in between.

The anterior tegumentary nerve (P2) arises from two roots, one of which emerges just dorsal to the anterior pedal nerve while the other emerges several millimeters more dorsally. The two roots anastomose and distribute branches to the muscles and skin along the ventral anterior body wall. From the right ganglion a branch goes to the muscles medial and anterior to the base of the penis and to the muscles along the shaft of the penis.

The anterolateral tegumentary nerve (P3) emerges from the dorsal surface of the ganglion, just lateral to the cerebropedal connective. It passes dorsal to the penis to distribute along the body wall. For its proximal course it runs parallel to a connective tissue band which connects the ganglion to the body wall. A branch of this nerve anastomoses with fine branchlets from the posterior tentacular nerve (C1) of the cerebral ganglion.

The posterolateral tegumentary nerve (P4) emerges lateral to the middle parapodial nerve and innervates the lateral body wall rostral to the anterior origin of the parapodium.

The posterior tegumentary nerve (P5) emerges from the surface of the ganglion near the pleuropedal connective. It has several important branches. The most anterior branch anastomoses with PL1, the pleural root of the posterior tegumentary nerve; the branches from this anastomosis innervate the anterior portions of the parapodia and the body wall rostral to them. A second branch, the lateral mantle branch, goes from the left ganglion to the lateral anterior edge of the mantle area. From the right ganglion this branch anastomoses with branch A6c of the siphon nerve from the abdominal ganglion and this anastomotic branch (P5a*A6c) innervates the opaline gland on its posterior side. From the left ganglion, the lateral mantle branch anastomoses with branch A1a of the vulvar nerve from the abdominal ganglion and this anastomotic branch (P5b*A1a) innervates the opaline gland on its anterior side.[10] Some fibers from the right lateral mantle branch usually run to the muscle of the body wall just lateral to the base of the penis.

The anterior parapodial nerve (P6) emerges from the caudal and lateral aspect of the ganglion. Its branches run under the longitudinal muscle band to innervate the anterior portion of the ipsilateral parapodium.

The middle parapodial nerve (P7) emerges from the ganglion as the most lateral and ventral nerve of the cluster. This nerve sends two or three branches under the longitudinal muscle bands to the middle section of the ipsilateral parapodium.

The middle pedal nerve (P8) emerges ventral to the posterior pedal nerve. It usually bifurcates very near the ganglion and innervates the middle portion of the foot. The distal branches of this nerve have a rosy or brownish color in adult animals.

The posterior pedal nerve (P9) is the most medial and dorsal of the cluster. This pigmented nerve innervates the posterior portion of the foot and sends a major branch to the posterior of the ipsilateral parapodium.

The unpaired nerve is

The parapedal commissure nerve (P10), which arises from the left pedal ganglion, separates from the commissure close to its midpoint, and innervates the middle to posterior midline of the foot.

PLEURAL GANGLIA

The pleural ganglia are situated next to the pedal ganglia (Figures 4-10, 4-12). They are connected to the cerebral ganglia, which are medial and rostral to them, by short, stout connectives. Each pleural ganglion is also joined to the ipsilateral pedal ganglion by a connective, which is not usually visible be-

---

[10]An asterisk in the name of a nerve indicates that it is an anastomotic branch, formed from the anastomosis between the nerves that make up the name.

cause the ganglia are usually so close together. Finally, the pleural ganglia give rise to the long pleuroabdominal (pleurovisceral) connectives (also called the visceral loop). The connective on the left side joins to the left abdominal hemiganglion; that on the right side connects to the right abdominal hemiganglion. Both connectives run dorsal to the anterior aorta.

The pleural ganglia give rise to two small nerves (Figure 4-12):

The pleural root of the posterior tegumentary nerve (PL1) is found bilaterally. It arises from the dorsal lateral aspect of the ganglion near the cerebropleural connective and joins the most anterior branch of the posterior tegumentary nerve (P5).

The pleural mantle nerve (PL2) is found only on the left, emerging medial and ventral to the pleural root of the posterior tegumentary, but still dorsal and lateral to the pleuroabdominal connective. It runs to the body wall muscle just anterior to the spermatheca at the edge of the subpallium. Its course and the general region of its distribution are symmetrical to that of the opaline branch (P5a) of the right posterior tegumentary.

The motor innervations and functions of the pleural ganglia are not known. According to Bottazzi and Enriques (1900), electrical stimulation of these ganglia at times initiates spontaneous gill contraction, while at other times it inhibits spontaneous gill movements. Most likely these actions are indirect and mediated by the abdominal ganglion.

The left pleural ganglion differs from the right in containing a very large cell (800 μm in diameter), the left (pleural) giant cell (identified cell LP1). In its electrophysiological properties (Hughes and Tauc, 1963) and its transmitter biochemistry (Giller and Schwartz, 1971a) this cell resembles cell R2 in the right abdominal hemiganglion (Figure 4-15). Hughes and Tauc (1963) believe this similarity indicates that the two giant cells are in symmetrical positions during early development, one in each of the larval pallial ganglia (see note 3). The left pallial ganglion is postulated to have fused with the left pleural and the right parietal with the supraintestinal (right abdominal hemiganglion), causing the asymmetrical distribution of the two giant cells. Hughes (1965) found one specimen in which cell R2 was actually located in the right pleural ganglion (Figure 4-15B).

As we shall see later (Chapter 7) this hypothesis has not proven to be the case, since there is no evidence for discrete symmetrical larval pallial ganglia. Cell R2 develops in the right hemiganglion asymmetrically from cell LP1. The two cells probably represent components of what in more primitive opis-

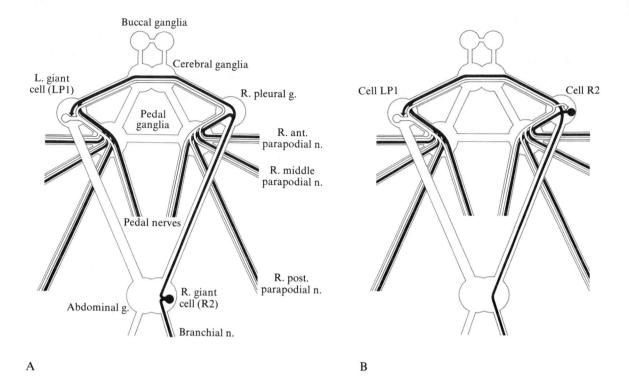

FIGURE 4-15.
The left (pleural) and right (abdominal) giant cells. [After Hughes and Chapple, 1967.]

A. Diagram of the typical branching of the axons of the two giant cells in *A. fasciata.*

B. Diagram of the branching of the giant cell axons in a unique preparation in which the cell bodies were symmetrically placed in the two pleural ganglia.

thobranchs are the pallial ganglia. But developmental processes recapitulate and foreshorten phylogenetic ones, and the two pallial ganglia develop from the outset as part of the left pleural and right abdominal ganglia. As a result, LP1 and the other cells that originally made up the left pallial ganglion develop within the left pleural ganglion, while R2 and the cells that originally made up the right pallial ganglion develop within the supraintestinal ganglion to form what is now the right abdominal hemiganglion.

Abdominal Ganglion

The abdominal or parietovisceral ganglion is situated at the edge of the visceral hump near the anterior aorta. Unlike the individual head ganglia of the circumesophageal ring, the abdominal ganglion is asymmetric and represents a fusion of several ganglionic masses. The right hemiganglion (thought to be a fusion of the right pallial and supraintestinal ganglia) lies next to and slightly anterodorsal to the left hemiganglion (thought to be a fusion of the subintestinal and visceral ganglia).[11]

The abdominal ganglion controls a variety of visceral functions (reproductive, respiratory, circulatory, and excretory) as well as the defensive (somatic) movements of the external organs of the mantle (Figure 4-16). The ganglion also controls several neuroendocrine processes (egg-laying and water balance) and a neuroglandular one (inking).

[11]The abdominal ganglion of *Aplysia* is often called the visceral ganglion. Alternatively, the left hemiganglion is sometimes referred to as the visceral ganglion and the right hemiganglion as the intestinal or parietal ganglion. The term "visceral ganglion" is not appropriate for the abdominal ganglion of *Aplysia* because it is thought to contain, in addition to the visceral homolog of more primitive opisthobranchs, three other ganglia (the subintestinal ganglion, supraintestinal ganglion, and the right pallial ganglion). The use of "visceral" and "intestinal" (or "parietal") for the two hemiganglia is also unsatisfactory because only the caudal part of the left hemiganglion is homologous to the visceral ganglion of primitive opisthobranchs. The noncommittal and commonly used term "abdominal ganglion" and the embryologically correct term "parietovisceral" (see Chapter 7) are clearly preferable.

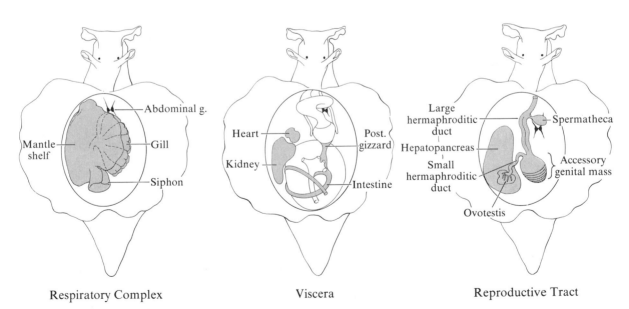

Respiratory Complex   Viscera   Reproductive Tract

FIGURE 4-16.
Motor innervation by the abdominal ganglion.

Reproductive activity is controlled in two ways: hormonal and neural. The bag-cell clusters release a hormone that triggers egg-laying and associated behavioral responses (Kupfermann, 1967; 1970; Strumwasser *et al.*, 1969; Arch, 1972a, b; Arch and Smock, 1977). In addition, the genital and vulvar nerves innervate and presumably control some phases of egg and sperm movement through the reproductive system. Thus electrical stimulation of the abdominal ganglion causes an increase in the peristalsis of the hermaphroditic duct and the vagina (Bottazzi and Enriques, 1900).

Respiratory function is also controlled in several ways. Activity of individual cells of the abdominal ganglion cause a variety of whole-gill movements (Kupfermann and Kandel, 1969; Peretz, 1969; Carew *et al.*, 1974, 1976). In addition, local reflexes mediate gill–pinnule retraction independent of the central ganglia (Peretz, 1970; Kupfermann *et al.*, 1971).

Individual cells in the ganglion control the circulation, heart rate, pericardial contraction, and the flow rate of hemolymph through branches of the arterial tree aorta (Mayeri *et al.*, 1974; Koester *et al.*, 1974; for earlier references see Dogiel, 1877; Schoenlein, 1894; Bottazzi and Enriques, 1900; Carlson, 1905).

Since ultrafiltration occurs across the walls of the heart into the pericardium, changes in pericardial size are likely to affect perfusion of the kidney and thereby excretion. Stimulation of the abdominal ganglion also results in movements of the posterior gizzard and intestine (Bottazzi and Enriques, 1900).

The ganglion also controls a variety of defensive responses: siphon and gill withdrawal as well as the secretion of ink from the purple gland (Carew and Kandel, 1977a; Kandel, 1976).

The main nerves of the right abdominal hemiganglion (Figure 4-17) are the following.

The vulvar nerve (A1) is a small nerve that branches to innervate the lower part of the mantle region and the vulvar orifice. One of its branches (A1a) anastomoses with a small branch of the posterior tegumentary nerve (P5b) from the right pedal ganglion, and one branch of this anastomosis (P5b*A1a) innervates the rostral surface of the opaline gland.

The branchial (*osphradial*) *nerve* (A2) is a stout nerve that goes to the base of the gill and splits into three branches. One branch goes to the osphradium. The second branch, which contains the branchial ganglion, enters the gill. The third branch extends to the purple gland and anterior portion of the mantle shelf.

The spermathecal (*bursa copulatrix*) *nerve* (A3) is a small nerve that goes to the spermatheca.

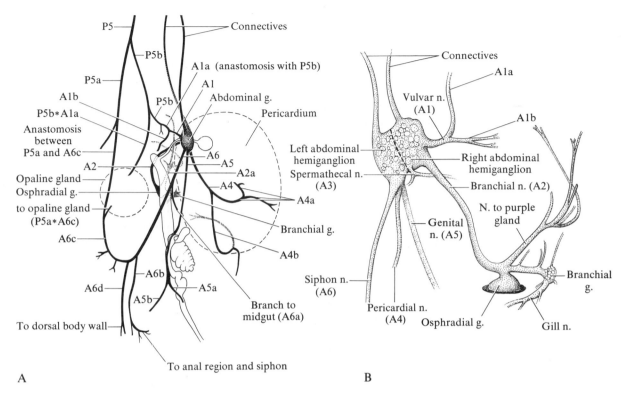

FIGURE 4-17.

The abdominal ganglion of *A. californica*. [Based on a drawing of *A. cervina* by MacFarland, 1909.]

A. Major nerves and anastomosis between the siphon nerve (A6) (a branch of the abdominal ganglion) and the posterior tegumentary nerve (P5) (a branch of the pedal ganglion).

B. Detail of the abdominal ganglia illustrating branches of the branchial nerve running to the branchial ganglion, osphradium, and purple gland. The dashed line indicates the division of the ganglion into hemiganglia.

Both ganglia give rise to two nerves (Figure 4-17):

The pericardial nerve (A4) innervates the heart and kidney, as well as the abdominal and gastroesophageal artery.

The genital nerve (A5) gives rise to the genital ganglion and then passes ventral to the duct of the spermatheca to innervate the genital duct and the accessory genital gland. A large branch passes to the rectum and innervates the small hermaphroditic duct and gland and also loops around the entrance of the afferent branchial vein.

The left abdominal hemiganglion gives rise to one nerve (Figure 4-17):

The siphon (anal or abdominal) nerve (A6) innervates the siphon, the muscular walls of the anus, the purple gland and the gill. One branch passes ventral to the rectum and innervates the genital duct and the accessory glands and adjacent body wall. Another recurrent branch (A6c) anastomoses with a branch of the posterior tegumentary nerve (P5a) from the pedal ganglion. One branch from this anastomosis (P5a*A6c) innervates the caudal surface of the opaline gland.

VARIATIONS IN THE *APLYSIA* CENTRAL NERVOUS STSTEM

The circumesophageal ring and the abdominal ganglion of the *Aplysia* species differ in the degree of fusion of their constituent ganglia. In what appears to be the most primitive species, *A. (Pruvotaplysia) punctata*, the cerebral, buccal, pedal, and pleural ganglia are round and distinct, and there are clearly visible commissures between the cerebral ganglia and the pleural and pedal ganglia. The two abdominal hemiganglia are also separate (Figure 4-18). Most species are more advanced, however, and the cerebral ganglia are fused and lie within the same connective tissue sheath. The pleurals and pedals also tend to lie close together and to fuse, as do the abdominal hemiganglia.

There are also a number of variations in the distribution of some peripheral nerves. The best studied instance is the innervation of the opaline gland and the anastomosis between the posterior tegumentary nerve (P5) and branches from the siphon and vulvar nerves of the abdominal ganglion (Figure 4-18). In two species of the subgenus *Varria*, *A. dactylomela* and *A. cervina*, MacFarland (1909) found that the opaline gland is innervated by nerves from both the pedal and the abdominal ganglia (Figure 4-18). A branch from the posterior tegumentary nerve (P5) of the pedal ganglion runs to the dorsolateral body wall musculature. It gives off a branch, P5b, which anastomoses with a branch of the vulvar nerve (A1a) of the right abdominal hemiganglion. This anastomotic nerve (A1a*P5b) in turn sends one branch to the opaline gland. The posterior tegumentary nerve gives off a branch that continues on as P5a; this branch anastomoses with a recurrent branch (A6c) of the siphon nerve (A6). At the junction of this anastomosis (P5a*A6c), another branch is sent to the opaline gland.

On the other hand, Mazarelli (1893) described the opaline innervation in *A. (Pruvotaplysia) punctata*, *A. (Aplysia) depilans*, and *A. (Varria) fasciata* as being supplied by the posterior tegumentary nerve (P5) and the vulvar nerve

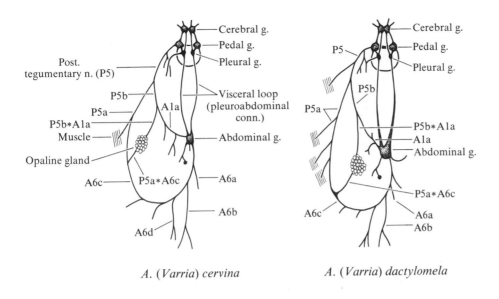

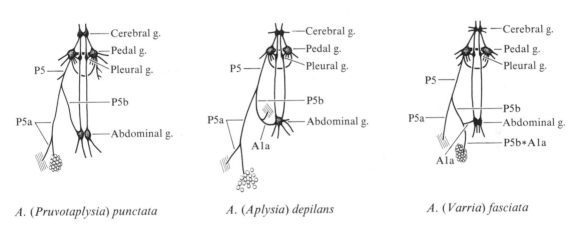

FIGURE 4-18.
Variations in innervation of the opaline gland among *Aplysia* species. [After MacFarland, 1909.]

(A1).[12] Blankenship (personal communication) found this to be true for *A. brasiliana* as well.

Hoffmann (1939) suggested that the innervation from the pedal ganglion was the primordial one since it innervates the body musculature of this region. The

[12]Eales (1921) also found this type of innervation in *A. punctata*.

abdominal ganglion innervation would then be secondary and would have developed as the ganglion grew closer to the opaline gland. Originally, the visceral loop was longer and the abdominal ganglion was further away from the gland. These suggestions may not be correct, however, because they attribute innervation to one or another ganglionic branch of an anastomosis on the basis of such insufficient criteria as geometry and proximity. With either arrangement it is possible that one or both ganglia innervate the opaline gland. Indeed, it would be interesting to determine with neurophysiological techniques whether in different species the same cells always innervate the gland but send their axons through different peripheral nerves.

VARIATIONS IN THE ANASPID NERVOUS SYSTEM

The order Anaspidea comprises the families Aplysiidae and Akeridae (genus *Akera*) (Figure 2-4, p. 34). In *Akera* (Akeridae) the connectives are long and torted and the ganglia of the abdominal complex are distinct and separated (Figure 4-19). This nervous system thus resembles that of the primitive opisthobranch *Acteon* and is therefore transitional between the more primitive cephalaspids and the slightly more advanced anaspid *Aplysia* (see Figure 2-2, p. 33).

Of the subfamilies of Aplysiidae (see Figure 2-4, p. 34), the nervous system of Dolabellinae most resembles that of *Aplysia*. The cerebral ganglia and left and right abdominal hemiganglia are also fused (Figures 4-19). As in *Aplysia*, the left hemiganglion is thought to represent a fusion of the left pallial, subintestinal, and visceral ganglia of more primitive opisthobranchs, whereas the right hemiganglion is thought to represent a fusion of the right pallial and supraintestinal ganglia (MacFarland, 1918; Hoffmann, 1939). In addition, the seventh pedal nerve of Dolabellinae, which corresponds to the fifth pedal nerve of *Aplysia*, anastomoses with a branch of the vulvar nerve of the abdominal ganglion and innervates the opaline gland. Moreover, unpublished observations on *Dolabella auricularis* by D. Willows indicate that the resemblance between *Dolabella* and *Aplysia* extends to the cellular level. The abdominal ganglion of the *Dolabella* species has cells R2, R14, R15, the bag cells, and other readily identifiable cells and clusters found in *Aplysia*.

In the subfamily Dolabriferinae (*Petalifera*) the pleurovisceral connectives are shortened to bring the abdominal ganglion complex to the periesophageal ring of ganglia. In the subfamily Notarchinae (*Notarchus*) the abdominal ganglia are concentrated and join the periesophageal ring (Figure 4-19).

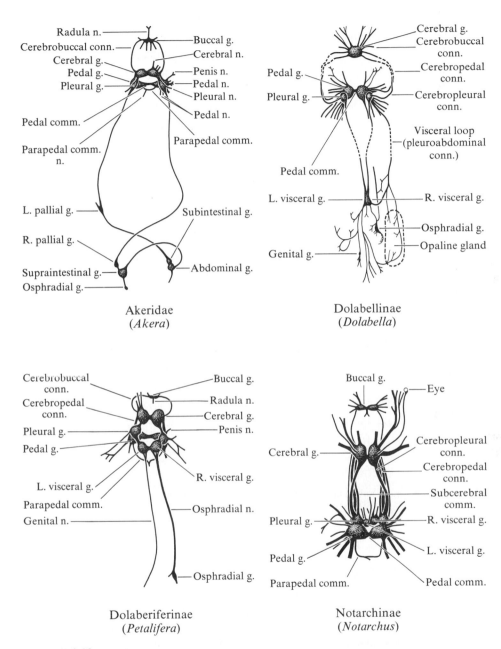

Radula n.
Cerebrobuccal conn.
Cerebral g.
Pedal g.
Pleural g.
Buccal g.
Cerebral n.
Penis n.
Pedal n.
Pleural n.
Pedal n.
Pedal comm.
Parapedal comm.
Parapedal comm. n.
L. pallial g.
R. pallial g.
Supraintestinal g.
Osphradial g.
Subintestinal g.
Abdominal g.

Akeridae
(*Akera*)

Cerebral g.
Cerebrobuccal conn.
Cerebropedal conn.
Cerebropleural conn.
Pedal g.
Pleural g.
Visceral loop
(pleuroabdominal conn.)
Pedal comm.
L. visceral g.
R. visceral g.
Osphradial g.
Opaline gland
Genital g.

Dolabellinae
(*Dolabella*)

Cerebrobuccal conn.
Cerebropedal conn.
Pleural g.
Pedal g.
Buccal g.
Radula n.
Cerebral g.
Penis n.
L. visceral g.
Parapedal comm.
Genital n.
R. visceral g.
Osphradial n.
Osphradial g.

Dolaberiferinae
(*Petalifera*)

Buccal g.
Eye
Cerebral g.
Cerebropleural conn.
Cerebropedal conn.
Subcerebral comm.
R. visceral g.
Pleural g.
L. visceral g.
Pedal g.
Parapedal comm.
Pedal comm.

Notarchinae
(*Notarchus*)

FIGURE 4-19.
Nervous systems of representative genera of three of the subfamilies of Aplysiidae, and the single genus *Akera* of the family Akeridae. [After Hoffmann, 1939.]

The subfamily Aplysiidae, to which the genus *Aplysia* belongs, also includes the genus *Syphonota,* which resembles *Aplysia* in having a periesophageal ring containing three pairs of ganglia, long pleurovisceral cords, and adjacent left (visceral) and right (supraintestinal) abdominal hemiganglia that are fused.

Nervous systems of two other neurobiologically interesting opisthobranch genera, the cephalaspid *Navanax* and the notaspid *Pleurobranchaea,* both shelled carnivores, are shown in Figure 4-20.

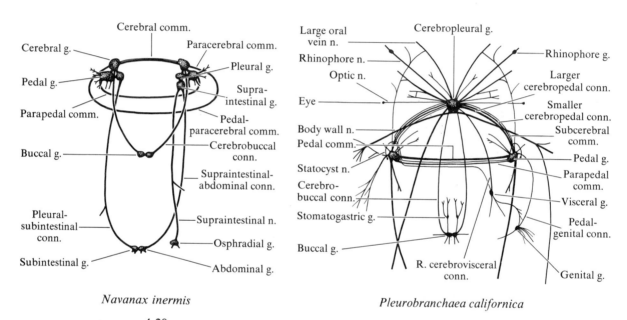

Navanax inermis   *Pleurobranchaea californica*

FIGURE 4-20.
Neurophysiologically interesting opisthobranchs. *Navanax inermis* [after Murray, 1971] and *Pleurobranchaea californica* [after Lee and Liegeois, 1974].

NEUROENDOCRINE SYSTEMS OF MOLLUSCS

As in other invertebrates and vertebrates, the molluscan central nervous system contains groups of nerve cells with endocrine properties (neuroendocrine cells). These neurons release substances into the blood stream to act at remote sites, thereby serving as a second coordinating system in parallel with the strictly neural one (Scharrer and Scharrer, 1963). In general, neuroendocrine systems tend to mediate long-term functions, such as metamorphosis, maturation, sexual development, color changes, and salt and water balance.

In gastropod molluscs, neurosecretory cells (as shown by morphological stains, such as the Gomori) are usually found in the cerebral and abdominal ganglia (Ortmann, 1960). In invertebrates in general and in molluscs in particular, the ratio of neurosecretory cells to ordinary neurons appears to be greater than that in the vertebrate brain. In the abdominal ganglion of *Aplysia,* for example, the bag cells, two clusters of neurosecretory cells involved in egg-laying, contain up to 800 relatively small (30–50 μm) neurons, approximately 30 percent of the total number of nerve cells in the adult ganglion. In the octopus, 650,000 of the animal's 10^8 nerve cells are concerned with color (chromatophore) changes (Wells, 1964, 1968; Young, 1971).

A particularly interesting and well-studied example of a molluscan neuroendocrine system is the optic gland of cephalopods, which controls sexual maturity. This gland sits on the optic tract, which connects the optic lobes with the central regions of the brain. In the immature female the gland is tonically inhibited by the subpedunculate lobes of the supraesophageal complex of the brain. But the optic gland can be stimulated to secrete its hormone either by cutting the connection from the subpedunculate lobes or by disrupting visual input to the brain (Figure 4-21). When this is done to an immature female she rapidly achieves precocious puberty and the size of the gonads increases 100-fold, from 0.5 g to 50 g. The hormone acts not only on the sex gland but also on the oviducts and the oviductal glands involved in depositing the shell around the egg. Moreover, the sexually precocious octopus will seek a suitable place to lay eggs and, having laid them, will remain with them until they hatch. She will stop eating during this period, blow water over the eggs to keep them clean, and protect them from predators. Thus, as with vertebrates (Lehrman, 1961), the gonadotrophic hormone affects not only the sexual apparatus but also behavior (Wells and Wells, 1959; Wells, 1964). Similar although less striking effects follow denervation of the optic glands of male octopods.

There are interesting parallels between the optic gland system of cephalopods and the pituitary system of vertebrates. In both cases the endocrine system that regulates sexual maturity is located in the central nervous system and is influenced by light; also in both there is normally a considerable delay in the onset of sexual maturity caused by a delay in the release of the gonadotrophic hormone. Finally, in both the trophic hormone produces endocrine as well as behavioral effects.

144

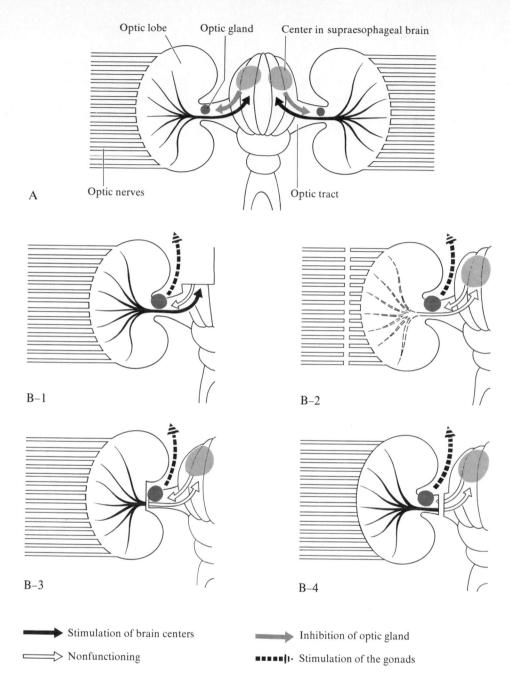

Optic lobe   Optic gland   Center in supraesophageal brain

A

Optic nerves   Optic tract

B–1

B–2

B–3

B–4

→ Stimulation of brain centers   → Inhibition of optic gland

⇨ Nonfunctioning   ■■■■|▸ Stimulation of the gonads

FIGURE 4-21.
The hormonal control of sexual maturity in octopods. The state of the gonads is determined by a secretion of the optic glands (A). Precocious secretion can be induced by any of the operations summarized in part B. These procedures abolish visual inputs to the brain or cut off control of the optic gland by a center in the hind part of the supraesophageal brain. [After Wells and Wells, 1959.]

SUMMARY AND PERSPECTIVE

The evolution of the molluscan nervous system has produced a great variety of forms, matched in diversity only by the nervous systems of arthropods and vertebrates. The molluscan nervous system ranges in complexity from the simple and primitive cord-like systems of the monoplacophorans, which are comparable in structure to the nervous system of flatworms, to the complex evolved brain of cephalopods, which compares favorably in size and capability to that of the lower vertebrates. Of all the molluscs, the gastropods have the greatest range and variety of structure in their nervous system.

The archetypal gastropod nervous system is thought to have been symmetrical (untorted), consisting of paired ganglia that were connected to each other by commissures, to other ganglia by connectives, and to the periphery by peripheral nerves. According to the most generally accepted model, the archetypal paired cerebral ganglia were interconnected by a commissure and each cerebral ganglion sent peripheral nerves to the eyes, tentacles, and esophagus. Three pairs of cerebral connectives led to the buccal ganglia rostrally and to the pleural and pedal ganglia caudally. The buccal ganglia innervated the muscles of the radula and the mouth parts; the pedal ganglia innervated the foot. On each side the pleural ganglion sent a connective to the intestinal (parietal) ganglion on the same side, which in turn sent a connective to the paired visceral ganglion. The visceral ganglion sent peripheral nerves to the internal organs of the visceral mass as well as to the mantle cavity and its organs. Following torsion, the pleurovisceral connectives became twisted into a figure eight, so that the left intestinal ganglion ended up below and on the right as the subintestinal (infraparietal) ganglion, and the right intestinal ganglion ended up above and on the left as the supraintestinal (supraparietal) ganglion. The remainder of the nervous system was not affected by torsion.

The ancestors of the opisthobranchs and pulmonates are thought to have evolved from the prosobranchs, the most primitive gastropods. But within each subclass there is a wide range of complexity and all subclasses share common mechanisms for achieving advanced levels of organization. Within each subclass there is the independent tendency toward a more concentrated nervous system. This is usually brought about by cephalization of the circumesophageal ring of ganglia resulting from an anterior migration of the posterior (parietal, pedal, and pleural) ganglia and their ultimate fusion with the cerebrals, and by fusion of the ganglia of the visceral loop.

The nervous system of the primitive torted opisthobranch—the cephalaspid *Acteon* or the anaspid *Akera*—consists of four paired ganglia (cerebrals, buccals, pleurals, and pedals) that form a ring around the esophagus. The cerebropleural ganglia give rise to the visceral (abdominal) nerve cords or connectives and are united posteriorly. Along their torted course lie the five abdominal ganglia: the two pallial ganglia, the left (supraintestinal) and right (subintestinal) hemiganglia, and the single visceral ganglion.

In the detorted *Aplysia* the circumesophageal ganglia are essentially in the same position as in *Akera*. Paired cerebrals, buccals, pleurals, and pedals are present and innervate roughly the same regions as in other gastropods. The cerebrals activate the tentacles, mouth, eyes, rhinophores, and penis. The buccals innervate the buccal musculature, the pharynx, the salivary gland, the esophagus, the esophageal crop, and the gizzard. The pedals innervate the head, foot, and parapodia. The pleurals often do not give rise to peripheral nerves but they do give rise to the visceral cords, now called the pleuroabdominal connectives.

The major difference between *Aplysia* and the most primitive opisthobranchs, such as *Acteon,* is in the abdominal ganglion complex. In *Aplysia* this complex also supplies the mantle and its glands, the gill, osphradium, genital organs, heart, kidney, and parts of the digestive apparatus. However, instead of five ganglia lying along the untorted visceral cords, there is only one large, asymmetrical, fused ganglion—the abdominal ganglion. This ganglion is made up of two hemiganglia that lie close together and are interconnected by a commissure. This asymmetrical ganglion is large and more complex than the single unpaired visceral ganglion of *Acteon* and is thought to be the result of fusion of the functions once carried out by four of the five ganglia that lie along the visceral cords of more primitive opisthobranchs. The right hemiganglion seems to be a fusion of the supraintestinal and the right pallial ganglia, and the left hemiganglion is a fusion of the subintestinal and the visceral ganglia.

As in other invertebrates and vertebrates, the molluscan central nervous system contains groups of nerve cells with endocrine properties. Neuroendocrine systems tend to be concerned with a variety of long-term functions, such as metamorphosis, maturation, sexual development, color changes, and salt and water metabolism.

SELECTED READING

Bullock, T. H., and G. A. Horridge. 1965. *Structure and Function in the Nervous Systems of Invertebrates.* Vol. 2. San Francisco: W. H. Freeman. An unrivaled source of information on the anatomy and function of the invertebrate nervous system.

Hoffmann, H. 1939. Opisthobranchia, Teil 1. In *Klassen und Ordnungen des Tierreichs, Vol. 3., Mollusca, Abteilung 2: Gastropoda, Buch 3: Opisthobranchia,* H. G. Bronn, ed. Leipzig: Akademische Verlagsgesellschaft M.B.H. The most detailed discussion available on the opisthobranch nervous system.

Wells, M. J. 1968. *Lower Animals.* New York: McGraw-Hill. A simple and good review of the neuroendocrine control and brain and behavior relationships in cephalopods.

CHAPTER 5

The Abdominal Ganglion of *Aplysia*: A Central Ganglion

The central neurons of molluscs and other invertebrates are grouped into ganglia, collections of interconnected nerve cells. In this chapter I shall review the structural organization of the central ganglia of *Aplysia,* using information on the best studied example, the abdominal ganglion. I also indicate the technical advances needed to improve our understanding of the morphological basis of the cellular interactions underlying behavior.

TYPES OF INVERTEBRATE GANGLIA

Invertebrate ganglia vary in size and complexity. At one extreme is the cardiac ganglion of the lobster, which contains only nine cells that initiate the rhythmic beat of the heart. At the other extreme are the complex fused ganglionic masses of octopods, which resemble in their complexity the brains of vertebrates.

The abdominal ganglion in *Aplysia* is of intermediate complexity and stands between these two extremes. It is a fused ganglionic mass (made up of three distinct embryonic ganglia) capable of mediating a number of somatic, visceromotor, and endocrine functions through a diversity of neurons. But it is relatively small in size and simple in organization.

GENERAL STRUCTURE OF A MOLLUSCAN GANGLION

The structure of the *Aplysia* abdominal ganglion resembles the ganglia of other higher invertebrates. The ganglion has a surrounding connective tissue sheath, a peripheral (cortical) region of cell bodies, and a central (medullary) region, or neuropil (Bullock and Horridge, 1965). The ganglion contains two classes of cellular elements: glial cells and neurons.

Connective Tissue Sheath

The sheath of the abdominal ganglion is made up of a fibrous connective tissue capsule that contains some muscle fibers. The sheath encloses not only the ganglion but also the fiber tracts running between ganglia (the connectives), the fiber tracts connecting symmetrical ganglia (the commissures), and the proximal part of the peripheral nerves (Coggeshall, 1967). The sheath is freely permeable to ions and other small molecules (as large as the dye methylene blue, MW 374), but it is a barrier to large molecules. The sheath also acts as a structural support for the neural tissue. Septa from the sheath penetrate into the ganglion, subdividing it into regions or compartments. As the animal grows, additional septa develop and further subdivide the ganglion (Figure 5-1).

In addition to its structural role the sheath is also a neurohemal organ: it contains the neurosecretory terminals that release neuroendocrine products. The *Aplysia* nervous system is avascular like that of all molluscs except cephalopods (Rosenbluth, 1963; Coggeshall, 1967). Blood vessels do not penetrate into the cellular components of the ganglion; nevertheless, the sheath of the abdominal ganglion is highly vascularized (Figure 5-2). The sheath also contains numerous fine terminal processes from the bag cells and the white cells (cells R3 to R14), the two types of neurosecretory cells in the ganglion that appear to release their secretory products into the sheath (Figure 5-2C).

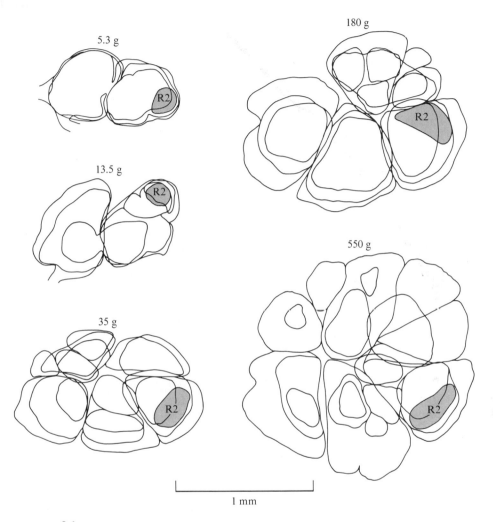

FIGURE 5-1.
Increase in connective tissue septa in *Aplysia* with age of the animal. The septa separate cellular groupings. Relative age is estimated from the animal's weight. The position of cell R2 is indicated. [After P. Skyles, in preparation.]

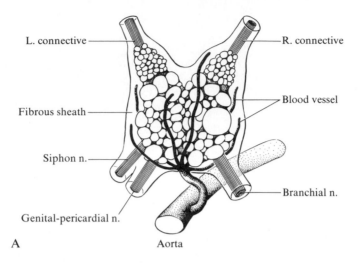

L. connective

R. connective

Fibrous sheath

Blood vessel

Siphon n.

Branchial n.

Genital-pericardial n.

A

Aorta

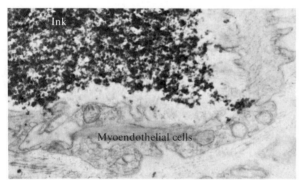

Ink

Myoendothelial cells

B-1

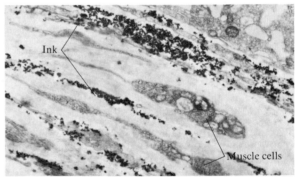

Ink

Muscle cells

B-2

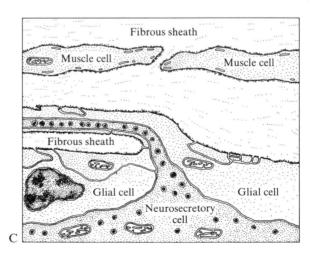

Fibrous sheath

Muscle cell

Muscle cell

Fibrous sheath

Glial cell

Glial cell

Neurosecretory
cell

C

FIGURE 5-2.

The·sheath of the abdominal ganglion of *Aplysia*.

A. The vascularization of the sheath by branches from the aorta. [After Coggeshall, 1967, and unpublished observations.]

B. Dye injection illustrating vascularity of the connective tissue sheath and avascularity of the abdominal ganglion of *Aplysia*. [From Coggeshall, 1967.] **(1)** Blood vessel in the sheath around the abdominal ganglion. India ink was injected into the vascular system of this animal shortly before fixation. The wall of the vessel is lined with myoendothelial cells. The luminal surface of the myoendothelial cells is lined with a layer of small fibers embedded in an amorphous matrix. This layer is emphasized in this picture because it separates the particles of ink from the myoendothelial layer. ($\times$ 9,720) **(2)** The injected ink, as well as being in vessels, is found free in the interstices of the sheath, as is illustrated in this picture. This finding shows that material in the vascular system can pass from the vascular lumen directly into the acellular parts of the sheath. Thus, the cellular components of the sheath—the muscle cells, fibroblasts, glial cells, axons, and outer neuronal cell bodies of the ganglion—are bathed in hemal fluid. ($\times$ 9,720)

C. Schematic illustration of the termination of a process from a neurosecretory cell in the sheath. [From Kandel, 1976.]

The Cell-Body Region

The monopolar cell bodies of neurons are gathered as a cortex around the periphery of the ganglion. Axons and synapses are not present in this cellular region in the ganglion (Rosenbluth, 1963; Coggeshall, 1967).[1] Each neuron sends its major axonal process into the neuropil, where it may branch extensively. The neuropil consists of the synaptic field of interconnections between processes of neurons in the ganglion and axons from other ganglia and from the periphery. The neuropil region also contains tracts and fibers that pass through the ganglionic core without making synaptic connections (Figure 5-3).

Invertebrate and vertebrate neurons share features of a common plan, although they differ in detail. One major difference is in the position of the receptive pole of the neuron, the dendritic tree. Vertebrate neurons are usually multipolar; the axon usually leaves the cell body at one pole and one or more

---

[1]Developing neurons in the abdominal ganglion do occasionally have synapses on the cell bodies (Schacher and Kandel, 1977). Although the cell bodies of adult animals are free of synapses, the membrane of some cell bodies contains receptor molecules that seem similar in pharmacological properties to those found in the neuropil region of the neuron (Tauc and Gerschenfeld, 1962). In view of the great concentration of receptors at the synaptic sites of the peripheral neurons of vertebrates (Harris, Kuffler, and Dennis, 1971) it is likely that a similarly high concentration of receptors is present in the neuropil region of *Aplysia*. Moreover, in some cells the receptors for transmitters are present only in the axonal and neuropil regions.

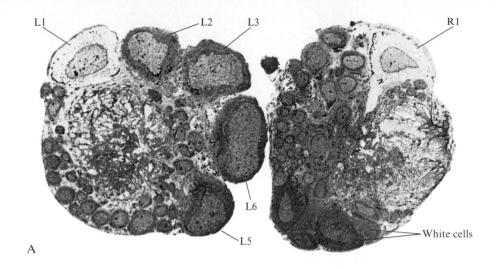

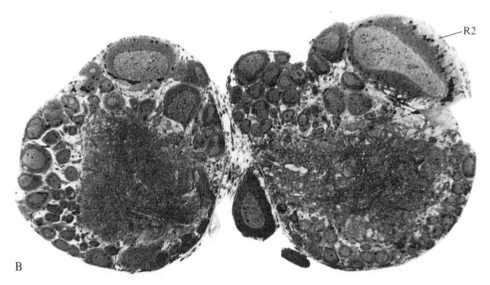

FIGURE 5-3.

Cellular rind and neuropil of the abdominal ganglion of *Aplysia*. [From Frazier *et al.*, 1967.]

A. Cross section of the left and right rostral quarter ganglion. There are six large cells in the left rostral quarter ganglion, five of which are illustrated here. Cell L6 is located midway between the dorsal and ventral surfaces of the ganglion in this plane of section. In more caudal parts of the ganglion cell L3 will have ended and L6 will be visible from the dorsal surface. Note that the cytoplasm of L1 is much lighter than that of the other identified cells. This is due in part to the absence of cytoplasmic granules and in part to the presence of cytoplasmic regions that are filled only with a fine filamentous material. Cell R1, which is located on the lateral aspect of the dorsal surface of the right rostral quarter ganglion, is symmetrical with L1 in both position and cytological appearance. None of the large dorsal white cells is visible in this section, but two smaller white cells are seen on the ventral surface of the ganglion. ($\times$ 37.5)

B. Cross section through the middle of the ganglion. Note the large and prominent "giant" cell R2 on the right dorsolateral margin of the ganglion. ($\times$ 37.5) dendritic trees emerge and branch out from there and other points. Synaptic input is distributed over the cell body and the dendritic trees. Invertebrate neurons are usually monopolar, as are vertebrate sensory cells; the cell body gives rise only to an axon, and the dendritic arborization usually emerges from the proximal part of the axonal region. The invertebrate cell body receives no synapses. The synaptic endings are distributed along the dendrites (Figure 5-4).

Another difference is in the relation of the receptive to the transmitting pole. In vertebrates these two poles are typically discrete and separated from one another. In invertebrates the receptive and transmitting sites are freely intermingled.

Each region of the ganglion contains characteristic glial or supporting cells.

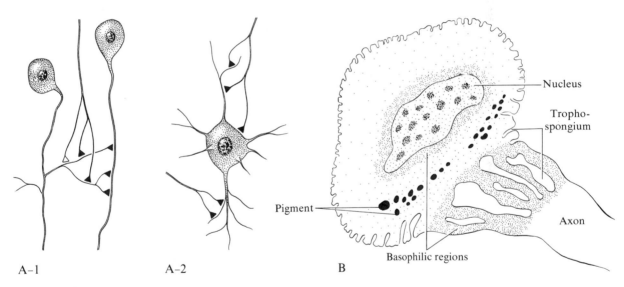

A–1   A–2   B

FIGURE 5-4.
The structure of invertebrate neurons.

A. Comparison of invertebrate **(1)** and vertebrate **(2)** nerve cells. Invertebrate neurons do not form synapses on the cell body; they typically form synapses only on the dendritic extensions of the main axon.

B. A diagram of a typical large ganglion cell of *Aplysia*. The cell body is broken up into several fascicles at the perikaryon–axon junction. This is an extreme example of the trophospongium or glial indentation characteristic of large ganglion cells. Ribosomes are concentrated in the region of the nucleus and in the perikaryon–axon junction, giving these regions a deep blue color when they are stained with a basic dye. [After Coggeshall, 1967.]

GLIAL CELLS

In *Aplysia,* as in most invertebrate nervous systems, different morphological types of glial cells are found in the connectives, in the cell-body layer, and in the neuropil. Little is as yet known about the function of glial cells (Kuffler and Nicholls, 1976). In animals in which they have been best studied, primarily in the leech and mudpuppy, glial cells have not been found to produce action potentials or synaptic potentials. However, they do undergo potential changes, which may have signaling functions. Glial cells are often electrically interconnected. As a result, a potential change in one glial cell is passively propagated to adjacent glial cells. Glial cells also have a resting potential that behaves as a very sensitive Nernst potassium electrode. They therefore undergo a depolarization (sometimes as large as 25 mV) as a result of the extrusion of potassium following the spike activity of neighboring nerve cells (for reviews see Kuffler and Nicholls, 1966, 1976; Orkand, 1977). The electrical properties of glial cells have not, as yet, been studied in *Aplysia.* The little that is known about glial function in *Aplysia* comes from a study of their position-dependent morphological differences and from autoradiographic and biochemical studies.

Connectives and Peripheral Nerves

The glial cells of the connectives and nerves envelop separate axon bundles. The exact relationship of glial cells to the axons they surround varies according to the size of the axon within the connectives and peripheral nerves. Small axons are frequently bundled together and enveloped by a single glial process. Two or three medium-sized axons are wrapped together by a single glial cell, whereas very large axons are often individually wrapped in a layer of processes from many glial cells, which often deeply indent the axon at several points (Coggeshall, 1967). Desmosome-like junctions unite the glial processes with one another (Figure 5-5; Coggeshall, 1967).

Glial cells of the connectives actively take up ^{3}H-leucine from the bathing solution and incorporate a significant amount of this labeled amino acid into protein (Thompson and Schwartz, unpublished observations). By contrast, the axons in the connective are only very slightly labeled and seem to have a low rate of protein synthesis, if any.

A possible function for these metabolically active glial cells is suggested by an experiment by Koester (personal communication). He examined the properties of the vasoconstrictor neurons in the abdominal ganglion that send their axons to the aortic sphincter and found that the distal part of the axon does not degenerate when separated from the cell body by cutting the connective. The axon continues to conduct action potentials to vasoconstrictor muscle for at least several weeks. A similar finding in the crayfish was previously reported (Hoy, Bittner, and Kennedy, 1967; Wine, 1973). What maintains the functioning of these cut axons in the absence of their cell bodies is not known. One possibility is that glial cells transfer proteins to the axon to maintain its viability. This seems to occur in the squid giant axon (Sarne *et al.*, 1976a, b).

Cell-Body Region and Neuropil

Glial cells are often found in a multi-layered capsule around large neurons and seem to define cell masses; they may possibly transfer nutrients from blood to neurons in this avascular nervous system (Coggeshall, 1967). Glial cells are relatively rare in the neuropil. Those that are present resemble the glial cells of the connectives and peripheral nerves.

NEURONS

In contrast to the paucity of information about the morphology and function of glial cells in the abdominal ganglion, there is a great deal of information about the properties of different nerve cells.

Fine Structure of the Neuron

The cytoplasm of a typical large neuron in the abdominal ganglion contains all the usual components for cellular function: mitochondria, granular and agranular endoplasmic reticulum, free and bound ribosomes, microtubules, neurofilaments, Golgi apparatus, granules and vesicles, multivesicular bodies, and round (1 to 5 μm) pigment granules that resemble lysosomes in their fine structure (Coggeshall, 1967). The cytoplasmic components within the cell body are not randomly distributed but seem to be organized. The free ribosomes and the cisternae of the granular endoplasmic reticulum are more concentrated near the nucleus and in the region where the axon emanates from the cell body (the axon hillock); pigment granules are usually found midway between the nucleus and the cell membrane (Figure 5-4B; Coggeshall, 1967).

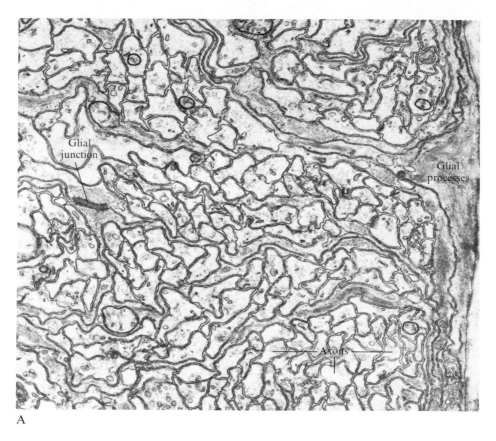

A

FIGURE 5-5.
Axon–glial interrelationships in *Aplysia.*

A. A cross section of the side of a connective. Note that the glial processes are oriented radially with respect to the side of the connective and contain large numbers of glial filaments. A characteristic glial junction can be seen uniting two of the glial processes. The axons are cut in cross section and have irregularly rounded outlines. Most of the axons in this picture have diameters from 0.8 to 2.0 μm. ($\times$ 22,000)

B. (1) A cross section of a peripheral nerve. The glial cells and their processes surround and presumably protect the axons. Note that two glial cells are united by a desmosome-like junction. The large axon in the lower center of the picture is deeply indented by a glial process. These indentations increase as the size of the axon increases, and they reach their acme in the giant axon in the right connective. ($\times$ 14,250) **(2)** A high-power view of a junction between two glial cells in a connective. Note that the extracellular space is filled with a plaque of electron-dense material that is approximately 300 Å wide. The tonofilaments in the glial processes on either side of the junction appear to insert in relatively electron-dense cytoplasm that borders the junction. ($\times$ 60,000) **(3)** A picture of the glial cells that surround the ganglion cell bodies. In contrast to the glial cells in the connectives and peripheral nerves, these glial cells do not possess a tonofilament-filled cytoplasm. The cytoplasm of these cells appears active, as evidenced by a large Golgi complex and the large numbers of mitochondria and vacuoles. Furthermore, in large animals there is a dense intercellular material that fills the extracellular space between these glial cells. ($\times$ 18,000) [From Coggeshall, 1967.]

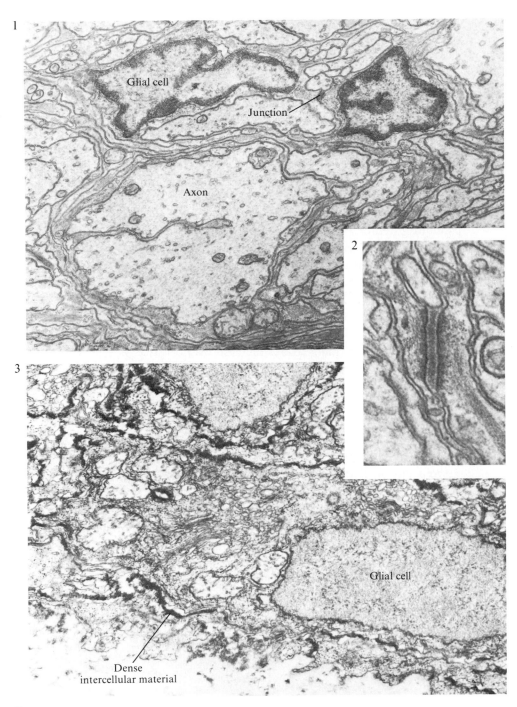

1

Glial cell

Junction

Axon

2

3

Glial cell

Dense
intercellular material

B

Some neurons also contain large granules similar to those found in neuro-secretory cells of vertebrates. Two groups of cells, the bag cells and white cells (R3 to R14) have features typical of neurosecretory cells—their processes end blindly in the sinuses of the sheath. But other granule-containing cells, L2, L3, L4, L5, L6 and R15 (a known neuroendocrine cell), do not send processes to the sheath (Frazier *et al.*, 1967). Although the various granules produced in different cells vary in size and shape from cell to cell (Coggeshall, 1967), in all cells granules appear to be finished in the Golgi complex (for a general discussion see Novikoff and Holtzman, 1976). The Golgi complex contains discrete clumps of material whose electron density is similar to that of the cores of the granules characteristic of the cell. Coggeshall (1967) has therefore suggested that granule formation in *Aplysia* neurons is similar to that in other granule-containing neurons—the clumps found in marginal buds attached to a Golgi cisterna break off to become granules (Palay, 1960).

Like other large invertebrate neurons, the neurons of the abdominal ganglion have prominent glial indentations called *the trophospongium*. In the largest cells these indentations are so extensive in the region of the axon hillock that the latter is broken up into many small intensely basophilic processes which reunite distally to form the axon (Figure 5-4). The number and prominence of these glial indentations increase with cell size.

Fine Structure of the Nucleus

Nuclei of smaller cells are round or oval, with a diameter approximately two-thirds that of the cell body. The nuclei of small neurons have only occasional nucleoli but the giant nuclei of large cells can have several thousand (Coggeshall, 1967).

The nucleus of a large cell, such as R2, may reach 500 μm in diameter. How this great size is achieved has been studied by Coggeshall and his colleagues (1967, 1970) and by Lasek and Dower (1971). Coggeshall (1967) examined sections of ganglia stained for DNA (Feulgen stain) and found that nuclei of large neurons were stained as deeply as those of smaller neurons, suggesting that the concentration of DNA is similar in large and small nuclei. Using the weight of the animal as an index of age, Coggeshall, Yaksta, and Swartz (1970) analyzed the DNA content in the nucleus of the giant cell R2 with a quantitative histochemical stain. They found that the DNA in R2 increases with animal weight, from approximately 2,000 times the haploid amount in small animals to over 75,000 times the haploid amount in large animals.[2] By graphing

[2]The haploid amount of DNA is that contained in a single sperm or egg cell.

the amount of DNA in R2 as a function of the animal's weight they found that the DNA did not increase continuously, but in multiples of two (Figure 5-6). This suggests that as the animal grows the DNA increases by repeated duplications of a large part of, perhaps all of, the genome, The smallest nucleus examined had 2,000 times the haploid DNA, indicating that at this stage the DNA had already undergone 11 duplications without cell division. The largest nucleus with 75,000 times the haploid DNA would therefore represent 15 duplications.

A similar conclusion was reached by Lasek and Dower (1971) using a spectrophotometric determination of the DNA following extraction of individual nuclei from cells by free-hand dissection. They examined a population of large animals and found that the nuclear DNA of R2 (and also of the left pleural giant cell LP1) was distributed in at least two populations: one containing 0.067 μg and the other containing twice that amount, 0.131 μg (Figure 5-6). Since the DNA content of the haploid *Aplysia* sperm cell is 1.0×10^{-6} μg, the data of Lasek and Dower suggest that the nuclei in the two populations had replicated 15 or 16 times. The number of nucleoli in the giant cell increased in fixed ratio with the DNA content of the nucleus (Lasek, Lee, and Przybylski, 1972), providing further evidence that the DNA increase represents synchronous replication of the entire genome.

Neurons with more than the diploid amount of nuclear DNA are not unique to *Aplysia*. Large neurons of the mammalian nervous system—the anterior horn cell of the spinal cord, the Purkinje cell, the sympathetic ganglion cell, and the Betz cell of the cerebral cortex—are tetraploid or octaploid. Neurons of the cerebral ganglia of the fruit fly *Drosophila* contain up to 16 times the haploid amount of DNA. Lasek and Dower (1971) suggest that large cells have additional DNA because they need extra copies of the genome to provide the transcription sites necessary to achieve their great size; beyond a certain point an increase in cell volume may require a concomitant increase in the number of transcription sites. (For a general discussion of evolution of genomic size by DNA doubling, see Sparrow and Nauman, 1976.)

Pigment Granules in the Neuronal Cell Body

The cell bodies of the neurons in the abdominal and other ganglia of *Aplysia* are bright yellow or orange. These colors are due to membrane-bound pigment granules, called lipochondria because of their lipid content (Arvanitaki and Chalazonitis, 1961). These granules resemble the end-stage lysosomes found in neuronal as well as nonneuronal cells (Novikoff and Holtzman, 1976). The pigment appears to be a carotenoid, β-carotene, a precursor of vitamin A, and

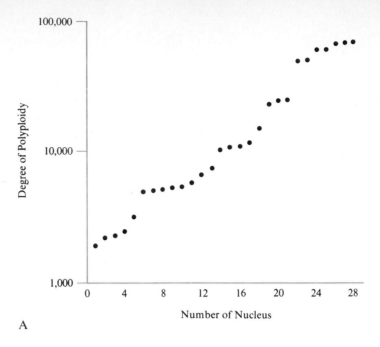

A

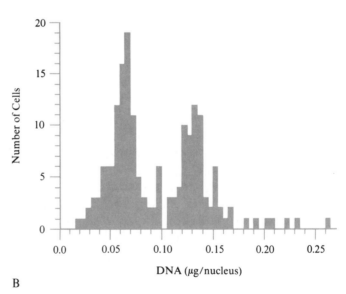

B

FIGURE 5-6.
Polyploidy in the identified cell R2 of *Aplysia*.

A. A graph of the degree of polyploidy plotted against the number of the nucleus. The ordinate is in a logarithmic scale. [From Coggeshall, Yaksta, and Swartz, 1970.]

B. The DNA content of individual nuclei from cells R2, L6 (the symmetrical cell of R2), and P1 is plotted in the form of a frequency histogram at intervals of 0.05 μg of DNA. The nuclei were extruded and DNA was analyzed by fluorometry. The cells fall into at least two classes; the means of the two groups are 0.067 and 0.131 g of DNA. [From Lasek and Dower, 1971. Copyright 1971 by the American Association for the Advancement of Science.]

may be used in the synthesis of a compound similar to rhodopsin (Krauhs, Sordahl, and Brown, 1977). Similar structures have also been described in retinal pigment epithelial cells of vertebrates (Spitznas and Hogan, 1970). In *Aplysia* neurons these pigments have their greatest absorption at 450 to 490 nm (Arvanitaki and Chalazonitis, 1961; Baur *et al.*, 1977). In cell R2 of *A. fasciata* or *A. californica* these pigments are involved in phototransduction. Monochromatic light at 490 nm is absorbed by the lipochondria and gives rise to hyperpolarizing inhibitory photo-responses due to an increase in the K^+ conductance of the membrane (Figure 5-7; Arvanitaki and Chalazonitis, 1961; Brown and Brown, 1972).[3]

There are several reasons for believing that the increased conductance to K^+ is in turn caused by an increase of free intracellular Ca^{++}. First, the effect of light can be simulated by injecting $CaCl_2$ into the cell. Second, the light response is abolished following injection into the cell of a calcium chelator (EGTA). Third, the pigment granules contain 5 nm globular particles that appear to bind Ca^{++}. Light absorption by the pigment changes the structure of the granules with the result that the globular particles change to membrane-like lamellae. It is postulated that as the granules change their conformation they release Ca^{++}, which triggers an increase in K^+ conductance and the hyperpolarizing response (Henkart, 1975; Brown, Baur, and Tuley, 1975; Baur *et al.*, 1977).

In addition, some cells contain the respiratory pigment myoglobin, which is red (see Chapter 3). The myoglobin found in the neurons of *A. californica* is similar to that found in radular muscle and has three absorption bands at 579, 542, and 418 nm (Arvanitaki and Chalazonitis, 1961; Wittenberg *et al.*, 1965).

THE NEUROPIL REGION

Three-Dimensional Architecture at the Light-Microscopic Level

Synaptic connections between neurons are made in the central neuropil region of the ganglion (Figures 5-3 and 5-4A). To understand the organization of the neuropil we must know the course of the axons within it. This can be determined by examining serial sections at the light microscope level (Coggeshall, 1967; King, 1976a, b) or by one of several recently developed techniques that

---

[3]In *A. fasciata* monochromatic light at 578 nm is absorbed by the heme-protein and causes a depolarization (Arvanitaki and Chalazonitis, 1961). In *A. californica* the same light produces only a weak hyperpolarization (Brown and Brown, 1972). The reason for this discrepancy is not clear; it might reflect a differential concentration of the two types of pigment.

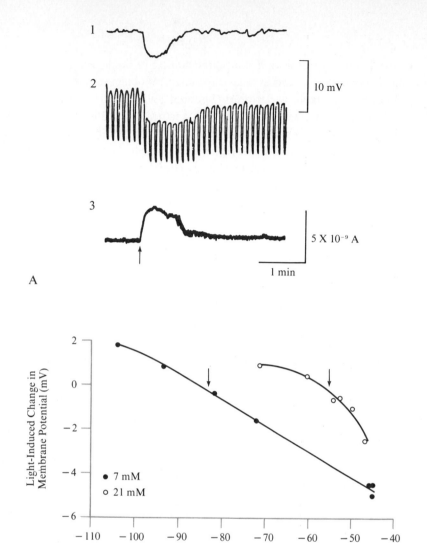

FIGURE 5-7.
Photoresponse in cell R2 of *Aplysia*. [From Brown and Brown, 1972. Copyright 1972 by the American Association for the Advancement of Science.]

A. (1) Membrane potential change produced by illumination of the *Aplysia* giant neuron (resting potential −45 mV). **(2)** Electrotonic potential produced across the membrane of R2 by passing inward-current pulses of 6 nA prior to, during, and after illumination. The decrease in electrotonic potentials indicates that the light produces a decrease in input resistance. **(3)** Membrane current produced by light when the giant cell membrane was voltage-clamped to the resting potential (−54 mV). The onset of illumination is indicated by the arrow at the bottom of the trace.

B. The relation between the light-induced membrane potential change (ΔV_L) and the membrane potential (V_m) prior to illumination, with the ganglion in two different concentrations of external K⁺-activity saline. The arrows indicate E_K in control and high-K⁺ solutions.

use the intracellular injection of a marker substance to label an individual cell and its processes. The substances used are:

1. One of several fluorescent Procion dyes (Stretton and Kravitz, 1968; 1973).

2. Cobalt ion and its subsequent precipitation in the cell as cobalt sulfide by incubation of the ganglion with ammonium sulfide (Pitman, Tweedle and Cohen, 1972).

3. The enzyme horseradish peroxidase, which can be localized histochemically (Muller and McMahan, 1976).

4. Labeled transmitter molecules, such as ^{3}H-serotonin (Goldman *et al.*, 1974; Pentreath and Berry, 1975), which are localized radioautographically.

5. Radio-labeled amino acids that are incorporated into proteins or radio-labeled sugar precursors (fucose or N-acetylgalactosamine), which in turn are incorporated into glycoproteins (Globus, Lux, and Schubert, 1968; Thompson, Schwartz, and Kandel, 1976), and which are localized radioautographically.

Proteins and glycoproteins are retained within the neuron, and some are transported down the axon to the terminals. These substances, although they may cross electrical synapses under some circumstances (Furshpan and Potter, 1968; Globus *et al.*, 1968; Bennett, 1973; Thompson, Schwartz, and Kandel, 1976), can be used to study the three-dimensional geometry of specific cells. Procion dyes, cobalt ions, and horseradish peroxidase can also be used to delineate the position within the ganglion of central sensory and motor neuron cell bodies that have peripheral axons. This is accomplished with all techniques by suspending the cut distal end of a peripheral nerve in the substance and allowing the cell bodies (whose axons run in the nerve) to be back-filled by either diffusion or current (Iles and Mulloney, 1971).

Studies of the crayfish and lobster, based upon intracellular injection of Procion or cobalt ion into neurons, have revealed that the three-dimensional structure of the cell body and main branches of a given identified neuron is fairly similar from animal to animal. In these animals the three-dimensional structure can serve as a fingerprint of these cells (Stretton and Kravitz, 1968, 1973; Selverston, 1973). In *Aplysia* the path of the primary axon is also fairly invariant, but there is sometimes considerable variability in the pattern of the secondary and tertiary branches (Figure 5-8; Winlow and Kandel, 1976). None-theless, this variability does not affect the functional pattern of interconnections as determined by electrophysiological methods, and suggests that the *exact* site of synaptic contact between two neurons need not be invariant.

The dendritic complexity of several well-studied identified cells in the ab-dominal ganglion was found to vary as a function of the extent of their synaptic connections. The white cells, a group of neurosecretory cells that make no known connections within the ganglion and receive few connections from other cells, have a modest dendritic tree. By contrast, L7, which receives many con-nections, and cell L10, which both makes and receives many connections, have rich and extensive dendritic trees. Cells with moderate synaptic intercon-nections (RB cells, R15, L2 to L6) have moderately complex dendritic trees (Figure 5-9).

By studying two (or more) interconnected cells at a time one can examine how specific axons relate to each other in the neuropil. Two types of relation-ships have been found in the few examples so far examined. In one type of relationship, a multibranched, presynaptic neuron, cell L10, was found not to send long processes to reach follower cells. Rather, the axons of the follower cells from different regions of the ganglion join to form fascicles that project to and penetrate the dendritic region of the presynaptic neuron (Figure 5-10; Winlow and Kandel, 1976). A similar pattern has been found in other inverte-brates (Krasne and Stirling, 1972; Kennedy and Davis, 1977). In another type of relationship, multibranched presynaptic neurons in the buccal ganglia of *Aplysia* project their processes to the dendritic region of the follower cells (Kupfermann and Weiss, unpublished). In each case the processes of the pre- and postsynaptic cells interdigitate so extensively that there seems to be much opportunity for multiple contacts (Figure 5-11).

Fine Structure

ELECTRON-MICROSCOPIC LABELS

In optimal instances where the ganglion is small, serial reconstruction at the electron-microscope level can be used to untangle the invertebrate neuropil (see for example King, 1975). However, for most ganglia conventional electron-microscopy usually has to be supplemented with techniques that can mark the central synapses of identified neurons. A number of the techniques for light-

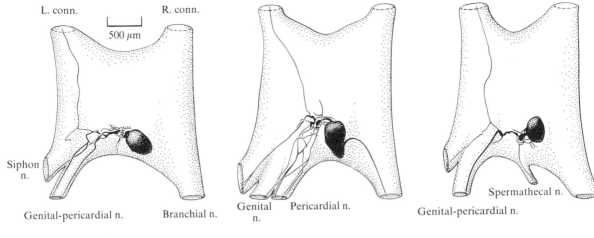

Cell R15

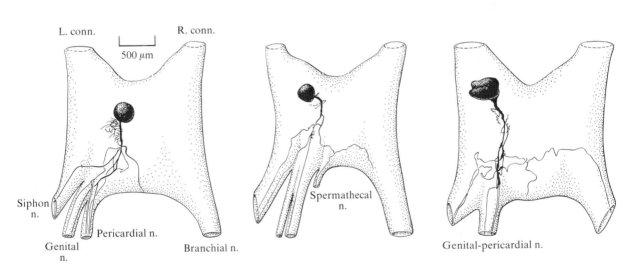

Cell L2

FIGURE 5-8.
The shape of identified cells in *Aplysia* as revealed by whole-mount preparations following intracellular injection of cobalt. Examples illustrate the variability of the structure of cells R15 and L2. [After Winlow and Kandel, 1976.]

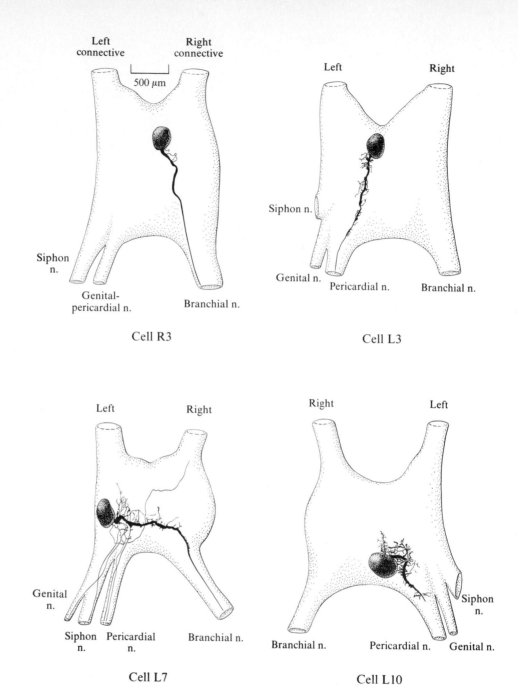

FIGURE 5-9.

Cobalt injections of various cell types in *Aplysia* to illustrate relation of size of dendritic arborization to extent of synaptic input and output. White cell (R3) has little synaptic input; cell L3 has moderate synaptic input; cell L7 has extensive synaptic input; cell L10 has extensive synaptic input and forms many connections to follower cells including L3 and L7. [After Winlow and Kandel, 1976.]

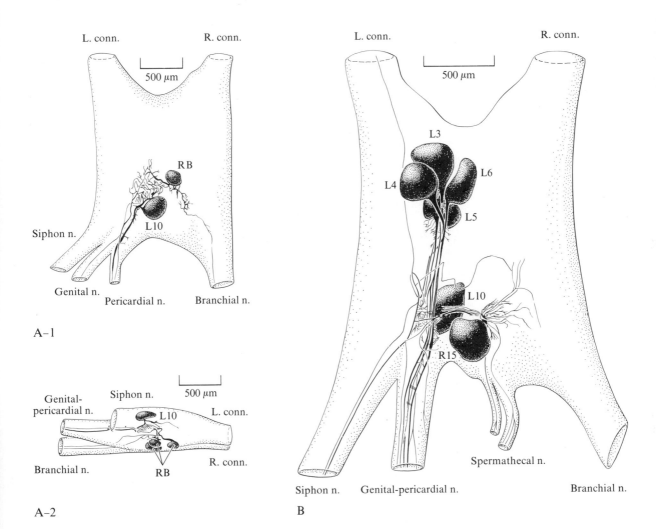

FIGURE 5-10.
Cell L10 and follower cells in *Aplysia*. [After Winlow and Kandel, 1976.]

A. Neuropil organization. **(1)** Dorsal view of one RB follower cell and L10. **(2)** Lateral view of three RB followers with L10.

B. Dorsal view of L10 and four inhibitory follower cells in the left rostral quarter ganglion and the excitatory follower R15.

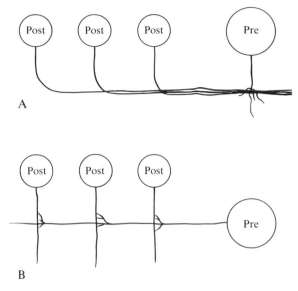

FIGURE 5-11.
Two patterns of interconnection. **A.** Presynaptic neurons receive projections of postsynaptic follower cells. **B.** Presynaptic neurons project to follower cells.

microscopic marking, considered above, can be applied to the electron microscope. With appropriate fixation, both Procion dyes and cobalt salts can be used as electron-microscopic markers (Purves and McMahan, 1972; Pitman *et al.*, 1972) but these two labels are not ideal because they cause cytological damage.

The most successful electron-opaque label is the enzyme horseradish peroxidase (Muller and McMahan, 1975; Bailey *et al.*, 1976; Thompson, Bailey *et al.*, 1976). This enzyme can be injected into a cell and is then transported along the axon. The ganglion is then reacted in the presence of diaminobenzidine and the substrate hydrogen peroxide. The enzyme reduces hydrogen peroxide and oxidizes diaminobenzidine, converting it into a polymer, which remains at all the positions within the cell where the enzyme is located. After postfixation with osmium tetroxide the diaminobenzidine polymer contains large numbers of osmium atoms and these render dense the regions that contain the polymer (Figures 5-12 and 5-13A).

Another approach is to inject intracellularly radioactive precursors of macromolecules or other substances (such as labeled transmitters) that are rapidly transported and can be fixed in the tissue (Globus, Lux, and Schubert, 1968; Pentreath and Berry, 1975; Thompson, Schwartz, and Kandel, 1976). The tissue is then examined by radioautography.

The sugars ^{3}H-L-fucose and ^{3}H N-acetylgalactosamine have been used in this way to study different cells in the abdominal ganglion (Thompson, Schwartz, and Kandel, 1976; Thompson, Bailey et al., 1976). Each of these sugars is incorporated into glycoproteins within the cell body. Labeled material is restricted to the soma and axonal and dendritic trees of the injected neurons and is transported along the axon to the terminals, making it possible to trace the regional anatomy of neurons in the light microscope (Figure 5-13B) and to identify processes and study synaptic morphology in the electron microscope (Figure 5-14).[4]

The availability of both radio-labeled and electron-dense markers makes it possible to mark both the presynaptic and the postsynaptic elements of synapses between identified cells. This type of double-labeling technique has now been used to identify regions of contact between L10 and L3 and the connections between sensory neurons and identified motor neuron L7 in the abdominal ganglion that mediate the gill-withdrawal reflex (Figure 5-15; Thompson, Bailey et al., 1976).

MORPHOLOGICAL BASIS OF SYNAPTIC ACTION

Using conventional thin-section electronmicroscopy in conjunction with single- and double-labeling techniques in the abdominal ganglion of *Aplysia*, a number of preliminary findings have emerged.

Fine structure of synaptic connections.   The cytology of *Aplysia* synapses (Rosenbluth, 1963; Coggeshall, 1967; Jourdan and Nicaise, 1971; Graubard, 1973; Thompson, Schwartz, and Kandel, 1976) has some features in common with that of vertebrates (Peters, Palay, and Webster, 1976). Presumed presynaptic terminals filled with granules or vesicles abut relatively clear, presumed postsynaptic processes. Three classes of synapses can be distinguished. In one

[4]Four cells, R2, L7, L3, and L10, have been studied in detail. In R2, L3, and L7 there is an uncomplicated distribution of label, which is restricted to the neuron that was injected. On the other hand, injection of label into L10, which makes some electrotonic synapses within the abdominal ganglion, always leads to the labeling of one other cell. Most likely, movement of the labeled precursor occurs across electrotonic synapses. A similar transfer of radioactivity from one injected Retzius cell in the leech to its coupled partner cell was reported by Globus, Lux, and Schubert (1973) after iontophoretic administration of ^{3}H-glycine.

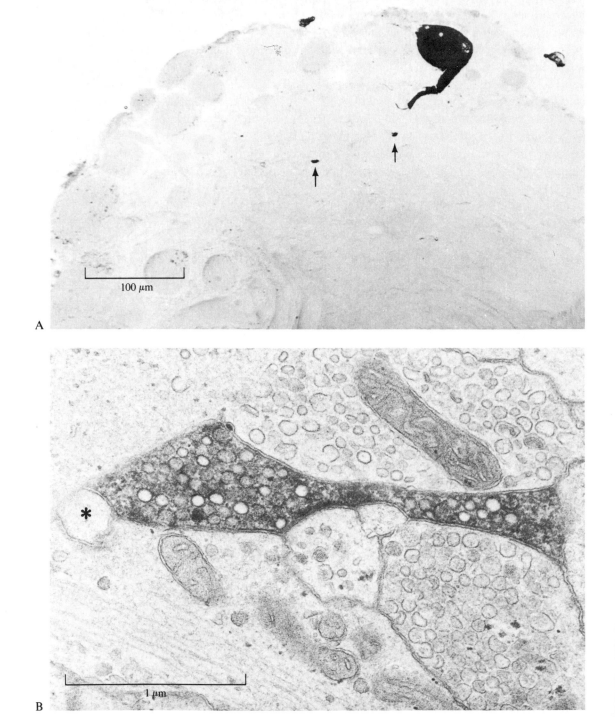

A

B

FIGURE 5-12.

A sensory neuron in *Aplysia*. [From Bailey, Thompson, Castellucci, and Kandel, unpubl.]

A. Light micrograph of a sensory neuron labeled with horseradish peroxidase. The dense, opaque reaction product fills the cell body, main process, and two short segments of processes within the neuropil (arrows), rendering all these areas clearly visible in this unstained 4 μm plastic section.

B. Electron micrograph of a sensory neuron varicosity labeled with horseradish peroxidase. The varicosity is filled with vesicles lying in a blanket of electron-dense reaction product. This varicosity is in close apposition to a small presumptive postsynaptic element (asterisk).

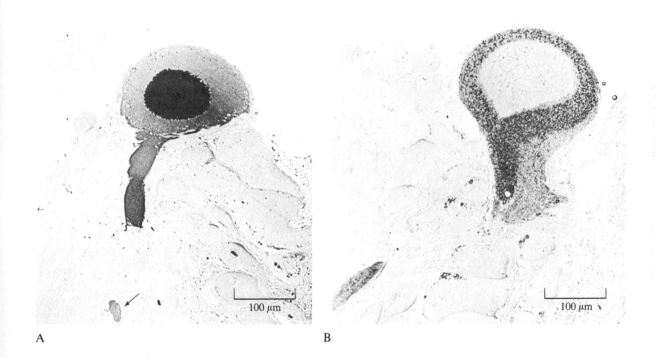

A   B

FIGURE 5-13.

The soma and portions of the principal axon of L7.

A. After injection with HRP. In this 4 μm thick section through the abdominal ganglion HRP reaction product delineates the soma, nucleus (darker), and portions of the principal axon of L7 cut both longitudinally, as it arises from the cell body, and transversely where it travels within the neuropil (arrow). A fine branch can be seen arising from the transversely sectioned profile. The overall morphology of the neuron appears undisturbed by the injection procedure.

B. After injection with ^{3}H-NAGA. The soma and portions of the major axon of L7 are marked by silver grains in this radioautogram of a 4 μm section through the abdominal ganglion. The label is highly localized to the injected neuron, and survey of sequential sections through the entire ganglion reveals no other cell bodies to be labeled. This radioautogram was exposed for four days. [From Thompson, Bailey, Castellucci, and Kandel, unpubl.]

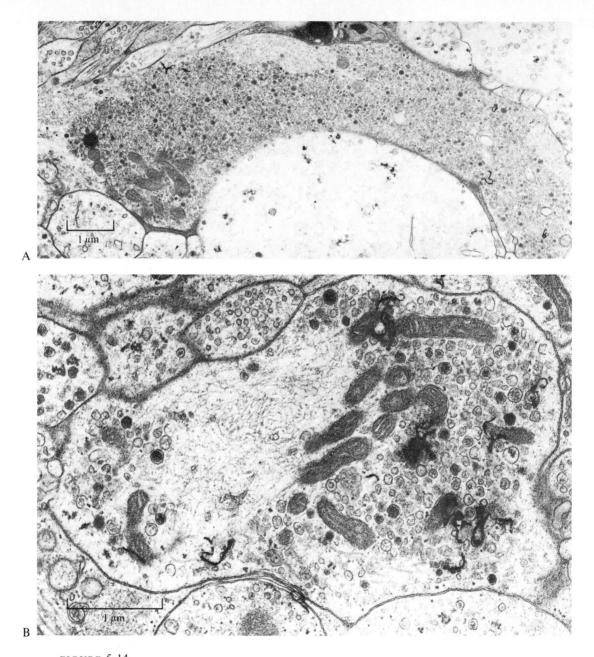

A

B

FIGURE 5-14.
Labeled terminals after injection of L10. [From Thompson, Schwartz, and Kandel, 1976.]

A. Thin section of a presumed synapse. This structure was found labeled in the neuropil of the left caudal quarter ganglion after injection of L10 from a 50-g animal with 5.5×10^4 CPM of fucose and three hours of incubation.

B. This presumed synapse in the neuropil of the left caudal quarter ganglion was labeled after injection of L10 from a 34-g animal with 1.5×10^5 CPM of fucose and 20 hours of incubation. The radioautograph was exposed for 40 days.

class there are no electron-dense bands of cytoplasm or widening of the intercellular gap between the pre- and postsynaptic processes at the zone of apposition, but vesicles cluster in the presynaptic element (Figure 5-16A). In the second class there is a widening of the intercellular gap and the gap is filled with granular material. Apposed membranes are relatively parallel and sometimes a narrow band of electron-dense cytoplasm seems to be closely applied to the cytoplasmic side of both pre- and postsynaptic elements. Vesicles are often clustered at focal sites in the presynaptic element (Figure 5-16B). Whereas these two classes of synapses both have flat types of appositions, the third class of synapse is novel in that the zone of apposition is indented. The postsynaptic member of the synapse protrudes finger-like into the vesicle-filled presynaptic process (Figure 5-16C). As is the case with the second class, there is a clearly defined active zone; the cleft is widened and the apposed membranes are parallel.

Presynaptic terminals are often large and can contain more than one type of vesicle.   Many presynaptic terminals are large (several micrometers in diameter) and varicose. These terminals can contain more than one type of vesicle (Table 5-1). The function of these several types is not known. Perhaps some vesicles simply represent different stages in the life cycle of a single synaptic vesicle type. Alternatively, some vesicles present at synaptic terminals may have functions other than the storing of chemical transmitter substances. Finally, it is possible that some presynaptic terminals may release more than

TABLE 5-1.
Variety of vesicle types found in the presynaptic terminals of identified neurons in Aplysia. (From Bailey et al., 1976; Thompson, Schwartz, and Kandel, 1976; Koester, Shapiro, and Thompson, in preparation.)

Cell	Transmitter	Size of vesicles (nm)	Electron density of vesicle core
LB_{VC}	ACh	1. 88 ± 5 SE (many) 2. 80–150 (few)	Clear Slightly dense
L10	ACh	1. 89 ± 4 SE (many) 2. 110 ± 5 SE (many)	Clear Slightly dense
Sensory neurons	?	80 ± 1.3 SE* (many)	Most clear

*Mean long dimension (axial ratio 1.2).

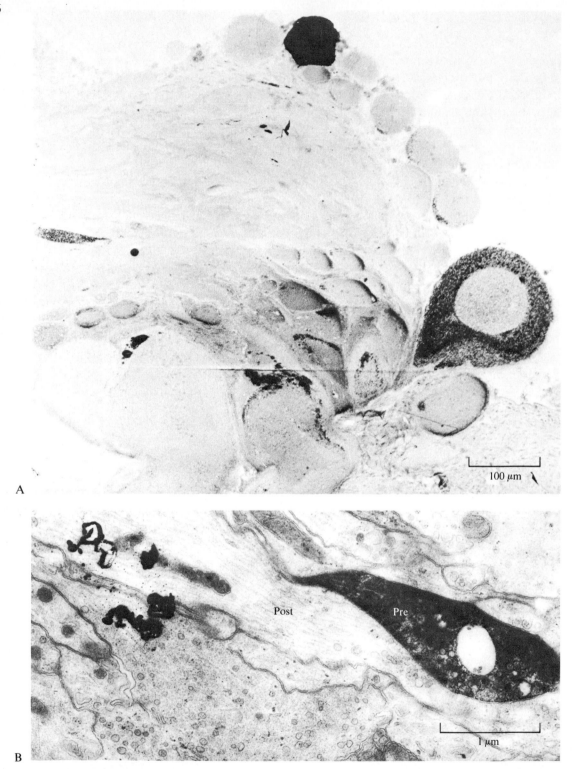

A

B

Post   Pre

100 μm

1 μm

one transmitter substance. But there is no compelling evidence to support this notion. Since individual identified cells can be dissected and their transmitter biochemistry determined, these problems can now be investigated by marking the synapses of cells with known transmitters and correlating vesicle morphology with transmitter biochemistry within a single identified neuron.

The postsynaptic receptive surface is often small and specialized.   The secondary and tertiary branches of the postsynaptic neurons contain many small (< 1 μm) specialized outpouchings called *spines,* which share synaptic specializations with the varicose synaptic terminal of presynaptic neurons, called *varicosities* (Figure 5-17). A single spine may receive the terminals of several different neurons. A similar relationship between varicose presynaptic terminals and spine-like postsynaptic sites is found in other invertebrate ganglia (Muller and McMahan, 1976; King, 1976a) and is characteristic of the vertebrate autonomic nervous system.

Presynaptic terminals can be postsynaptic sites for other presynaptic terminals. In invertebrate neurons the presynaptic releasing regions (usually varicosities) are often located adjacent to postsynaptic receptive areas. In some cases the varicosity itself seems to serve as the receiving region for synaptic input from other neurons. These morphological arrangements may be the anatomical bases for presynaptic facilitation and inhibition (see *Cellular Basis of Behavior,* Fig. 6-21 and Chapter 11).

Presynaptic neurons make multiple postsynaptic contacts.   A single presynaptic neuron often makes multiple contacts with a single postsynaptic cell. Moreover, after contacting a given postsynaptic process the presynaptic process

FIGURE 5-15.
Double label injection of a sensory cell and the gill motor neuron L7. [From Thompson, Bailey, Castellucci, and Kandel, unpubl.]

A. In this radioautogram of a 4 μm section through the abdominal ganglion, L7 is marked by silver grains over its soma and axon and a sensory neuron cell body containing HRP appears uniformly and darkly stained. Fine branches of the sensory neuron are indicated by the arrow.

B. This contact between a vesicle-filled sensory neuron varicosity (*Pre*) and a fine process of L7 (*Post*) exhibits many features characteristic of a sensory synapse, though it lacks a synaptic specialization. Serial reconstruction of such a sensory neuron varicosity would be expected to reveal at least two synaptic specializations onto adjacent fine processes. Such regions of contact between HRP-labeled sensory neurons and a radioautographically labeled L7 are being examined in serial sections to determine if all of the morphological criteria for a synaptic contact are met.

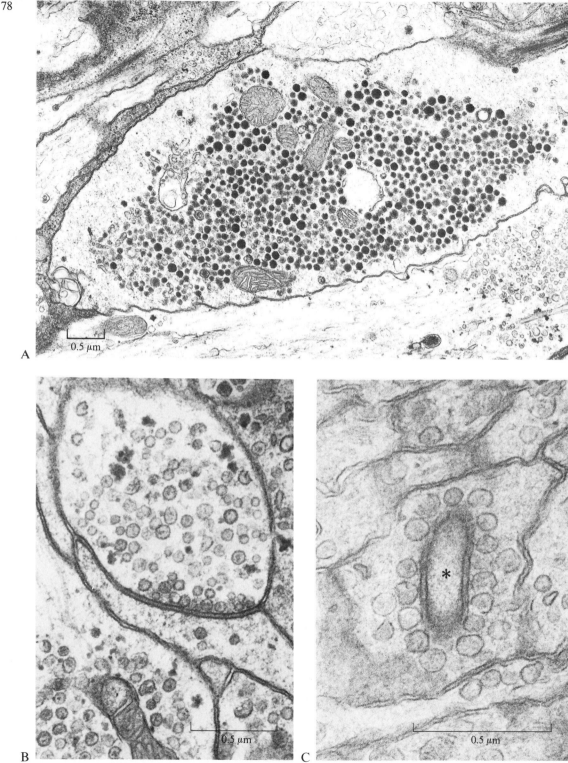

A

0.5 μm

B

0.5 μm

C

0.5 μm

FIGURE 5-16.

Three types of contact in the neuropil.

A. Flat type of synaptic apposition without prominent specialization. These are seen commonly and consist of a vesicle and/or a granule-filled profile abutting a relatively clear neuronal process and seem to represent random sections that miss the active zones of the varicosity. Two such regions of specialized membrane are illustrated in parts B and C. [From Thompson, unpubl.]

B and **C.** Clearly defined synaptic contacts **B.** Flat type of synaptic apposition. The vesicles are closely applied to the presynaptic membrane; the widened intercellular space is filled with a granular material; the membranes are parallel; and the very narrow layers of electron-dense cytoplasm are closely applied to the cytoplasmic leaflets of the apposed synaptic membranes. [From Thompson and Bailey, unpubl.] **C.** Indented type of synaptic apposition. Here a small postsynaptic finger (*) invades a vesicle-filled presynaptic terminal. At regions of a widened cleft a collection of thin filaments appears to connect vesicles with presynaptic membrane. [From Bailey and Thompson, unpubl.]

often goes on beyond it to contact other processes forming *en passant* synapses (Figure 5-18).

A single postsynaptic cell receives synaptic input from a variety of synaptic terminals. A variety of synaptic boutons have been found to end on axons of the large identified cells in the abdominal ganglion (Coggeshall, 1967; Frazier *et al.,* 1967; Graubard, 1973; Thompson, Bailey *et al.,* 1976). Coggeshall's studies, summarized in Table 5-2, show that a given neuron receives terminals containing different types of vesicles and granules. Since different types of

TABLE 5-2.

Types of vesicles in the presynaptic terminals contacting the axonal processes of some identified cells in Aplysia. *(From Frazier* et al., *1967.)*

Cell	Number of terminals	Size of vesicles in different terminals (nm)	Electron density of vesicle core
R3–R13	Few	1. 100 2. 40–50 3. 40–50 4. 120	Small, dense core Clear Clear Moderately dense core
R15	Numerous	1. 120 2. 100–120 3. 120	Clear Slightly dense core Dense core
L2–L6	Numerous	1. 100–120 2. 80–100	Diffusely dense Clear
R2	Numerous	1. 80 2. 100 3. 100	Clear Small, dense core Diffusely dense
L1 and R1	Sparse	1. 80–100 2. 100–120	Diffuse Diffuse, not very electron dense

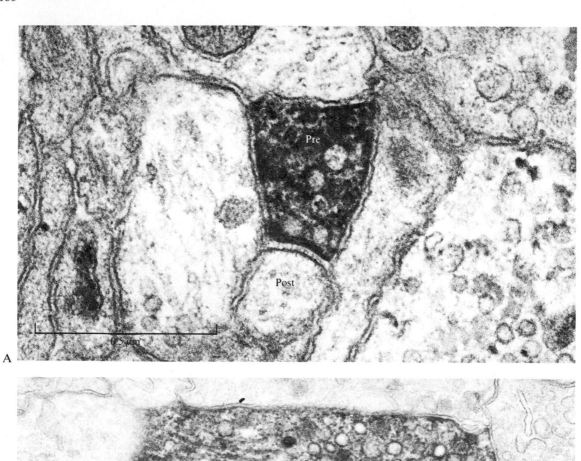

A

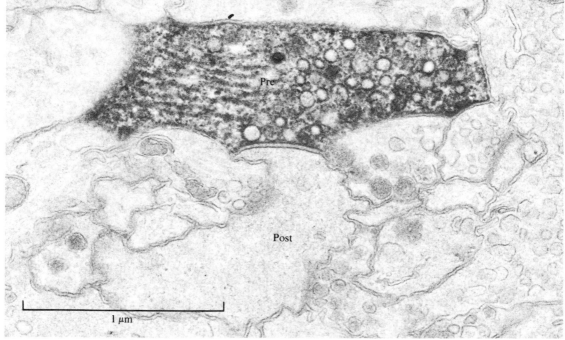

B

FIGURE 5-17.

Postsynaptic spines. [From Bailey, Thompson, Castellucci, and Kandel, unpubl.]

A. Small horseradish peroxidase-filled profile of a sensory neuron in synaptic register with a single unlabeled postsynaptic element. Note the slightly widened synaptic cleft containing thin, dense striations running perpendicular to the apposed synaptic membranes. Serial reconstruction has shown that these small postsynaptic elements have a spine-like geometry. (Scale = 0.5 μm)

B. In this micrograph a greater portion of the presynaptic sensory varicosity is included in the plane of section and the unlabeled postsynaptic element has been cut more parallel to its long axis (unlike the spine transversely sectioned in part A). Vesicle apposition, a slight widening of the cleft (in this case containing a faint intermediate density), and parallel alignment of pre- and postsynaptic membranes are characteristic of the active zones at these contacts.

granules may package different chemical transmitters, the evidence is in accord with pharmacological findings that suggest that a given neutron may receive input from presynaptic terminals of neurons using a number of different transmitters. For example, cell R15 appears to receive some synaptic connections from cells that utilize acetylcholine, whereas others utilize serotonin (Gerschenfeld, Ascher, and Tauc, 1967) or dopamine (Ascher, 1972).[4]

RELATIONSHIP BETWEEN CELL BODY AND AXON TERMINALS

The ability to mark the synapses of identified cells makes it possible to study the relationship of the cell body, where transmitter vesicles and enzymes are synthesized, to the axon terminals, the site of transmitter release.

As the stout axonal process of the neuron leaves the region of the cell body, the character of the cytoplasm changes abruptly (see Figure 5-4). The cytoplasm of the cell body is deeply basophilic and filled with a larger variety of organelles than is the axon. There is no obvious structure that could act as a mechanical boundary separating the two regions (Coggeshall, 1967). Several organelles, such as vesicles, mitochondria, smooth endoplasmic reticulum,

[4]Scanning electron-microscopy has also been used to study the neuropil of the abdominal ganglion. Lewis, Everhart, and Zeevi (1969) found that they could trace neuronal fibers over considerable distances from the cell body to their presumed synaptic terminations. Their preliminary studies suggest that the terminal processes of neurons exhibit extensive anastomotic branching and terminate synaptically on each other as well as upon a layered structure. The nature of these layered structures is unclear, but Lewis and his colleagues suggest that they represent nodal points in the neuropil where groups of interconnecting processes converge. The nodal points are made up of three elements: (1) the presynaptic elements: (2) the postsynaptic processes they synapse upon; and (3) interstitial material. Zeevi (1972) defined these nodal points as being "superspines" that serve as receptive and integrative loci for neurons. To identify the nature of these structures will require a combination of scanning and transmission electro-microscopy.

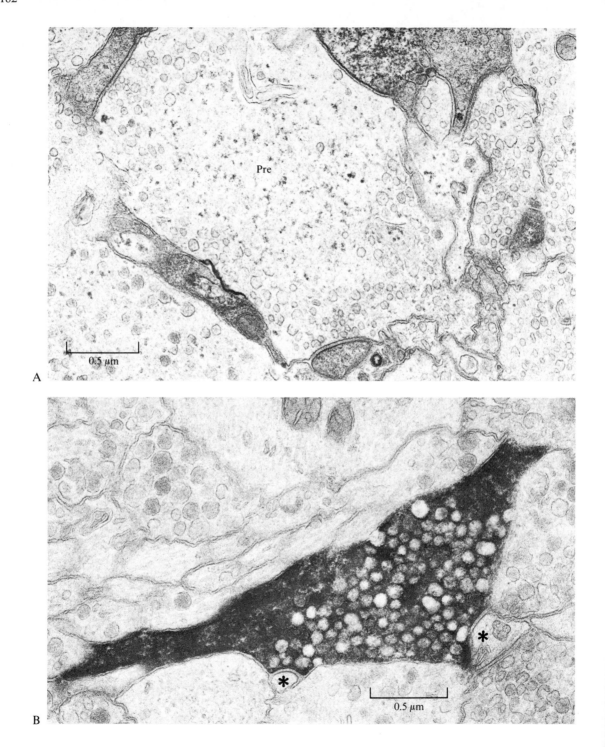

A

B

FIGURE 5-18.

Multiple contacts made by a single labeled neuron.

A. An unlabeled presynaptic process (*Pre*) contacting more than one postsynaptic HRP-labeled L7 profile. [From Thompson, Bailey, Castellucci, and Kandel, in preparation.]

B. Portion of a vesicle-laden HRP-labeled sensory varicosity forming two distinct synaptic contacts (*) with postsynaptic elements. [From Bailey, Thompson, Castellucci, and Kandel, unpubl.]

neurofilaments, and microtubules, are transported from the cell body to the axon. Other organelles, such as rough endoplasmic reticulum and free ribosomes, are absent from the axon. Perhaps one obstacle to the movement of some of these organelles beyond the axon hillock and down the axon is lack of a transport mechanism. However, since these organelles could diffuse slowly down the axon, it is likely that they may be tied into the cell body by some system of fibrils.

Movement of organelles from the cell body down the axon to the terminal region can be studied following injection of the cell body with ^{3}H-fucose (Thompson, Schwartz, and Kandel, 1976). Electron-microscopic examination revealed that areas containing heavy labeling coincide with accumulations of vesicles, mitochondria, and smooth endoplasmic reticulum. This distribution suggests that these organelles move through special areas or channels in the axons (for review see Schwartz, 1979).

The cell body and axon of cholinergic neurons R2 and L10 contain the same two populations of vesicles found in the terminals, namely, many moderately electron-lucent (60–130 nm) and some dense-core (80–150 nm). This indicates that the vesicles are assembled in the cell body and transported down the axon to the terminals. Indeed, statistical studies reveal that much of the radioactivity that leaves the cell body is incorporated into vesicles for transport down the axons to the terminals (Thompson, Schwartz, and Kandel, 1976).

SUMMARY AND PERSPECTIVE

Invertebrate ganglia vary in size and complexity. Each ganglion is usually covered with a connective tissue sheath, which in some cases also serves as a neurohemal organ. The cell bodies form a cortex on the outside of the ganglion surrounding the central neuropil.

The cell bodies are usually large, monopolar, and free of synapses. They commonly also have a large nucleus. For example, in cell R2, which has a diameter of 800 μm, the nucleus may reach 500 μm in diameter and contain 0.13 μgm of DNA. Individual cells and even their nuclei can be dissected by hand, so that one can compare the molecular composition (of the DNA and other macromolecules) of functionally different neurons and of different parts of an individual neuron. In addition, one can correlate biochemical properties of a neuron with its fine structure, and both of these aspects of cellular function can in turn be correlated with the behavioral role of the cell. Lacking, however, are comparable data about the neuropil. This ignorance about the architecture of the neuropil in *Aplysia* has been a major deterrent to a more detailed study of central ganglionic function.

The recent development of techniques for marking the synaptic regions of individual neurons should soon allow the unraveling of the three-dimensional architecture of the neuropil and the correlation of synaptic structure with synaptic function and biochemistry. A number of radioautographic labels (such as fucose or n-acetylgalactosamine) and electron-opaque markers (such as horseradish peroxidase) are now available that allow recognition of both the pre- and postsynaptic elements of the synapses of identified cells. As a result, much progress is to be expected. Indeed, several findings have already emerged that form the basis for exploring the principles underlying the morphological basis of synaptic actions. For example, by using double-label techniques, presumed synaptic contacts between neurons have been shown. This provides the first direct evidence supporting the electrophysiological inferences about monosynaptic connections between certain identified cells. Moreover, these techniques have delineated a number of features of the presynaptic terminals, including: (1) that the terminals are often relatively large and bulbous (varicose) and can contain more than one vesicle type; (2) that they end on relatively small spines given off by secondary processes of the postsynaptic cells; and (3) that a single presynaptic neuron makes multiple synaptic contacts with a single postsynaptic cell.

Since labeled sugars get incorporated into the glycoproteins of synaptic vesicles, one can also study the transport of transmitter vesicles from their site of synthesis in the cell body to their site of release at the synaptic terminals. These types of studies are an essential prerequisite for analyzing the relationship of transmitter biochemistry to synaptic physiology and for examining the molecular mechanisms of different forms of synaptic plasticity.

SELECTED READING

Kater, S. B., and C. Nicholson, eds. 1973. *Intracellular Staining in Neurobiology.* New York: Springer-Verlag. A review of intracellular marking methods for light microscopy.

Kennedy, D., A. I. Selverston, and M. P. Remler. 1969. Analysis of restricted neural networks. *Science,* 164:1488–1496. Application of Procion dye technology to the crayfish escape system.

King, D. G. 1976. Organization of crustacean neuropil. I: Patterns of synaptic connections in lobster stomatogastric ganglion. *J. Neurocytology,* 5:207–237. II: Distribution of synaptic contacts on identified motor neurons in lobster stomatogastric ganglion. *Ibid,* pp. 239–266. An important study of the synaptic organization of an invertebrate ganglion (the stomatogastric ganglion of the lobster) by means of serial reconstruction.

Stretton, A. O. W., and E. A. Kravitz, 1968. Neuronal geometry: Determination with a technique of intracellular dye injection. *Science,* 162:132–134. The first description of a reliable dye (Procion yellow) for intracellular staining of individual nerve cells at the light-microscopic level.

CHAPTER 6

Interaction Between the Peripheral and Central Nervous Systems in *Aplysia*

The nervous system of molluscs, like that of vertebrates, contains both central and peripheral neurons. In vertebrates the peripheral motor neurons are restricted to certain viscera, such as the stomach and intestine. In molluscs the peripheral nervous system is more extensive. For example, peripheral motor pathways course under the skin to innervate both somatic muscles (foot and appendages) and visceral muscles. These pathways can mediate some responses in the absence of central ganglia.[1] Analysis of a given behavior therefore requires a detailed understanding of the relative contribution of the peripheral plexus and central nervous system.

The existence of central and peripheral pathways can confound the analysis of behavioral modifications in molluscs (Kandel and Spencer, 1968; Jacklet and Lukowiak, 1974). But the recent development of experimental controls for distinguishing between the two components of a dually innervated effector system allows one now, in certain favorable cases, to analyze the roles of both components in behavior. In this chapter I shall review the types of innervation found in several organ systems in *Aplysia* and the relative contribution of central and peripheral pathways to various behaviors.

[1]Bullock and Horridge (1965), Prosser (1946), Jordan (1901), Kandel and Spencer (1968), Peretz (1970), Newby (1972), Kupfermann *et al.,* (1971); Carew *et al.* (1972), Lukowiak and Jacklet (1972), Jacklet and Lukowiak (1974), Perlman (1975), Bailey *et al.* (1975 and unpubl.).

THREE PATTERNS OF INNERVATION
IN A COMMON EFFECTOR SYSTEM

In 1903 Bethe suggested that the peripheral and central nervous systems of *Aplysia* mediate different types of effector responses. He suggested that the peripheral nervous system mediated local responses that are largely restricted to the site of stimulation, and that the central nervous system controlled rapid actions over distances that extend greatly beyond the site of stimulation. Although Bethe's work was supported by several subsequent studies (e.g., Hoffmann, 1910), his proposal left certain questions unresolved. Can any response be mediated exclusively by central or peripheral systems or do all responses involve concomitant activation of both systems? Do the central ganglia innervate effector organs directly or do they invariably control effector structures via the peripheral plexus?

To answer these questions it has been necessary to isolate and examine independently the central and peripheral components of a given movement. This type of study has now been carried out on several effector systems in *Aplysia:* the gill, siphon, parapodia, foot, heart, and buccal muscles. In general, the results confirm Bethe's distinction between *relayed* (or *remote*) *reflexes,* which are largely mediated by the central nervous system with minimal or no contribution from the peripheral nervous system, and *local responses,* which invariably involve the peripheral nervous system but not always the central nervous system. This distinction, however, is not absolute and often depends upon the strength of stimulation. Weak stimuli tend to produce much more selective activation of one or the other pathway than do strong or noxious stimuli.[2]

Relayed Reflexes: The Gill-Withdrawal Reflex

The best studied case of a relayed reflex in *Aplysia* is the gill-withdrawal reflex to stimulation of the siphon. A weak tactile stimulus applied to the siphon—either water jets of 125 g/cm^2 applied to 300 mm^2 of siphon skin or a behaviorally comparable mechanical probe stimulus of 400 g/cm^2 applied to 0.25 mm^2 of skin—elicits a conjoined response at the site of stimulation, the siphon (to be considered below), and a relayed reflex response at a remote site, the gill (Figure 6-1A). Central pathways, originating within the abdominal

---

[2]Kupfermann *et al.* (1971, 1974), Jacklet and Lukowiak (1974), Perlman (1975), Peretz *et al.* (1976), Carew *et al.* (1976).

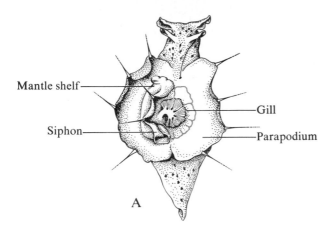

A

Gill-Withdrawal Reflex   Pinnule Response

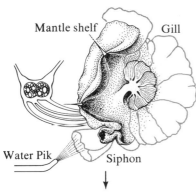 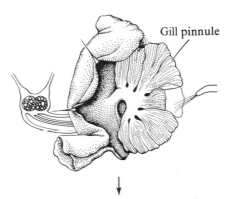

Gill-Withdrawal Reflex After
Removal of Abdominal Ganglion

Pinnule Response After
Removal of Abdominal Ganglion

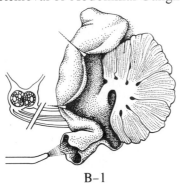 

B-1   B-2

FIGURE 6-1.
Peripheral and central nervous contributions to the gill-withdrawal reflex and the pinnule response. [After Kupfermann *et al.,* 1971.]

A. Dorsal view of an intact animal showing the relaxed siphon, mantle shelf, and gill.

B. Comparison of gill-withdrawal reflex and pinnule response. The mantle shelf has been retracted to expose the mantle cavity and gill. **(1)** The gill-withdrawal reflex to siphon stimulation before and after removal of the abdominal ganglion. **(2)** The pinnule response before and after removal of the abdominal ganglion. Whereas removal of the abdominal ganglion eliminates the gill-withdrawal reflex to weak stimuli, it does not block the pinnule response.

ganglion, are both necessary and sufficient for mediating the gill-withdrawal component of this defensive response to weak and moderate intensity stimuli.[3]

The abdominal ganglion contains at least six motor cells that produce movements of the gill (L7, LD_{G1}, LD_{G2}, RD_G, $L9_{G1}$, $L9_{G2}$). The three major motor neurons (L7, LD_{G1}, LD_{G2}) are known to make direct connections with gill muscles (Figure 6-2). The three minor motor cells are presumed to make direct connections but the evidence for them is incomplete.

The abdominal ganglion also contains a cluster of about 24 sensory neurons that innervate the siphon skin (Figure 6-3). These cells have a low threshold

---

[3]There has been general agreement that the gill-withdrawal reflex to weak tactile stimuli is entirely mediated by the abdominal ganglion, whereas the reflex to strong or noxious stimuli involves peripheral pathways as well. However, there were, until recently, conflicting findings on the neural control of the reflex response to *moderate-intensity* stimuli (water jet 125–150 g/cm²; mechanical probe 400–1800 g/cm²). Kupfermann and co-workers (1971) reported that the abdominal ganglion mediates 94 percent of the response to moderate tactile stimuli, whereas Peretz and co-workers (1976) reported that it mediates little or none of the response.

In joint experiments the two groups found that the differences seem to result from the type of preparation and stimuli used, and the criteria for accepting responses. Kupfermann and co-workers studied intact animals that exhibited near maximal responses to moderate-intensity jets of seawater and used only preparations that showed responses with a minimal criterion of at least 35 percent of maximal spontaneous pumping movements. Peretz and co-workers used an isolated gill-mantle preparation and did not exclude small reflex responses.

To see if these differences explained the discrepancy, Carew and co-workers (1976) videotaped the reflex elicited by water jets of moderate intensity (250 g/cm²) in intact animals. Whereas sham operation (N = 10) reduced the reflex to 98.3 percent of control, deganglionation (N = 10) reduced it to 10.1 percent. Using a protocol developed jointly with Peretz and co-workers, they next examined the reflex with a strain gauge in isolated preparations, and alternated a controlled-force probe to the rim of the siphon (Byrne *et al.*, 1974) with a "tapper" stimulus to the base (Peretz *et al.*, 1976). A single force of 4.0 g was used (probe diameter 0.56 mm, $p = 1600$ g/cm₂; tapper diameter 1.5 mm, $p = 228$ g/cm²). Preparations were accepted only if probe responses were at the outset at least 35 percent of maximal spontaneous contractions. Deganglionation (N = 10) reduced the responses to probe and tapper to 5 and 2 percent of control, respectively.

In additional experiments (N = 7) no criterion on the probe response was imposed. Now the results were clearly different. Whereas the mean probe response after deganglionation was again small (14 percent of control), the mean tapper response was 121 percent of control.

Thus, the CNS mediates at least 90 percent of the reflex to moderate-intensity stimuli in the intact animal and at least 95 percent in the surgically isolated preparation when minimal response criteria are used. Although the CNS still mediates at least 86 percent when small reflex responses to a feedback-controlled probe are examined, the CNS does not, on the average, contribute to small responses to the tapper used by Peretz and co-workers.

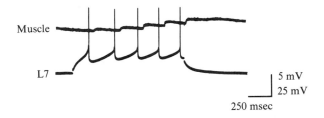

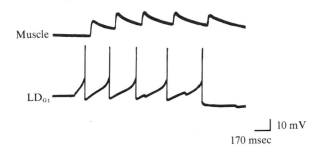

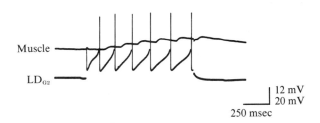

FIGURE 6-2.
Intracellular recordings from gill motor neurons L7, LD_{G1}, and LD_{G2} and from gill muscle cells. [From Carew *et al.*, 1974.]

to mechanical stimulation (100 g/cm²) that is identical to the threshold for the reflex (Byrne, Castellucci, and Kandel, 1974; 1978). The sensory cells make monosynaptic connections with the gill motor cells and to several types of inter-neurons (Figure 6-3).

This neural circuit seems to account for most of the relayed reflex. For example, a weak tactile stimulus to the siphon activates about eight sensory

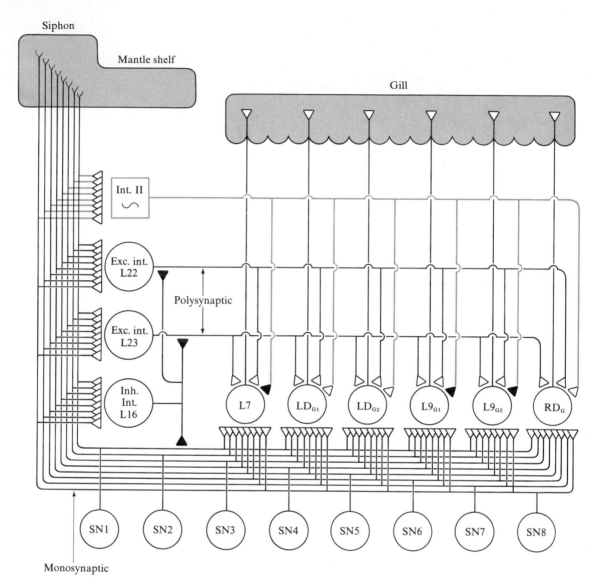

FIGURE 6-3.
The neuronal circuit of the gill-withdrawal reflex. Simplified wiring diagram of the reflex to stimulation of a point on the siphon skin. The entire siphon skin is innervated by approximately 24 mechanoreceptors. Stimulating a 0.5 mm point on the skin activates about eight sensory neurons. The sensory neurons (SN) make direct (monosynaptic) connections to six identified gill motor neurons and to at least one inhibitory interneuron (L16) and two excitatory interneurons (L22, L23). In this and all subsequent circuit diagrams, light lines indicate unidentified cells; black lines indicate verified cells and connections. Squares indicate cell populations; circles indicate single identified cells. Black triangles indicate inhibitory connections; white triangles indicate excitatory connections. [After Byrne et al., 1974.]

cells, causing each to fire an average of one spike per stimulus (Figure 6-4A). Byrne, Castellucci, and Kandel (1978) have modeled the reflex by stimulating a single sensory cell eight times so as to simulate the activation of a population of eight sensory cells (Figure 6-4).

Surgical removal of the abdominal ganglion or blocking synaptic transmission with high-Mg^{++} solutions abolishes the reflex response to weak stimuli. Even removing a single major motor cell (L7 or LD_{G1}) from the reflex by hyperpolarization reduces reflex response by 30–40 percent (Figure 6-5). The two

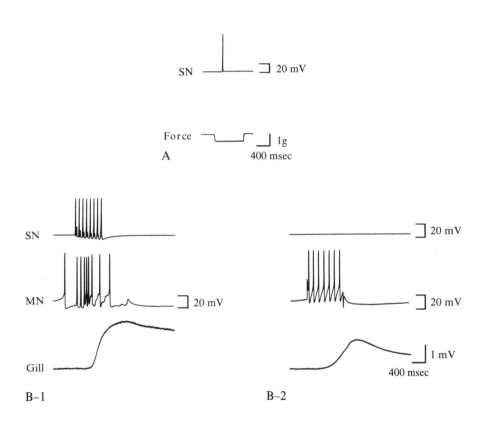

FIGURE 6-4.
A single sensory cell elicits gill withdrawal. [From Kandel *et al.,* 1976.]

A. A controlled-force punctate stimulus (800 msec in duration) was delivered to the siphon skin while intracellular recordings were obtained from an individual sensory neuron.

B. (1) The sensory neuron (SN) was fired intracellularly with a burst of eight spikes, producing a series of action potentials in the motor neuron (MN) and a large reflex response. **(2)** Intracellular stimulation of the motor neuron with seven spikes produced a contraction that is about 40 percent of the full reflex response in part B-1.

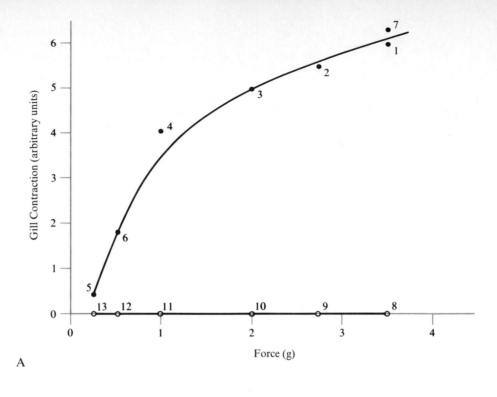

A

1. Seawater

Gill

L7

2. High Mg++

3. Seawater

20 mV

2 sec

B

1. Hyperpolarized

Gill

L7

2. Nonhyperpolarized

3. Hyperpolarized

4. Direct firing

80 mV

1 sec

C

cells alone thus account for 70 percent of the reflex. Conversely, surgical isolation of the siphon from the gill and mantle shelf—so that the siphon is connected to the abdominal ganglion only by the siphon nerve, thereby disrupting all possible peripheral pathways—does not interfere with the ability of a stimulus to the siphon to elicit gill withdrawal (Figure 6-6).

At moderate-intensity stimuli (125–500 g/cm^2 water jet; 400–1800 g/cm^2 mechanical probe stimulus) a small peripheral component begins to be recruited. This component contributes 10 percent to the total response (Kupfermann et al., 1974; Carew et al., 1976). At even greater pressures (500 g/cm^2 water jet; 1800 g/cm^2 probe) the peripheral contribution may become larger still and serve as an amplifying mechanism to enhance the central response to noxious stimuli (Kupfermann et al., 1971, 1974; Peretz et al., 1976; Carew et al., 1976). Thus one can examine the reflex as mediated predominantly by the central nervous system or as mediated conjointly by the central and peripheral nervous systems, depending on stimulus intensity. Other factors that preferentially elicit central as opposed to conjoint mediation are a minimal criterion of reflex responsiveness and type of mechanical stimulus used (Carew et al., 1976, and unpubl.)

FIGURE 6-5.

A. Graded gill response to tactile stimuli of different intensities. In a simplified preparation constant-force stimuli of different amplitudes were applied to the siphon skin every 25 min. The evoked gill contractions were recorded by means of a photocell; 6 arbitrary units (au) equals about 65 percent of a maximal gill contraction. The numbers on the data points indicate the sequence of stimuli. After stimulus 7 the abdominal ganglion was removed, abolishing the reflex (stimuli 8 to 13). In some preparations a small component (about 5 percent) of the reflex remained in response to stimulation of 4 g after the ganglion was removed. The diameter of the probe was 0.5 mm, so that 1 g produced a pressure of 400 g/cm^2. [From Kandel, Brunelli, Byrne, and Castellucci, 1976.]

B. Effects on the gill-withdrawal reflex of functional removal of the abdominal ganglion by blockade of synaptic transmission. In each part of the experiment the gill reflex was elicited by a jet of seawater (indicated by the short bar) applied to the siphon. In the first pair of traces the ganglion was bathed in normal seawater. In the second pair it was bathed in a high-Mg^{++} solution, which blocks synaptic transmission in the ganglion. The third pair of traces shows the gill response and the synaptic output to cell L7 after the ganglion was returned to normal seawater. [After Kupfermann et al., 1971.]

C. Effects on the gill-withdrawal reflex of functional removal of cell L7 from the neural circuit by hyperpolarization. The gill-withdrawal reflex was elicited by a jet of seawater (indicated by the short bar) applied to the siphon every 5 min. On alternate trials cell L7 was hyperpolarized so that the excitatory input could not discharge it. In the last set of traces L7 was directly fired by a long depolarizing pulse that was adjusted to fire L7 in a pattern comparable to that seen during gill contraction. The size of the gill contraction due to direct firing of L7 was approximately 30 to 40% and equaled the reduction of gill contraction caused by removing L7 from the reflex. [After Kupfermann et al., 1971.]

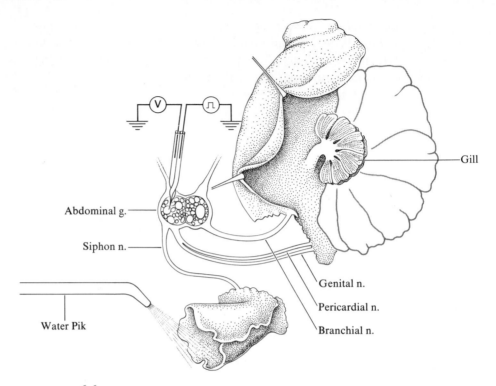

FIGURE 6-6.
Isolated siphon preparation to eliminate possible peripheral pathways between the siphon and the gill. The siphon can be isolated except for its connection to the abdominal ganglion via the siphon nerve. The abdominal ganglion remains connected to the gill via the branchial and pericardial-genital nerves. [After Kupfermann *et al.*, 1971.]

Local Responses: The Pinnule Response

The best-studied case of a purely peripherally mediated behavior is the response of the gill pinnules to direct stimulation. Peretz (1970) has found that a weak tactile stimulus applied directly to a gill pinnule elicits a local response, the *pinnule response,* which is restricted to the stimulated gill pinnule (Figures 6-1B and 6-7A).[4] The central gill motor neurons are not activated by weak

[4]The pinnule response to pinnule stimulation has been referred to as the "gill-withdrawal reflex" by Peretz (1970; and see Peretz and Lukowiak, 1975). But in view of the earlier use of that term for the gill movements produced by siphon stimulation (Kupfermann and Kandel, 1969; Kupfermann *et al.,* 1971), the use of the same term for a gill response initiated in a different way and mediated in a different manner seems confusing. The term "pinnule response" or "pinnule reflex" is therefore preferable.

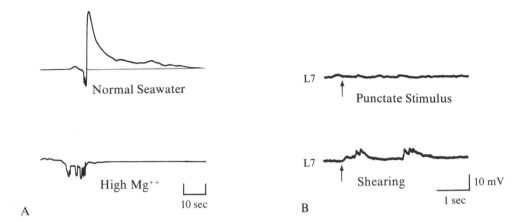

FIGURE 6-7.

A. Effects of a high-Mg^{++} solution on the peripherally mediated pinnule response. In normal seawater a local pinnule response (recorded by a photocell) was produced by gently stimulating a single pinnule with a glass probe. When the gill was perfused with seawater containing four times the normal Mg^{++} (200 mM) the pinnule response was completely abolished, even to vigorous tactile stimulation. [From Carew *et al.*, 1974.]

B. Synaptic input to gill motor neuron L7 from the gill and the siphon (L7 was hyperpolarized to prevent background spike activity). A brief punctate stimulus (sable brush) applied to a pinnule resulted in no synaptic input to L7. A shearing stimulus (brush stroke) to the gill resulted in a small synaptic input to L7. (The same stimulus applied to the siphon resulted in a massive synaptic input eliciting several spikes in L7.) [From Kupfermann *et al.*, 1971. Copyright 1971 by the American Association for the Advancement of Science.]

stimuli to the gill pinnules; peripheral pathways are both necessary and sufficient for its mediation. With stronger stimuli the response spreads to other pinnules. At these strengths central motor neurons are synaptically excited and may contribute to the responses (Figure 6-7B).

Preliminary evidence suggests that these peripheral pathways operate by means of chemically mediated synapses (Carew *et al.*, 1974). The local responses are blocked in high-Mg^{++} solutions (Figure 6-7A). Much further work is required for a better understanding of the peripheral neural circuitry mediating the pinnule response. But a beginning has been made by Peretz and his colleagues, who have found a large number of neurons within the gill itself (Peretz, 1970; Peretz and Moller, 1974).[5]

[5]In a series of studies Peretz and his colleagues have defined the morphology and distribution of the peripheral neurons. Particularly interesting is the finding by fluorescence that some of the peripheral neurons contain biogenic amines.

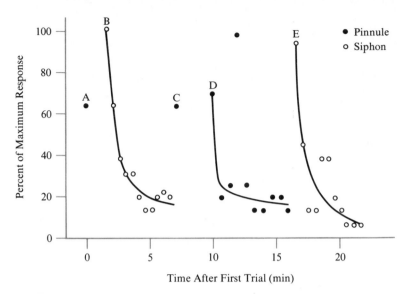

FIGURE 6-8.

Independence of habituation of the gill-withdrawal reflex and the pinnule response. Gill or pinnule contractions were measured by a photocell. Siphon stimuli consisted of jets of water from a Water Pik. The pinnule stimuli consisted of brush strokes across the gill. This stimulus produced a relatively large contraction involving several pinnules. Similar results were obtained with a weak Water Pik stimulus applied directly to the pinnules. Response magnitudes are represented as a percent of the maximum response obtained (point *B,* 100 percent). At point *A* a single shearing stimulus was presented to several pinnules. At *B* the gill-withdrawal reflex was habituated by presenting a series of stimuli to the siphon every 30 sec. After habituation of the gill-withdrawal reflex, a stimulus to the pinnules similar to that given at *A* resulted in a contraction (point *C*) that was identical to that of *A* before habituation of the gill-withdrawal reflex. At *D* the pinnule response was habituated. A test stimulus presented to the siphon (point *E*) showed that during habituation of the pinnule response the gill-withdrawal reflex had recovered almost completely. The gill-withdrawal reflex then was again habituated. [After Kupfermann *et al.,* 1971.]

Because of the relative independence of central and peripheral pathways at weak stimulus intensities, stimulation of one pathway does not affect the responsiveness of the other. For example, habituation of the gill-withdrawal reflex does not affect the pinnule response nor does habituation of the pinnule response affect the gill-withdrawal reflex (Figure 6-8). However, moderate and strong stimuli can recruit activity in both the central and peripheral pathways (Kupfermann *et al.,* 1971, 1974; Peretz and Howieson, 1973; Peretz *et al.,* 1976; Carew *et al.,* 1976).

Conjoint Responses: Central and Peripheral
Mediation of Siphon Withdrawal

Not all responses elicited by stimulating an organ directly involve only peripheral pathways. In some cases central and peripheral pathways are activated concomitantly. This is the case in the siphon component of the defensive-withdrawal reflex. As we have seen, weak stimulation of the siphon causes a centrally mediated withdrawal of the gill; it also causes withdrawal of the siphon. Siphon withdrawal can be elicited in the isolated siphon without the presence of central ganglia; however, central pathways also contribute to this response (Kandel and Spencer, 1968; Newby, 1972; Lukowiak and Jacklet, 1972; Jacklet and Lukowiak, 1974; Perlman, 1975; Bailey *et al.*, 1975).

The abdominal ganglion contains eight motor neurons that produce siphon contraction. These motor cells receive excitatory connections from the same centrally located mechanoreceptor sensory neurons that innervate the gill (Figure 6-9). To estimate the relative contribution in the intact animal of the central and peripheral pathways to siphon contraction with stimuli of moderate intensity (400 g/cm^2 water jet), Perlman used three different techniques: acute reversible deganglionation (using high-Mg^{++} solution), chronic surgical deganglionation, and reversible removal of individual cells with hyperpolarization. With each of these methods the central component contributed about 55 percent; the remaining 45 percent was mediated by peripheral motor cells (Figures 6-10 and 6-11).

Although the peripheral contribution to the behavior is slightly less than the central contribution, Perlman (1975) and Jacklet and Lukowiak (1974) found that the two pathways share several features. For example, both the conjoined responses (produced by stimulating both the peripheral and central pathways together) and the peripheral response (produced in the deganglionated preparation) undergo habituation, and the kinetics of habituation in the peripheral system are quite similar to those in the central one (Figure 6-11; for a detailed discussion of other similarities see Jacklet and Lukowiak, 1974). Moreover, the kinetics for the centrally and peripherally mediated siphon components resemble those of the centrally mediated gill component. This similarity originally suggested that the peripheral and central contribution to siphon withdrawal may involve some common elements. Thus, the similarity in the kinetics of habituation could be explained if the peripheral and central siphon motor neurons and the central gill motor neurons all received innervation from the same centrally located sensory neurons (Figure 6-12).

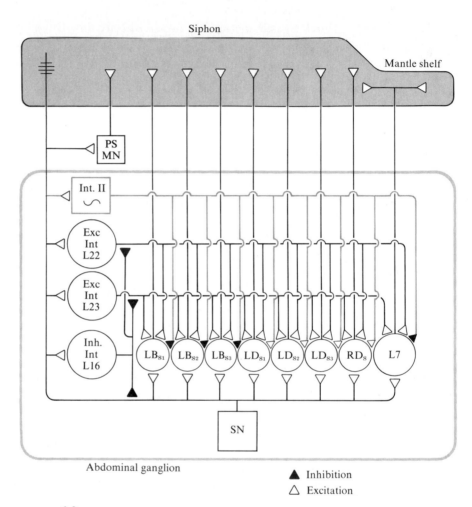

FIGURE 6-9.

The neural circuit for siphon withdrawal. The circuit consists of eight central motor cells and a group of 25–30 peripheral motor neurons. These are in turn innervated by a population of sensory neurons (SN) within the abdominal ganglion that make direct connections to both the central and peripheral motor neurons (PSMN) as well as a group of interneurons. [After Perlman, 1975; Bailey *et al.*, unpubl.]

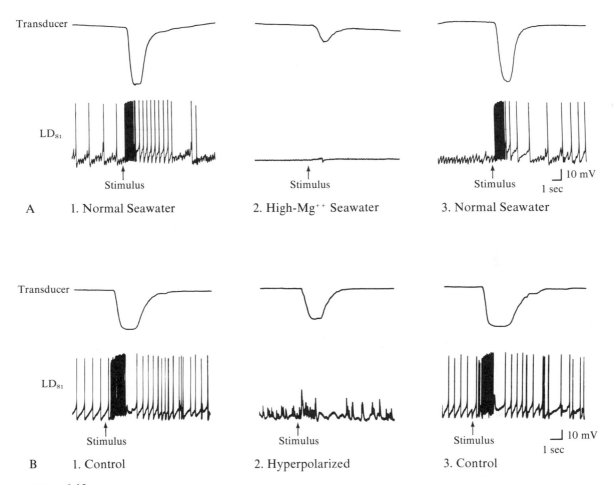

FIGURE 6-10.

Contribution of central motor neurons to siphon-withdrawal reflex. [From Perlman, 1975.]

A. The effect of blockade of synaptic transmission on the siphon-withdrawal reflex elicited by tactile stimulation (800 msec jets of seawater) to the siphon skin. **(1)** The gill-withdrawal reflex recorded with a transducer and the intracellular response of motor neuron LD_{S1} to the tactile stimulus. **(2)** In a high-Mg^{++} solution that blocks synaptic transmission as indicated by the absence of activity in LD_{S1}, the siphon-withdrawal reflex is significantly reduced by more than 50 percent. **(3)** Return of normal seawater. Both the responses of the siphon motor neuron and the siphon-withdrawal reflex are restored.

B. The effects of hyperpolarizing an individual motor neuron, LD_{S1}, on the siphon-withdrawal reflex. **(1)** Control. Tactile stimulus applied to the siphon elicits a burst response in LD_{S1} and a concomitant siphon-withdrawal reflex. **(2)** Cell LD_{S1} was hyperpolarized so that tactile stimulation no longer elicited firing in the cell. The transducer response was now reduced by about 30 percent. **(3)** Hyperpolarization of LD_{S1} was removed and the reflex was restored to its initial height.

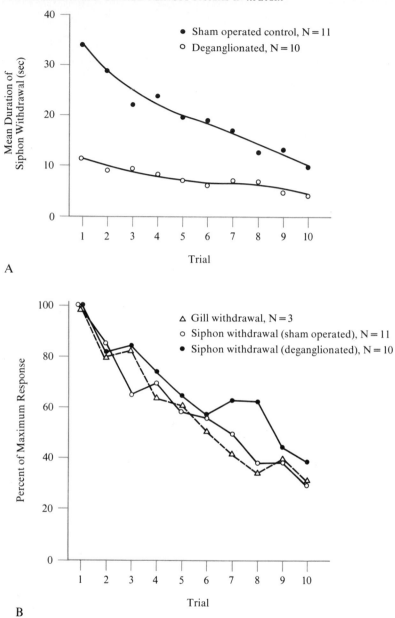

FIGURE 6-11.
Kinetics of habituation of siphon and gill withdrawal. **A.** A comparison of habituation of
the siphon response in the intact animal (sham operated control) and the deganglionated
animal. **B.** The two responses are normalized so that they can be more directly compared
to each other and to the gill-withdrawal reflex. [From Perlman, 1975.]

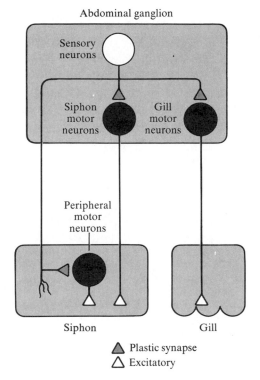

Abdominal ganglion

Sensory
neurons

Siphon
motor
neurons

Gill
motor
neurons

Peripheral
motor
neurons

Siphon

Gill

▲ Plastic synapse
△ Excitatory

FIGURE 6-12.
A hypothetical circuit to explain the similarity of the
kinetics of habituation illustrated in Figure 6-11. The
peripheral siphon motor neurons and the central siphon
and gill motor neurons are innervated by common
sensory cells. [After Bailey *et al.*, unpubl.]

PERIPHERAL MOTOR CELLS IN THE SIPHON OF *APLYSIA*

Lukowiak and Jacklet (1972) and Byrne and co-workers (1974) had earlier
described about 30 cells distributed along the course of the siphon nerve, extending from the base of the siphon to its tip. Bailey and co-workers found that
these peripheral cells fell into two categories (Figure 6-13). One category, consisting of a single cell—typically the largest, with a diameter of 120 to 190 μm—
has the morphological appearance of neurosecretory cells. This cell appears
white on direct illumination, has a small dendritic tree, and, like neurosecretory
cells, contains many large electron-dense granules in its cytoplasm (Figure

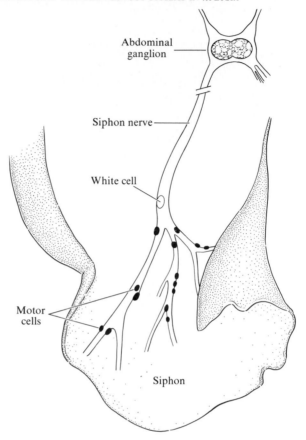

FIGURE 6-13.
The location of the peripheral neurons along the course
of the siphon nerve. There is one large white cell and
25–30 smaller cells. [After Bailey *et al.*, unpubl.]

6-14). It receives little sensory input following stimulation of the siphon; it also
receives few spontaneously occurring synaptic potentials (Figure 6-15A).
Intracellular stimulation of these cells does not produce movements of the
siphon. The second group consists of about 25–30 smaller cells, with cell bodies
of about 50–100 μm. These cells are pigmented and do not have the granules
characteristic of neurosecretory cells; they have a larger dendritic field than the
white cell (Figure 6-14B), and receive extensive spontaneous synaptic input
similar to that received by the central siphon motor cells (Figure 6-15A). Direct
stimulation of each of these cells produces a restricted movement of the siphon,
most commonly of its base (Figure 6-15B; Bailey *et al.*, 1976, and unpubl.).

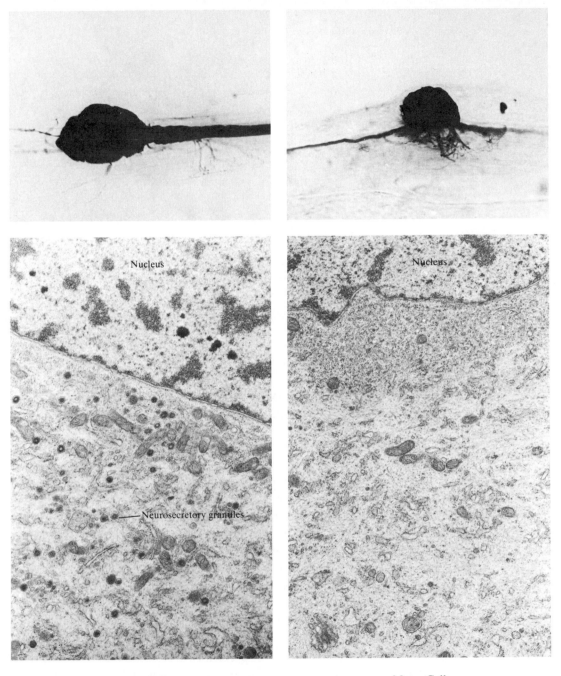

White Cell   Motor Cell

FIGURE 6-14.
Comparison of the white cell and the peripheral siphon motor cell. Both cells have been injected with cobalt chloride. The white cell has a very small dendritic tree while the motor cell
has a more extensive dendritic tree (top). Electron-microscopic comparison of the cytoplasms of
the two types of cells reveals many electron-dense, presumably neurosecretory granules, in the
white cell. These granules are not present in the motor cell. [From Bailey *et al.,* unpubl.]

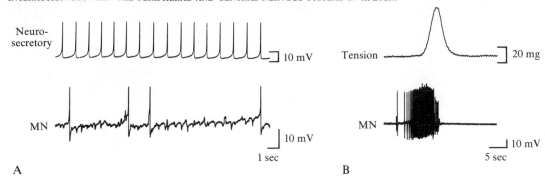

FIGURE 6-15.
Electrical properties of the peripheral white cell and the peripheral siphon motor neuron.

A. The neurosecretory cell has a regular firing pattern and very little synaptic input. The motor neuron has an irregular firing pattern modulated by extensive excitatory and inhibitory synaptic potentials.

B. Siphon movement produced by the stimulation of a single peripheral motor cell. [From Bailey *et al.*, unpubl.]

Adjacent peripheral siphon motor cells are electrically coupled to one another (distant ones, however, appear not to be coupled). Each motor cell receives monosynaptic excitatory connections from centrally located mechanoreceptor neurons (Figure 6-16). The connections that these peripheral motor cells receive are similar to those the sensory neurons make on the central motor cells. Moreover, as suggested by the hypothesis illustrated in Figure 6-12, the peripheral motor cells are activated before the central motor cells (Figure 6-16).

Repeated firing of a sensory neuron produces depression of the excitatory postsynaptic potential in the peripheral siphon motor cells similar to the depression it produces in central siphon and gill motor cells (Figure 6-17). In all three systems habituation of the behavioral response is related to the depression of the excitatory postsynaptic potential produced by the common mechanoreceptor neuron on the central and peripheral motor cells. The central and peripheral systems have similar kinetics of habituation because both systems share the same (central) sensory neurons and the different branches of the sensory neuron show similar response depression.

The studies of the peripheral siphon motor neurons illustrate that the peripheral nervous system is made up of discrete neurons with specific interconnections that are organized in a manner similar to that of the central nervous system. The *neuron doctrine*, which states that the individual neuron is the independent trophic and functional unit of the nervous system (Palay and Chan-Palay, 1977) also applies to peripheral neuronal systems. The peripheral nervous system of the siphon is not a syncytium or random nerve net. What

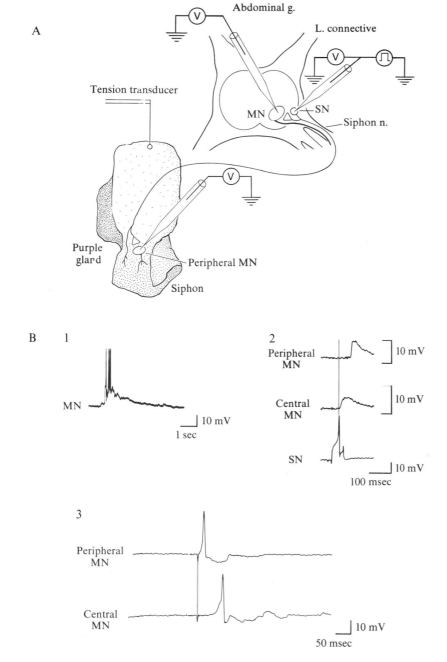

FIGURE 6-16.
Connections between central sensory neurons and peripheral siphon motor cells. [After Bailey *et al.*, unpubl.]

A. Experimental set up for studying the electrophysiological properties and interconnections of the peripheral and central motor neurons and the central sensory cells.

B. Connections. **(1)** The repetitive burst of activity in the peripheral motor neuron produced by stimulating the siphon skin. **(2)** Direct, presumably monosynaptic, connection in the peripheral motor neuron produced by a single action potential in a central sensory neuron. **(3)** Comparison of the synaptic responses to siphon stimulation of peripheral and central motor neurons. The EPSP produced in the peripheral motor neuron precedes that of the central one.

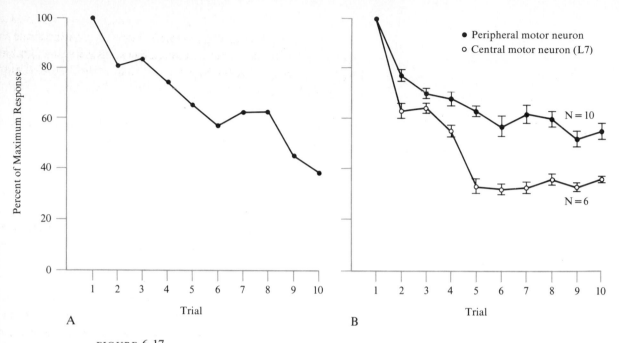

FIGURE 6-17.
Comparison of the kinetics of habituation of siphon withdrawal **(A)** and the synaptic depression produced in the peripheral motor cells by stimulation of a single sensory neuron **(B)**. [From Bailey *et al.*, unpubl.]

gives the peripheral nervous system of the siphon its local character is the presence of *axon reflexes* mediated by the collateral branches from central sensory cells. An implication of the presence of axonal reflexes is that ablation of a central ganglion need not eliminate all central influences. A behavior that persists after ablation of a central ganglion need not be entirely mediated by peripheral neurons. In the case of siphon-withdrawal, the sensory component of the reflex remaining after removal of the abdominal ganglion is *not* mediated by peripheral sensory neurons but by the peripheral axon collaterals of central sensory neurons.

PERIPHERAL–CENTRAL INTERACTIONS IN THE SIPHON OF *SPISULA*

Prior (1972a, b) has studied peripheral and central interactions in the siphon of the clam *Spisula*. This clam has paired siphons: one is incurrent, the other excurrent. A weak tactile stimulus to the siphon results in a local muscular contraction consisting of closure of the siphon and local contraction of the siphon wall in the area contacted by the stimulus. A slightly stronger stimulus causes, in addition to the local contraction, an activation of the siphon retractor muscles, resulting in a withdrawal of the siphon into the mantle cavity. The local contraction of the siphon wall musculature is mediated by peripheral pathways and persists when the visceral ganglion is removed. By contrast, the withdrawal of the siphon is abolished by removing the ganglion, suggesting that the siphon retractor muscle is innervated by the central ganglion.

Prior (1972a) found a cluster of cells on the point of branching of the peripheral siphon nerve that send their axons peripherally to the siphon wall musculature. Although stimulation of these cells does not produce movement, Prior proposed that they may nonetheless be peripheral motor neurons because they are strongly activated by tactile stimulation of the siphon and the mantle. They are thought to be activated by collateral axons of centrally located sensory neurons (Prior, 1972a). As is the case in *Aplysia,* this synaptic input undergoes depression with repeated stimulation, paralleling the habituation.

Although the data for *Spisula* are less complete than for *Aplysia,* the indirect evidence suggests a similar model for local reflex responses: activation of peripheral motor cells by axon collateral branches of central sensory neurons (Prior, 1972a). The findings in *Spisula* thus suggest that innervation of peripheral motor cells by axons of central sensory neurons may be fairly general. It will therefore be interesting to examine the development of peripheral motor neurons in *Aplysia* to determine whether they first arise in the abdominal ganglion and only then migrate to the periphery or whether they arise from peripherally located stem cells.

In *Spisula* the presumed peripheral motor cells have a higher input resistance and lower threshold than the presumed central motor cells. This difference provides a mechanism for intensity discrimination for motor systems sharing a common sensory input and may explain why weak stimuli selectively activate peripheral motor cells and stronger stimuli also activate central ones (Prior, 1972b).

INNERVATION OF THE PARAPODIA AND FOOT OF *APLYSIA*

A less complete analysis of the relationship between central and peripheral systems in *Aplysia* has come from early studies of the control of the parapodia and the foot. Bethe (1903) found that after removal of the entire central nervous system the parapodia could still undergo peristaltic movements similar to those that occur in swimming. This finding was not confirmed, however, by either Fröhlich (1910c) or Hoffmann (1910). Fröhlich found spontaneous contractions only in isolated parapodia that had been badly damaged during the surgical procedure or by strong and prolonged stimulation. In freshly denervated, undamaged parapodia, stimulation produced only localized contraction; the swimming movements were centrally controlled by the pedal ganglia (Fröhlich, 1910c; Hoffmann, 1910). The spread of contraction in damaged parapodia may be muscular rather than neural (Bullock and Horridge, 1965).

Central mediation of parapodial movement has also been confirmed by Jahan-Parwar and Fredman (1978a, b). They found that stretching one of the parapodia produces coordinated contraction of both parapodia. This reflex requires that the parapodial nerves that connect the parapodia to the pedal ganglion be intact. But the reflex can still be obtained even when the parapodia are isolated from the foot so as to sever any peripheral pathways.

A beginning in the analysis of the central control of parapodial movement has recently been made. Several central motor cells have been found in the cerebral and the pedal ganglia that control flaring and antiflaring of the parapodia (Hening, Carew, and Kandel, 1976; Jahan-Parwar and Fredman, 1978a, b, and Chapter 8 below).

The isolated foot of *Aplysia* is also incapable of undulating after the pedal nerves to the foot are severed (Jordan, 1929). To study the mechanisms of pedal wave propagation Jahan-Parwar and Fredman (1978a) have used a split-foot preparation first developed by Jordan (1901) and Herter (1931). The foot is cut longitudinally, thereby eliminating coordination between the two halves of the foot by means of peripheral pathways. In this preparation Jahan-Parwar and Fredman found normal, highly coordinated waves of contraction that move synchronously along both halves of the foot, from the anterior to the posterior margin. Similarly, stretching of one side of a split foot produced contractions of both halves. Finally, tactile stimulation of one side of the head caused both sides of the foot to contract. These three experiments indicate that the bilaterally coordinated movements of the pedal waves and their reflex contractions are mediated by the central nervous system.

To further examine the central control of locomotor activity, Jahan-Parwar and Fredman subdivided one-half of a foot into three segments. Stretching the middle segment caused contraction of the other ipsilateral segments as well as contraction of the entire contralateral half-foot. These experiments suggest that the mechanism for the propagation of the pedal wave of contraction resides centrally and does not require for its expression the integrity of peripheral pathways.

Further evidence for central control of locomotion comes from the experiments illustrating that neurons in the pedal ganglia (Hening, Carew, and

Kandel, 1976, 1977) and cerebral ganglia (Jahan-Parwar and Fredman, un-publ.) can control longitudinal contraction of the foot. As we will see in Chapter 8, the central program for locomotion resides within the head ganglia and is expressed in the isolated central nerve ring lacking any peripheral feed-back (Hening *et al.,* 1977). As is the case with the gill and the siphon, the periph-eral neurons in the foot seem primarily to mediate local reflexes, such as local suction of particular parts of the sole during locomotion (Jahan-Parwar and Fredman, unpubl. observ.).

CENTRAL INNERVATION WITHOUT PERIPHERAL MOTOR CELLS

A few muscular organs of *Aplysia,* such as the heart and the buccal mass, lack peripheral motor cells. Their neural control appears to reside completely in the central nervous system. For example, the heart of *Aplysia* is myogenic and beats in the absence of peripheral neural control. However, this myogenic beat is modulated by central neurons: two types of excitors (LD_{HE} and RB_{HE}), and a group of three inhibitors (LD_{HI1}, LD_{HI2}, LD_{HI3}) all of which directly inner-vate the heart muscle. There is no evidence for other peripherally located motor cells (Mayeri *et al.,* 1974).

The buccal musculature is also centrally innervated by both excitor and inhibitor motor cells (Banks, 1975; Orkand and Orkand, 1975; Cohen, Weiss, and Kupfermann, 1978). Denervation leaves the muscle unresponsive to local stimulation. A similar situation seems to apply for the heart and buccal mass of other molluscs (Heyer *et al.,* 1973; Rosza, 1974; Spray, Spira, and Bennett, 1976).

CAPABILITY FOR HABITUATION IN CENTRAL AND PERIPHERAL PATHWAYS

One of the key findings in the study of both central and peripheral pathways of the mantle organs in *Aplysia* and in *Spisula* is that both types of pathways are capable of mediating habituation (Kandel and Spencer, 1968; Peretz, 1970; Kupfermann *et al.,* 1971; Prior, 1972a; for a detailed review see Jacklet and Lukowiak, 1974). For example, repeated weak tactile stimuli to the isolated siphon produce marked habituation of the siphon-withdrawal reflex with kinetics that resemble the habituation of the central pathways (Figure 6-12).

As another example, the centrally mediated (relayed) gill-withdrawal reflex undergoes habituation, as does the peripherally mediated pinnule response.

The existence of dual pathways in the relayed reflex, each capable of independent habituation, illustrates an *extensive redundancy* in the neural control of behavior. This dual capability is interesting from a phylogenetic point of view and suggests some speculations on the evolution of this behavioral modification.

Even single-celled organisms show response depression. Behavior modifications resembling habituation have been thoroughly studied in protozoa, particularly in *Paramecium* (Thorpe, 1963). Because organisms that do not have a nervous system undergo response depression, it is not surprising that behaviors controlled by the nerve nets of simple metazoa, such as the coelenterates, habituate. Most students of comparative neuroanatomy argue that there is an intimate relation between the development of independent receptors and sensory systems and the development of a central nervous system. Thus, both Cajal (1911) and Parker (1919) have argued that central neurons may have evolved from a primitive receptor-effector cell. Most receptor cells show some accommodation (adaptation) to a constant stimulus or to one that is rapidly repeated, responding less effectively as the stimulus is maintained or repeated at high rates. For such organisms as coelenterates, whose nervous system is made up of peripheral nerve nets supplemented by pacemaker centers and epithelial conducting systems, well-developed receptor adaptation can be an effective means for generating a simple type of habituation. In the more differentiated nervous systems of higher invertebrates a more flexible mechanism for habituation develops, such as homosynaptic depression in the synaptic terminals of mechanoreceptor cells. In higher invertebrates homosynaptic depression (and perhaps other types of plastic changes) appears to supersede sensory adaptation as a mechanism for response decrement. Homosynaptic depression provides greater behavioral flexibility than adaptation because it can be overridden by a process of opposite sign, presynaptic facilitation, resulting from the activity of a heterosynaptic input. In vertebrates homosynaptic depression seems also to be the mechanism for at least some forms of response depression, but here the site of plasticity is located still further centrally by one or more synapses. The locus for habituation of the flexion reflex of the cat seems to be the synapse between second- or third-order interneurons and the motor neuron (Spencer *et al.,* 1966b). This postulated "centralization" of the neural locus for habituation is illustrated in Figure 6-18.

According to this speculative model, part of the evolution of habituation involves not an increase in neuronal plasticity, but a restriction to a progres-

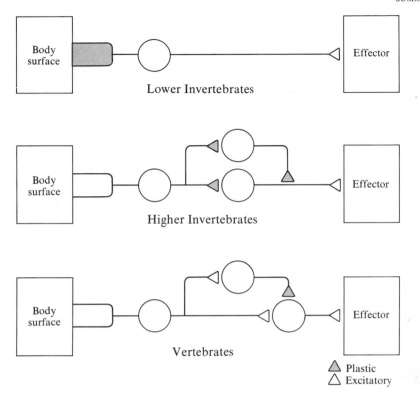

FIGURE 6-18.
A model for the evolution of habituation, comparing possible mechanisms for habituation in lower invertebrates, higher invertebrates, and vertebrates. See text for details.

sively more central site. A central locus permits far greater generalization of habituation, as one indeed sees in vertebrates. According to this view, *Aplysia* and other molluscs stand at a crossroad in evolution and therefore have available to them both primitive peripheral and more advanced central mechanisms for response depression.

SUMMARY AND PERSPECTIVE

The heart and buccal mass in *Aplysia* lack peripheral motor systems. Other effector structures possess some peripheral motor mechanisms that provide local control. Analysis of the mantle organs in *Aplysia* suggests that there are three patterns of central–peripheral interactions: (1) *local responses,* e.g., pinnule withdrawal of the gill, which seem primarily to involve peripheral mediation, although this can be modulated by central influence; (2) *relayed reflex responses,* e.g., the gill-withdrawal reflex, which primarily involve central pathways; and (3) *conjoined responses,* in which most stimuli elicit concomitant activation of both central and peripheral pathways.

In *Aplysia* and probably in other molluscs the peripheral nervous system does not seem to be a functional nerve net that can conduct information over a distance and between organs, as is the case in coelenterates (Bullock and Horridge, 1965). Some peripheral pathways in *Aplysia* consist of discrete neurons that primarily mediate local responses to weak stimuli. However, peripheral pathways can also be recruited as an amplifying mechanism to increase the responses to very strong stimuli. Little is known about the organization of peripheral neuronal systems. In the two cases that have been analyzed (siphon withdrawal of *Aplysia* and the clam *Spisula*), the peripheral system was found to be similar in its organization to the central system. For example, the peripheral and central siphon motor cells have similar properties and both cells share common central sensory cells.

The gill of *Aplysia* also contains peripheral neurons of unknown function (the branchial ganglion) and isolated nerve cells at various points along the peripheral nerve. It is not known, however, whether the pinnule response involves any of these peripheral neurons, or whether it is due to axon reflexes of central cells or even to direct stimulation of muscle. This latter possibility must be considered in view of the evidence that gill muscle epithelium has both septate and gap junctions and might be electrically coupled (Gilula and Satir, 1971). By intracellular recordings from peripheral neurons and muscle, it should also be possible to specify here—as was done in the siphon—how peripheral neurons participate in the pinnule response (for a beginning see Peretz and Lukowiak, 1975; Carew *et al.,* 1974).

On the basis of the electrical properties of the peripheral neurons in the siphon skin, it would appear that most of them are specialized in their motor (or sensory) functions, much as central neurons are. However, some peripheral cells may be less specialized than central neurons. Coggeshall (1971) has found one type of peripheral neuron in the secretory epithelia of the accessory genital mass (see Chapter 3, p. 99) that may be both sensory and motor. The apical processes of these cells are ciliated and protrude into the oviduct. These apical cilia resemble those found in chemo- or mechanoreceptor neurons and may therefore have a sensory transducer function. The basal part of the cell contains granules and is drawn out into a fine process that resembles an axon.

These processes end directly on large secretory cells. Coggeshall suggests that these neurons serve a sensory function at their apical end and motor functions at their basal end. The existence of receptor–effector cells was postulated by Parker (1919) to be an early stage in the evolution of the nervous system prior to the development of differentiated sensory, interneurons, and motor neurons. Coggeshall's work suggests that some such cells may exist in some parts of the peripheral nervous system of *Aplysia*.

SELECTED READING

Jacklet, J. W., and K. Lukowiak. 1974. Neural processes in habituation and sensitization in model systems. *Progress in Neurobiology,* 4:1–56. Compares the capability for habituation of central and peripheral systems in *Aplysia.*

Prior, D. J. 1972. Electrophysiological analysis of peripheral neurones and their possible role in the local reflexes of a mollusc. *Journal of Experimental Biology,* 57:133–145. An early attempt to analyze a peripherally mediated response in molluscs.

CHAPTER 7

Development of *Aplysia* and Related Opisthobranchs

Our understanding of an animal's behavior is considerably enhanced by knowing how that behavior developed. Developmental studies can elucidate the sequence and pattern of emergence of an organism's behavioral repertory and can therefore address the following questions: Does the development of a behavior require a number of preliminary steps or does the behavior emerge in an all-or-none form? How does the time course of development of the neural circuit of a behavior relate to the time course for the expression of that behavior? Can a behavior be modified as soon as it is developed?

Studies of the development of particular ganglia in *Aplysia,* in addition to their relevance for behavior, can also produce insights into the program of differentiation for nerve cells. It may be possible to learn about the developmental origins of various identified cells, and to see how the development of one type of cell in the ganglion relates to that of another. The problem of cell lineage has scarcely been explored in nervous systems and could be profitably examined in *Aplysia.*[1] Specific questions are: Do clonally related cells—cells that are linearly descended from a common neuroblast—have some functional relationship to each other in the differentiated nervous system? Do

---

[1] The early studies of the embryonic development of *Aplysia* (see in particular Blochmann, 1883; Carazzi, 1900, 1905; and Saunders and Poole, 1910) have only recently been followed up (Kriegstein *et al.*, 1974; Kriegstein, 1977a, b; Switzer-Dunlap and Hadfield, 1977).

these cells release the same transmitter substance? Do they form functional sensory or motor clusters? This chapter describes the development of *Aplysia,* and particularly the development of the nervous system, in comparison with what is known about other gastropods.

PHASES IN THE LIFE CYCLE OF *APLYSIA CALIFORNICA*

The development of *Aplysia californica* occurs in four phases: (1) embryonic development, which extends from fertilized egg to hatching; (2) larval development, which extends from hatching to metamorphosis; (3) metamorphic development; and (4) postmetamorphic or juvenile development, which extends from metamorphosis to reproductive maturity (Figure 7-1).

The first phase, *embryonic development,* takes approximately 14 days at room temperature. Four to five days after the fertilized eggs are laid, movements of the embryo are first seen inside the egg capsule. Spontaneous hatching occurs six to seven days later.[2] When the egg hatches, a free-swimming larval form, the *veliger,* emerges.

The second phase, *larval (veliger) development,* takes about 34 days at room temperature. The veliger larvae swim as plankton in the ocean and feed on unicellular algae (and possibly bacteria and organic detritus). The veliger is a complete and independent organism that performs all the necessary life functions *except* reproduction. To achieve reproductive maturity, the veligers must metamorphose and undergo a radical change in form, diet, behavior patterns, and life style.

The third phase, *metamorphosis,* takes two to three days (days 48 to 51 postfertilization, or days 34 to 37 posthatching). During this period, the animals do not eat or increase in overall size. Rather, they undergo preparatory changes that allow them to switch their diet from unicellular alga to seaweed (macrophytic algae) and their feeding habits from ciliary to radular feeding. They also stop ciliary swimming and begin to crawl.

The final phase, *postmetamorphic (juvenile) development,* takes about 35 days. It begins with the completion of metamorphosis and ends with reproductive maturity, about 85 days after metamorphosis or about 133 days after deposition of the fertilized egg.

---

[2]The *Aplysia* embryo itself undergoes several developmental stages: cleavage, blastula, gastrula, trochophore, and early veliger. The trochophore stage is analogous to the trochophore larva of annelids and primitive molluscs, but in *Aplysia* the trochophore is not a functional (free-living) larval stage.

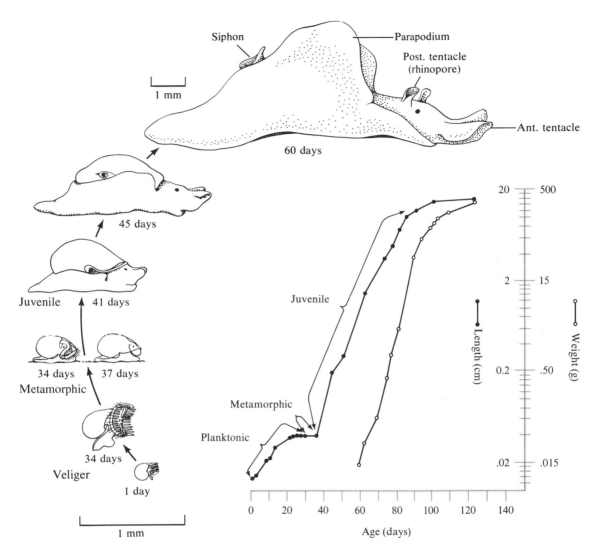

FIGURE 7-1.
Life cycle of *Aplysia californica*.

Graph of animals' weight and length as a function of age (time after hatching). Each point of the graph is an average of five measurements. Maximum shell diameter was used to measure the length of veligers and extended body length to measure juveniles. The data are based on an environment of 22°C seawater. External features of animals from each of these phases are illustrated in the drawings on the left. Drawings are to the same scale, with the exception of the 60-day juvenile. The rapidity of growth following metamorphosis is highly dependent on diet. [Data from Kriegstein *et al.*, 1974.]

TABLE 7-1.

Features of the embryonic and larval periods of five Hawaiian aplysiids. (From Switzer-Dunlap and Hadfield, 1977.)

	Zygote diameter (μm)	Embryonic period (days)	Shell length at hatching (μm)	Larval period (days)	Shell length at metamorphosis (μm)
Aplysia dactylomela	92	8–9	143	30	310–315
Aplysia juliana	82	7–8	125	28	315–325
Aplysia parvula	75	6–7	107	?	520–525*
Dolabella auricularia	93	9–10	148	31	290–300
Stylocheilus longicauda	55	6–7	105	33	325–340

*Shell lengths of larvae collected from plankton and metamorphosed in the laboratory.

A similar developmental timetable has been obtained for a number of Hawaiian aplysiids: *A. dactylomela, A. juliana, A. parvula, Dolabella auricularis* and *Stylocheilus longicauda* (Table 7-1).

Sperm, Egg, and Fertilization

The sperm cell of *Aplysia* arises in the ovotestis and undergoes a number of transitions to achieve the elongated structure of the mature spermatozoon that is capable of swimming and penetrating the ovum (Figure 7-2A). The spermatozoon consists of a nucleus and the necessary machinery for delivering itself to and penetrating the egg: the flagellum (or swimming organ) and mitochondria (Figure 7-2B). The spermatozoa of three European species of *Aplysia* range from 150 to 225 μm in length (Thompson and Bebbington, 1969).

Each fertilized egg is approximately 100 μm in diameter and contains a large amount of yolk, which is concentrated at one end (Figure 7-3).[3] The eggs are

[3]Typically, one end of the egg is fairly free of yolk; this region is the *animal* pole. It contains the nucleus, and it is here that the *mitotic spindles* appear and the *polar bodies,* the discarded daughter nuclei of meiosis, are formed. The opposite end of the egg, where the yolk concentration is highest, is the *vegetal* pole. Eggs such as those of *Aplysia* that tend to concentrate yolk at one end are moderately "telolecithal." Those of cephalopods are strongly telolecithal. The elongated yolk-rich eggs of arthropods and coelenterates are called "centrolecithal" because the large yolk mass is centrally located.

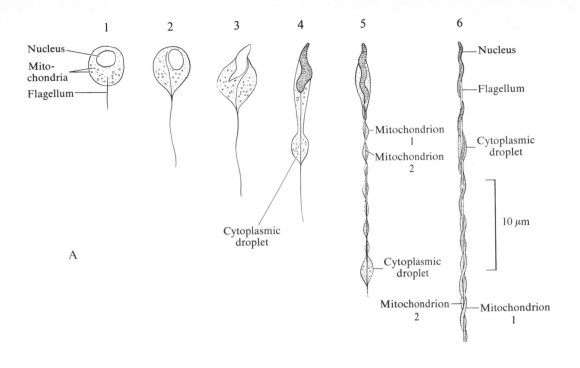

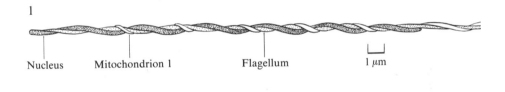

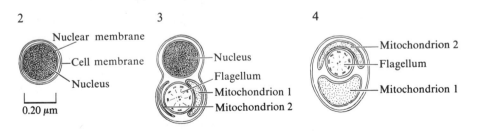

FIGURE 7-2.

The sperm of *Aplysia*. [After Thompson and Bebbington, 1969.]

A. Stages in spermiogenesis of *Aplysia punctata*. Camera lucida drawings from living sperm suspended in saline and stained with intravital dyes. **(1)** Early spermatid. **(2)** Spermatid with elongating flagellum. **(3)** Spermatid showing the beginning of the nuclear helix. **(4)** Spermatid with elongating nucleus and formation of cytoplasmic droplets. **(5)** Spermatid with dual mitochondrial helices. **(6)** Spermatozoon with elongated nuclear helix, flagellum reaching almost to the anterior extremity, and dual mitochondrial helices.

B. The fine structure of *Aplysia* spermatozoon based upon electronmicrographs. **(1)** Anterior end of a spermatozoon. **(2)** Section through the head tip. **(3)** Section through the mid-region of the head. **(4)** Section through the principal piece.

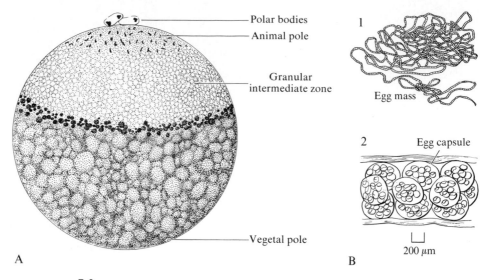

FIGURE 7-3.
The egg of *Aplysia.*

A. Fertilized egg stained with neutral red showing animal pole, polar bodies, vegetal pole, and intermediate zones. [After Ries and Gersch, 1936.]

B. The egg mass of *Aplysia.* **(1)** Several egg capsules, each containing several eggs, are encased in a gelatinous ribbon (cordon) and deposited in a long string in an egg mass. **(2)** A segment of a cordon.

packaged in capsules that hold 4–100 eggs, depending on the size of the animal (P. Kandel and T. Capo, personal communication). The egg capsules are encased in a gelatinous ribbon (cordon) and deposited as a long string containing from 100,000 to 3,000,000 eggs (Figure 7-3B and Table 7-2; MacGinitie, 1934).

TABLE 7-2.
Spawn of four species of Aplysia.

	Egg capsules (average no. per cm)	Ova (average no. per capsule)	Total no. of ova	Embryonic period (days)
A. depilans[a]	160	25	3,308,000	14–16 at 25°C
A. fasciata[a]	118	43	25,877,400	14–16 at 25°C
A. punctata[a]	532	4	135,180	20–22 at 15°C
A. californica[b]	100	4–100	1,500,000	8–11 at 25°C

[a]From Thompson and Bebbington, 1969.
[b]From Kriegstein, Castellucci, and Kandel, 1974; and P. Kandel and T. Capo, personal communication.

Embryonic Development

CLEAVAGE

Cleavage (or segmentation) refers to the first few cell divisions of the fertilized egg. Cleavage is generally considered to end when the cells formed by division have arranged themselves around a central cavity, or blastocoel. At this point, the embryo is referred to as a *blastula*. The blastula then undergoes an invagination (gastrulation) into the gastrula and becomes a two-layered structure. Like the eggs of other animals, cleavage of the fertilized *Aplysia* egg does not lead to an increase in the total mass of the embryo, but partitions the cytoplasm and cell membrane of one large cell into many smaller cells that can serve as building blocks of a convenient size. A major biochemical event during cleavage is DNA assembly and chromosome replication necessary to supply each cell with a *complete* set of chromosomes. In anaspid and cephalaspid opisthobranchs the diploid number of chromosomes is usually 34 (Inaba, 1959).

The eggs of *Aplysia*, like those of other molluscs, undergo spiral cleavage. The first and second cell divisions are at right angles to each other (meridional), forming four cells of nearly equal size (Figure 7-4). But subsequent divisions are oblique, resulting in alternatively right and left displacements of the cleaving cells (see the 8-, 16-, and 32-cell stages in Figure 7-6C). Therefore, daughter cells remain in the original quadrant of the parent cells.

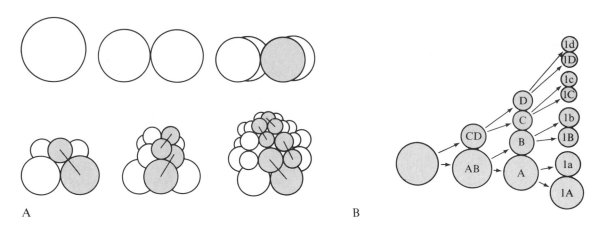

A   B

FIGURE 7-4.
General features of spiral cleavage. **A.** The progeny of one cell in the four-cell stage are shaded to illustrate that the daughter cells remain in the same quadrant. Lines indicate the axes of the preceding mitoses. **B.** The fates and terminology of cells during three divisions (A, B, C, D refer to the four macromeres). [After Villee *et al.*, 1963.]

During the first cleavage, a lobe of protoplasm is pinched off at the vegetal pole of the egg cell. The lobe (the *polar* lobe) remains attached to one of the daughter blastomeres by a narrow neck, and once the separation of the cell is completed the material flows back into this daughter blastomere, making it larger than the other blastomere. During the next cleavage a lobe again forms and when division is completed flows back into a daughter cell. The polar lobe is incorporated into the D macromere (see below) and is thought to contain the morphogenetic determinants for the mesoderm (Saunders, 1970; for a general discussion of spiral cleavage see Raven, 1966; Kume and Dan, 1957; Cather, 1971).

The development of molluscan eggs, including those of *Aplysia,* is of the *determinate* type. At the very earliest division each blastomere receives a partic-ular part of the original egg cytoplasm and membrane that endows its lineage with a specific ability to give rise to certain organs. These *morphogenetic* or *determining substances* (determinants) are present in the egg before cleavage begins (see Raven, 1966; Davidson, 1976).[4] As a result of the selective dis-tribution of the determining substances the fate of each blastomere—the form and timing of cleavage as well as the capability for differentiation in some molluscan forms—is independent of the presence of the other blastomeres (Crampton, 1896; Wilson, 1904). Thus, Wilson (1904) found that isolated blastomeres of the prosobranch *Patella coerules* destined to have cilia in the embryo become ciliated at the same time and after carrying out the same number of divisions as they do in the intact embryo.[5]

Modern studies have shown that the molecules that selectively specify the patterns of gene activation in each mother cell are differentially distributed to

[4]In the *mosaic* or *determinate eggs* of annelids and molluscs the fate of each cell is determined at the very beginning of the blastula stage. Injury or removal of certain cells in the blastula or earlier embryo leads to a defective embryo. By contrast, the *regulative* or *indeterminate eggs* of echinoderms and chordates produce a normal (dwarf) embryo after injury to a given cell. The morphogenetic determinants of mosaic eggs are thought to be located in and near mem-brane of the egg and to have assumed their determining positions *prior to* the onset of cleavage. The mechanical division pattern of the cell is determined by the already distributed determinants rather than the other way around. Thus, the difference between determinate (mosaic) cleavage and indeterminate (regulative) cleavage seems to lie in the *relative* times of fixation of the cortical patterns and cleavage. If the cortical determinants become fixed before cleavage, cleavage is deter-minate. If cleavage begins before the final distribution of cortical determinants is completed, the early cleavages are indeterminate. The association between spiral cleavage and determinate devel-opment may, however, be fortuitous. (See Berrill, 1971; Cather, 1971).

[5]In more detailed experiments Clement (1968) has examined the embryonic value of the first quartet of cells in *Illyanassa* by surgically deleting single cells. He found, for example, that deletion of micromere 1a resulted in the loss of the left eye, whereas deletion of 1c resulted in loss of the

the blastomeres (Davidson, 1976; Cather, 1971). The morphogenetic determinants are not, however, distributed in the cytoplasm of the blastomeres but seem to be localized in the cortex, presumably the cell membrane (Cather, 1971). The determinants are synthesized during oogenesis and stored in and near the egg membrane until just before or just after fertilization. The distinct patterns of gene activity necessary for differentiation in a given cell are thought to result from the interaction of the totipotential genome of each blastomere and the specific determinants characteristic of that cell.

In its early stages the structure of the molluscan embryo is, cell for cell, almost identical to that of annelids (Figure 7-5). In both groups mesoderm arises from a cell that buds from primitive endoderm: gastrulation begins after the sixth or seventh division and rapidly produces a ciliated larva or embryo called *trochophore* that has the beginning of a coelomic cavity. The fundamental difference between molluscs and annelids appears only later in development (Figure 7-5). The molluscan veliger larva shows no sign of the segmentation present in the annelid post-trochophore larva.

Because the molluscan embryo develops asymmetrically and has a relatively small number of relatively large cells laid down in a determinable manner, the cells of the blastula have recognizable shapes and positions. Carazzi (1900) and Saunders and Poole (1910) described a complete cell lineage for *Aplysia punctata* to at least the 300-cell stage (three days after fertilization).

The cell divisions are unequal. The first division in *A. punctata* passes through opposite poles of the egg and splits it into two cells, the *macromeres* AB and CD (Carazzi, 1900). Both cells contain part of the vegetal pole, but cell AB is much larger than cell CD.[6] In the second division, which also passes through opposite poles (meridional), AB and CD divide to form the four parent

---

right eye. These results are consistent with the probable cell of origins for the eyes from specific micromeres in the embryo and is therefore consistent with classic mosaicism. However, even though the eyes do not originate from macromere D (originating instead from micromeres 1a and 1c), removal of macromere D leads to the loss of both eyes (Clement, 1956). (For discussion of the labeling system of micro- and macromeres see p. 228). These and other results indicate that macromere D exerts an influence comparable to the primary organizer in vertebrates. This macromere is important in establishing the anterior–posterior axis of the embryo, and participates throughout cleavage in setting aside cells for specific fates and establishing the inductive interactions required for organ formation. Cather (1971) suggests that the D quadrant has this unique role because it possesses a unique portion of the cortex of the vegetal pole of the embryo. Thus, although the cells of origin for various organs are highly determined, their developmental sequence requires a sequence of inductive interactions.

[6]This arrangement, where AB is greater than CD, is found in some other opisthobranchs; but in most molluscs CD is greater than AB (Raven, 1966).

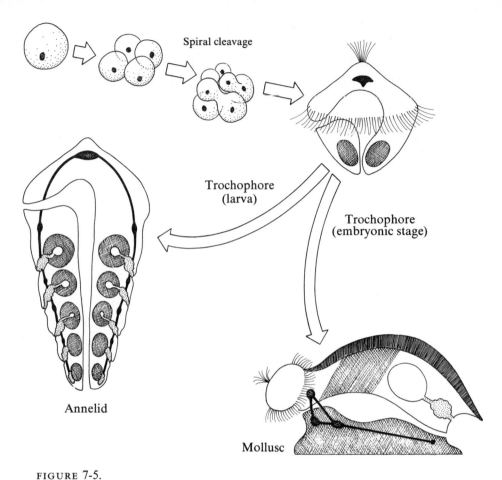

Spiral cleavage

Trochophore
(larva)

Trochophore
(embryonic stage)

Annelid

Mollusc

FIGURE 7-5.
Both gastropods and annelids have spiral cleavage and a trochophore stage. However, the details of the trochophore stage vary. *Aplysia* passes the trochophore stage within the egg capsule. By contrast, a few gastropods, most other molluscs, and most annelids have a true free-living trochophore larva stage. [After Wells, 1968.]

macromeres, A, B, C, and D. At the third division a small cell or *micromere* is pinched off from the animal pole of each macromere (Figure 7-6A, B, C). Following the nomenclature introduced by Conklin (1897), this first quartet of micromeres is labeled 1a, 1b, 1c, and 1d; the residual macromeres are labeled 1A, 1B, 1C, and 1D. Prior to the third division the cleavage spindle of each micromere is in an oblique position with respect to the egg axis and is in a right-hand spiral (viewed from the animal pole). As a result, each micromere is displaced to the right, or clockwise, with respect to its macromere (Figure 7-6C). With each successive division, the spindles become oriented approxi-

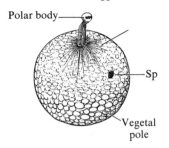

Polar body

Sp

Vegetal pole

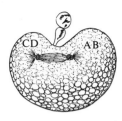

CD AB

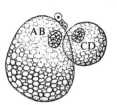

AB CD

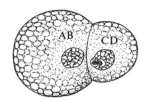

AB CD

A. First Division

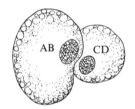

AB CD

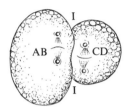

AB CD

I
I

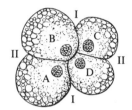

B C
A D

I
I
II
II

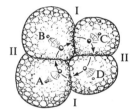

B C
A D

I
I
II
II

B. Second Division

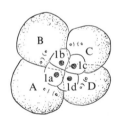

B C
1b 1c
1a 1d
A D

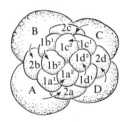

B 2c C
1b¹ 1c² 1c¹
2b 1b² 1d² 2d
1a¹ 1a² 1d¹
A 2a D

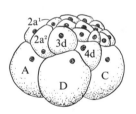

2a¹
2a² 3d
A 4d
D C

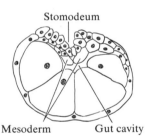

Stomodeum

Mesoderm Gut cavity

C. Third Division   D. Fourth Division   E. Development of Mesoderm

FIGURE 7-6.

Early division in *Aplysia punctata*. [After Carazzi, 1905.]

A. In the first division the zygote divides into two unequal cells, AB and CD. Both cells have components of animal and vegetal poles.

B. In the second division AB splits into two equal cells, A and B, and CD splits into two equal small cells, C and D. Roman numerals I and II indicate the axes of the first and second division.

C. In the third division the four parent cells of macromeres (A, B, C, D) divide to form macromeres (1A, 1B, 1C, 1D) and descendant cells or micromeres (1a, 1b, 1c, 1d).

D. In the fourth division the second generation of micromeres that arises from the micromeres are called 2a, 2b, 2c, 2d. Those that will arise from the second generation of micromeres are called 1a^1, 1b^1, 1c^1, 1d^1 and 1a^2 to 1d^2.

E. Gastrulation by epiboly. An outgrowth of micromeres forms a cap of cells that give rise to the mesoderm.

mately at right angles to those of the preceding division so that a succession of clockwise and counterclockwise divisions occur, and each successive quartet of micromeres is alternately displaced clockwise and counterclockwise. This alternation is called *alternating spiral cleavage* (Raven, 1966).

In the fourth division (Figure 7-6D), which is counterclockwise, the macromeres 1A, 1B, 1C, and 1D pinch off a second quartet of micromeres, 2a, 2b, 2c, and 2d. The parent macromeres are designated 2A, 2B, 2C, and 2D. The first quartet of micromeres, 1a, 1b, 1c, and 1d, also divide counterclockwise, giving rise to two series, $1a^1$, $1b^1$, $1c^1$, $1d^1$ and $1a^2$, $1b^2$, $1c^2$, $1d^2$, where cells closer to the animal pole are given the superscript 1. At the fifth division another quartet of micromeres (3a to 3d) is formed from a clockwise splitting of the macromeres 2A to 2D (which then become designated 3A to 3D). The earlier generations of micromeres also divide. Thus, $1a^1$ splits into $1a^{11}$ and $1a^{12}$, $1a^2$ into $1a^{21}$ and $1a^{22}$, and 2a into $2a^1$ and $2a^2$. Figure 7-7 illustrates the first five cleavages, showing in detail the cleavages of macromere A.

After the fifth division the regular features of spiral cleavage disappear and bilateral symmetry begins to appear as the cells of the D quadrant start behav-

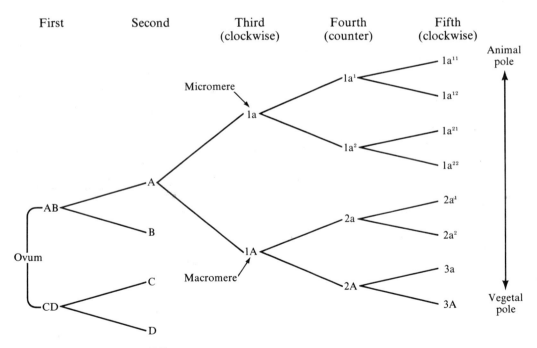

FIGURE 7-7.
Sequence of spiral divisions in *Aplysia punctata* and nomenclature of daughter cells.

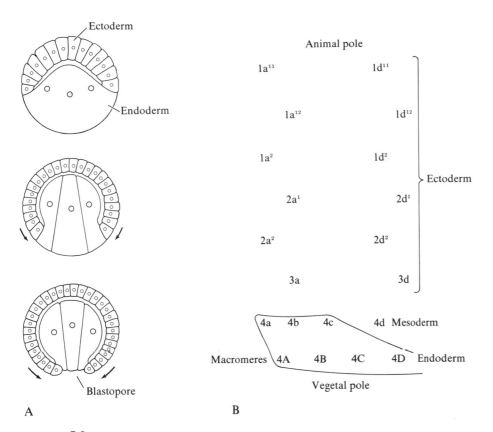

FIGURE 7-8.
Origin of the three germ layers in the sixth division in *Aplysia*. [After Wilmoth, 1967.] **A.** Diagram showing gastrulation by epiboly. **B.** Scheme to indicate quartet orientation, labeling system, and origin of the three germ layers. The 2a–2d and 1a–1d quartets have both divided. Cell 4d forms coelomic mesoderm; cells 4a, 4b, 4c, and the macromere quartet form endoderm; the remainder form ectoderm and ectomesoderm.

ing differently from those of the other quadrants (Raven, 1966). As cleavage progresses, the macromeres become covered with a cap of micromeres at the animal pole (Figures 7-6E and 7-8A).

The future *ectoderm* arises from the first three micromere quartets. The first quartet gives rise to the ectoderm for the head, and the second and third give rise to the rest of the ectoderm. The future *mesoderm* arises at the sixth division from cell 4d (the micromere produced from macromere 3D). The future *endoderm* arises at the sixth division from macromeres 4A, 4B, 4C, and 4D, and from micromeres 4a, 4b, and 4c (Figure 7-8B; Raven, 1966; Saunders and Poole, 1910).

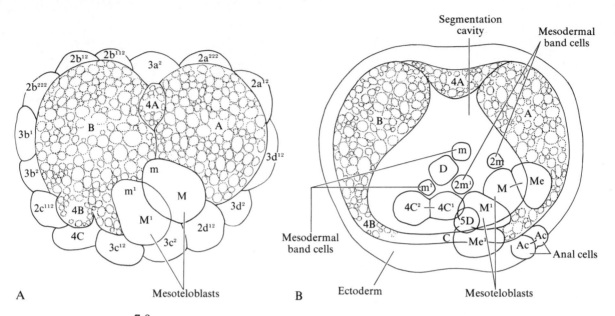

FIGURE 7-9.

Onset of gastrulation in *Aplysia.* [After Saunders and Poole, 1910.]

A. Section of an early embryo seen from the vegetal pole. The two cells M and M^1 have each given rise to small anterior mesoderm cells m and m^1, respectively.

B. Diagrammatic sagittal section seen from the vegetal pole of an egg in the 170-cell stage. The cells labeled m, m^1, 2m, and 2m^1 are derivatives of the mesoteloblasts M and M^1 and are seen spreading in an anterior direction from the region of the anal cells, Ac. The cells labeled Me are other derivatives of M. The macromeres A and B have diverged from each other to form a segmentation cavity (the precursor of the gut cavity).

LATE EMBRYONIC DEVELOPMENT

With the formation of the precursors of the three germ layers at the sixth division, the small ectoderm cells from the animal pole overgrow and enclose the macromeres of the vegetal pole, thus forming a thin layer of cells covering the entire embryo (Figures 7-8A and 7-9A). This process is called *epiboly,* the overgrowth of one part by another, and it is this enveloping process that leads to gastrulation in *Aplysia* (Figures 7-6E and 7-8A; Carazzi, 1905; Saunders and Poole, 1910; Raven, 1966; for discussion of epiboly see Karsznia *et al.,* 1969; Saunders, 1970).

After three days of development, the embryo (now at the trochophore stage) begins to rotate within the egg capsule. Rotation is accomplished by means of ciliary action of the velar cells. This ring of cells, called the *prototroch,* or *pre-*

velum, develops anteriorly from a simple ring of cells derived from macromere cell B and will ultimately become the velum, the feeding and swimming organ of the larva. Caudally, a pair of conspicuous anal cells (derived from two ecto-dermal cells, $2d^{22221}$ and $2d^{22222}$) increase greatly in size until they project from the surface (Figure 7-10A). They are thought to function as excretory organs in the embryo (Saunders and Poole, 1910). The interior of the embryo is largely occupied by the two large macromeres A and B. At the vegetal pole, the cells A and B begin to diverge from one another (Figure 7-9B) thereby creating an irregular cavity called the *archenteron* or *segmentation cavity,* which grows into the stomach and rest of the midgut (Figure 7-10A).[7]

Now organ development begins. The prevelum continues to develop at the anterior end of the embryo, and the nervous system, secondary kidney, primi-tive kidney, gut, digestive glands, muscles, shell gland, and shell also start to differentiate. In addition, two independent transformations occur: (1) oral-anal (or ventral) flexion and (2) torsion (Figure 7-10). In *Aplysia* both occur almost simultaneously, but they are quite distinct transformations (Figure 7-11, Saun-ders and Poole, 1910). Oral-anal flexion occurs in other molluscan classes, but torsion occurs only in gastropods.

Oral-anal flexion consists of a more rapid growth of the dorsal part of the embryo relative to the ventral part so that the anus (originally posterior) be-comes relatively closer to the mouth, which is anterior (Figures 7-10A and 7-14). This process can be traced in Figure 7-10 (parts A3 to B4) by focusing on the anal cells.

Torsion, the most dramatic event in gastropod embryogenesis, consists of the rotation of the anus and visceral mass in relation to the head and foot so that the intestine, right digestive gland, and kidney are carried from the ventral sur-face to the right side of the body. In most prosobranchs this involves a 180° rotation (Figures 1-5, 1-6, and 1-7, p. 18ff.), but in *Aplysia* it is only about 120° (Saunders and Poole, 1910). It can be followed in Figure 7-10 (parts A2 to B4) by focusing on the secondary kidney. As a result of torsion and oral-anal flex-ion, the digestive system resembles a U-shaped tube.

According to Crofts' hypothesis, based on studies of prosobranch larvae (see Chapter 1), torsion results from a rapid initial rotation of the mantle region due to the contraction of unopposed muscle fibers attached to the shell on the left side. This rapid phase is followed by slower rotation due to differential growth.

---

[7]The segmentation cavity is not homologous to a blastocoel. In an embryo with a hollow blastula the blastocoel is the internal cavity, lined by putative ectoderm and endoderm, and into which the mesodern cells bud. In *Aplysia* a hollow blastula is not formed.

A. Oral-Anal Flexion

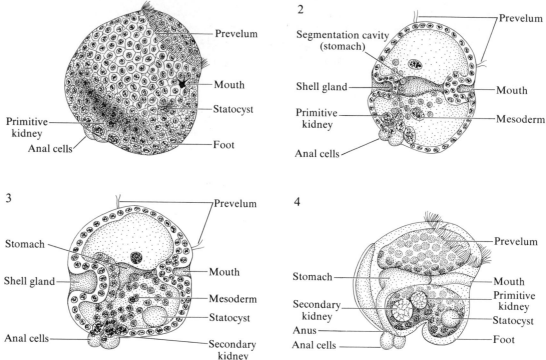

1

- Prevelum
- Mouth
- Statocyst
- Primitive kidney
- Anal cells
- Foot

2

- Prevelum
- Segmentation cavity (stomach)
- Shell gland
- Primitive kidney
- Anal cells
- Mouth
- Mesoderm

3

- Prevelum
- Stomach
- Shell gland
- Mouth
- Mesoderm
- Statocyst
- Anal cells
- Secondary kidney

4

- Prevelum
- Stomach
- Secondary kidney
- Anus
- Anal cells
- Mouth
- Primitive kidney
- Statocyst
- Foot

B. Torsion

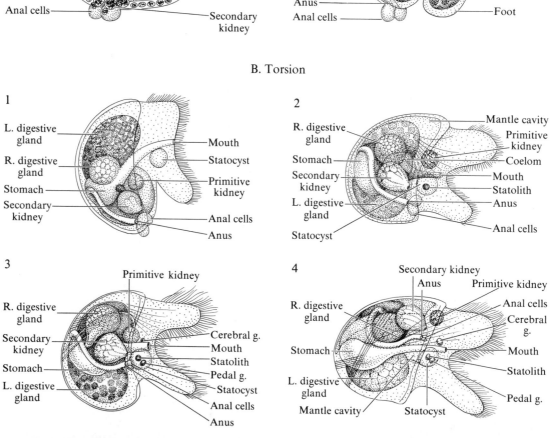

1

- L. digestive gland
- R. digestive gland
- Stomach
- Secondary kidney
- Mouth
- Statocyst
- Primitive kidney
- Anal cells
- Anus

2

- R. digestive gland
- Stomach
- Secondary kidney
- L. digestive gland
- Statocyst
- Mantle cavity
- Primitive kidney
- Coelom
- Mouth
- Statolith
- Anus
- Anal cells

3

- Primitive kidney
- R. digestive gland
- Secondary kidney
- Stomach
- L. digestive gland
- Cerebral g.
- Mouth
- Statolith
- Pedal g.
- Statocyst
- Anal cells
- Anus

4

- Secondary kidney
- Anus
- Primitive kidney
- Anal cells
- Cerebral g.
- R. digestive gland
- Stomach
- L. digestive gland
- Mantle cavity
- Statocyst
- Mouth
- Statolith
- Pedal g.

C. Free-Swimming Veliger

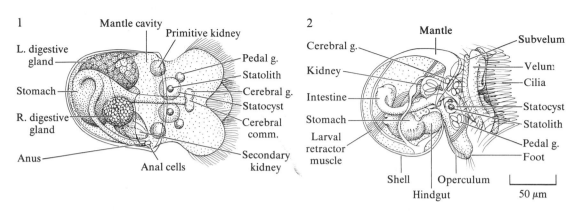

FIGURE 7-10.
Embryonic development in *Aplysia:* transition from trochophore to veliger larva.

A. Oral-anal flexion. **(1)** Two- to three-day embryo (300-cell stage) immediately before rotation (right side). **(2)** Embryo immediately before rotation (as in 1) seen in light-microscopic section. The ectoderm is represented as peeled off from the right half of the embryo and thus seen in section. The right half of the ectoderm is omitted. The greatest part of the interior is occupied by the macromeres A and B, which have diverged from each other and thus enclose the segmentation cavity. **(3)** Slightly older embryo than in 2. **(4)** View of an embryo about 24 hours older than that shown in 3, seen from the right side. [After Saunders and Poole, 1910.]

B. Torsion as viewed from the right side of a free-swimming veliger. **(1)** Slightly older embryo showing first appearance of coelom (pericardial cavity) and beginning of torsion. **(2)** Assumption of characteristic veliger form. Anal-oral flexion and torsion continue. (Note movement of anal cells in figures 2 to 4.) **(3)** Anal-oral flexion and torsion continue. **(4)** Anal-oral flexion and torsion completed. [After Saunders and Poole, 1910.]

C. Free-swimming veliger. **(1)** Dorsal view. [After Saunders and Poole, 1910.] **(2)** Lateral view. [After Kriegstein, 1977b.]

Whether a similar mechanism operates in *Aplysia* is not known. As is the case in several prosobranchs (see Chapter 1), the embryo of *Aplysia punctata* has two muscles, a large and a small velum retractor (Figure 7-12). Only the large retractor develops prior to torsion. Saunders and Poole (1910) suggest that its unopposed action as well as differential growth cause torsion. However, the development of the large muscle seems to proceed while torsion is in progress (Figure 7-12). To contribute to torsion the muscle should have to develop before torsion begins as is the case in *Haliotis,* studied by Crofts. Moreover, to determine if muscular contractions are involved one needs to study living, not preserved, specimens as was done by Saunders and Poole.

While flexion and torsion are occurring the foot increases in size and extends further below the mouth. During days 5 to 10 after fertilization the prototroch

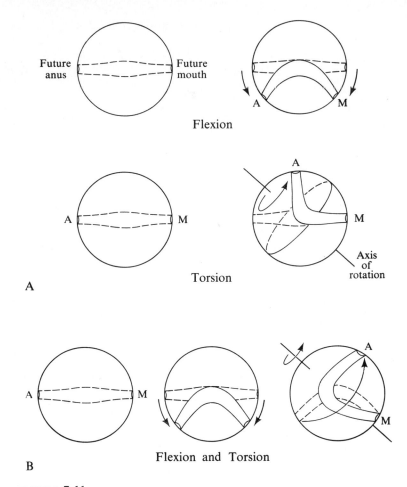

FIGURE 7-11.
Oral-anal flexion, torsion, and detorsion. **A.** Flexion and torsion illustrated as independent movements. Flexion is a bending along the anterior-posterior axis of the embryo. Torsion is a separate process that involves a rotation of the gut, bringing the anus anterior. **B.** Flexion and torsion in sequence.

(the precursor of the velum) is transformed into a bilobed velum, the organ used for food capturing and locomotion, and the embryo enters the *veliger stage*. The expansion of the anterior portion of the body places the mouth ventrally between the lobes of the velum. The anal cells are still present in the posterior part of the body but have begun to migrate rostrally (Figure 7-10B4). The operculum appears on the surface of the foot. The principal growth during this prehatching stage involves the shell, the operculum, and the velar lobes.

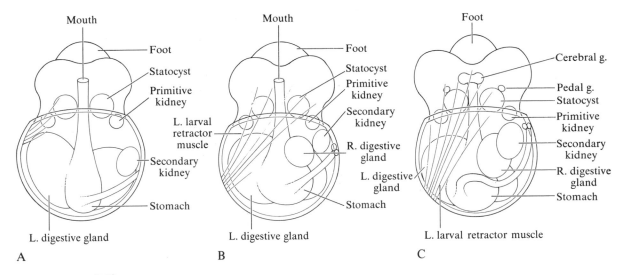

FIGURE 7-12.
Muscle development in *Aplysia punctata*. **A.** Dorsal view of an embryo at a stage when the large retractor muscle is first formed. **B.** Similar view of the embryo at a slightly later stage. **C.** Similar view of a free-swimming larva. [After Saunders and Poole, 1910.]

Beneath the shell, the anus moves further dorsally to the lateral margin of the foot in its migration toward the position normally seen in gastropod larvae (Figure 7-10B4). The larva is now ready for swimming in the ocean. The velum and foot can be completely withdrawn into the shell, with the operculum protecting the larva from small predators. Between 10 to 14 days after fertilization, the veligers break out of the egg capsules into the ocean as free-swimming larvae.

Larval Development

Kriegstein (1977b) divides larval development into six stages (Figure 7-13).

Stage 1. Newly hatched veligers (0 to 10 days posthatching). Newly hatched veligers have a maximum diameter of 125 μm. They already have paired statocysts at the base of the foot. An opaque, white grain 5 μm in diameter marks the site of the larval kidney near the right side of the shell aperture. The mantle margin is not fixed to the shell and can be withdrawn to accommodate the head-foot when the animal closes its operculum.

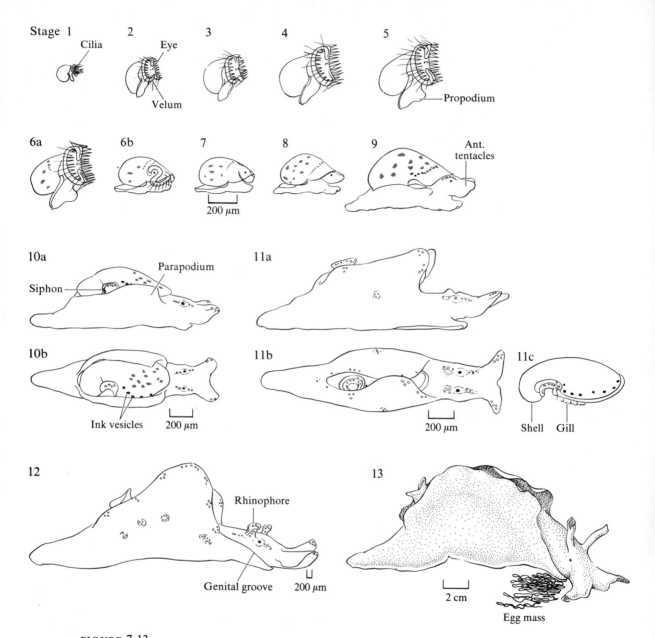

FIGURE 7-13.
Thirteen stages in the posthatching life cycle of *Aplysia californica.* Stage 6a is the animal swimming and 6b is the animal at the same stage crawling. The solid spots appearing at Stage 6 are red in nature; the open spots appearing at Stage 10 are white. Starting at Stage 10 the entire skin becomes red. See text for details. [After Kriegstein, 1977b.]

Stage 2. Appearance of eyes (2 to 3 weeks). The eyes are the most prominent feature of animals at this stage. They develop from velar ectoderm and consist of lenses surrounded by darkly pigmented granules. They first appear as two small black spots, one on either side of the midline slightly above the center of the velum. The shell begins to coil sinistrally at this stage.

Stage 3. Development of the larval heart (3 to 4 weeks). At this stage the larval heart appears and beats about 50 times per minute (at 22°C). The stomach and digestive gland also contract, but only about seven times per minute. The animal can inhibit both heart beat and digestive contracture. The opening of the kidney pore into the intestine is marked by two to four opaque grains.

Stage 4. Largest shell diameter (4 to 5 weeks). At this stage animals measure 400 μm across the largest shell diameter and 450 μm across the fully extended velum. The shell grows no further until after metamorphosis. The mantle, which lays down the shell matrix, extends the full length of the shell. The white pigments marking the larval kidney now disappear and the kidney becomes less visible in intact specimens.

Stage 5. The development of the propodium (5 weeks). The propodium, starting as a swollen area in the middle of the foot, grows out to form a stalk-like appendage densely ciliated at the tip and capable of stretching 200 μm. When the propodium is retracted, as when the animal swims, it forms a bump in the middle of the foot. Once the propodium develops the veliger can crawl, but veligers do not settle and are not competent to metamorphose until they have reached the next stage.

Stage 6. Appearance of red spots (5½ weeks). Four to six irregularly shaped red spots now appear at the right side of the outer perivisceral membrane and a red band appears overlying the mantle margin just beneath the shell. The appearance of these pigments is the most reliable indicator that veligers are competent to metamorphose. An active band of cilia can be seen in a U-shaped extension of the right lateral mantle margin. These cilia mark the anus and the site of the siphon primordium. Metapodial and propodial glands become prominent and are visible in the foot of intact larvae. Three to four ink vesicles are also present in the mantle below the right dorsal shell margin, but these are not visible in intact larvae.

Metamorphic Development

The essential condition for settlement and crawling is the development of the propodium. Once it develops (Stage 5) the veliger can be coaxed into crawling, provided that a suitable substrate is present. Larvae at Stage 6 spend proportionately more time at the bottom of the culture chamber. This appears to be a preadaptation for settling. Animals are now competent to metamorphose but will undergo no further development unless presented with a specific triggering substance.

For *A. californica* the triggering substance is the seaweed *Laurencia* (Kriegstein *et al.*, 1974). When presented with a choice of six algae at once (*Plocamium, Laurencia, Polysiphonia, Daisia, Chondrus,* and *Ulva*) animals preferentially crawl upon *Laurencia* and very rarely upon the others, with *Daisia* being a distant second choice. Larvae of *A. juliana* will metamorphose on the green seaweed *Ulva* (*U. fasciculata, U. reticulata,* and *U. lactuca*). *A. dactylomela* is less discriminating and will metamorphose on a variety of red seaweeds; *Laurencia* is particularly effective but others (*Martensia, Chandroccus, Gelidium, Polysiphonia,* and *Spyridea*) can also serve as triggering substances (Switzer-Dunlap and Hadfield, 1977).

For the first hour or so after contacting *Laurencia, A. californica* will typically alternate periods of crawling and periods of sitting still. Occasionally during this period the larvae will extend their velum and swim off. Usually within one hour the larvae stop crawling and attach themselves by secretions from the metapodial glands. The velum and propodium are then both retracted and the operculum partly closed. Unless disturbed the larvae remain in this position during the early stages of metamorphosis.

Within three hours of settlement the cells bearing the long velar cilia are shed. This signals the beginning of metamorphosis (Stage 7). Within hours the velar lobes decrease in size. Once the velar cilia are shed the larvae cannot swim. The pelagic (swimming) phase ends and the benthic (crawling) phase begins. One day after settlement the radula begins to extend and retract in movements that anticipate feeding. Kriegstein divides metamorphic development into two stages (Figure 7-13).

Stage 7.   Shedding of velar cilia (metamorphosis day 1). The velar cilia are shed and the velar lobes are reduced to stubby rudiments. The definitive two-chambered heart starts to beat and intraventricular and aortic valves begin to function.

Stage 8.   Fusion of the velar lobes (metamorphosis day 3). The velar lobe rudiments become tightly apposed medially and fused at their bases. They begin to grow anteriorly and laterally and gradually extend forward from beneath the shell to form the anterior tentacles. The metapodium grows and its distal tip projects 30 μm beyond the limits of the operculum. The larvae now have pink, semicircular eyebrow markings above their eyes. The larval heart stops beating. At the end of this stage animals begin to graze on seaweed.

Juvenile Development

Kriegstein divides juvenile development into four stages (Figure 7-13).

Stage 9.   Development of pink color (7 days postmetamorphosis). The juveniles begin to turn pink. The color spreads through the skin and becomes progressively darker. Parapodial rudiments develop as outgrowths of the lateral margins of the metapodium. The metapodium continues to grow posteriorly and the operculum is cast off. Shell growth, which was halted during late pelagic life, resumes, but the newly secreted shell is flat and broad and involves growth of the dorsal margin of the shell aperture so that the aperture itself becomes long, oval, and flattened.

The siphon primordium is visible as a curved ciliated ridge of mantle tissue protruding from beneath the shell margin and flaring back against the shell. At the right anterior edge four to six purple ink-filled vesicles appear in a row. The largest one is nearest the siphon and the others are progressively smaller. As the mantle grows beneath the shell the red spots, which were located on the right side in earlier stages, become distributed more evenly over the newly enlarged mantle. Approximately nine spots are visible.

Stage 10.   Appearance of the parapodia (10 days postmetamorphosis). At this stage the juveniles are completely pink, with the exception of the mantle, which is not pigmented but has the red spot previously described. The parapodia now overlay half of the shell; the siphon is visible as it projects between the parapodia and the shell on the right side. White crystalline spots appear in small clusters at the tips of the anterior tentacles above the eyes and on the left margin of the siphon. The number of red spots on the mantle increases to 15–20. The purple ink vesicles also increase in number. A small gill becomes apparent beneath the right dorsal region of the mantle. The heart, which beats 200 times per minute at 22°C, has grown and moved forward and to the left, where it is visible in intact animals beneath the surface of the shell.

Stage 11. Appearance of the rhinophores (17 days postmetamorphosis). The rhinophores begin to develop on either side above the eyes. The parapodia increase in size and contact one another anteriorly. The mantle margin grows over the edges of the shell.

Stage 12. Genital groove (late juvenile appearance). A genital groove becomes visible on the right side of the head. The anterior edges of the two parapodia form a prominent anterior siphon. The rhinophores reach their definitive proportions. Red spots resembling those on the head become distributed over the body epidermis. White spots also appear in great numbers in well-defined clusters on the anterior tentacles, rhinophores, parapodia, siphon, and head. The mantle has overgrown the shell on all sides, leaving an oval-shaped bare spot in the middle of the shell.

Stage 13. Reproductive maturity. Animals begin to copulate and lay eggs. Their skin color takes on brown and gray coloration. The edges of the anterior tentacles, rhinophores, and parapodia become convoluted in fleshy folds.

DEVELOPMENT OF THE NERVOUS SYSTEM

Stages in Development

The nervous system of *Aplysia* develops gradually (Figure 7-14). The first step begins in the trochophore, prior to hatching, with the differentiation of the statocysts and the cerebral and pedal ganglia. Development of the nervous system continues throughout the veliger and juvenile stages, lasting a total of about 140 days. Near the position of each future ganglion the ectoderm first thickens by cell division, followed by separation of cells from the base of the thickening.

The statocysts are the first neural structures to develop. When the embryo rotates at the 300-cell stage, the two statocysts form from a small ectodermal invagination on each side, just ventral to a line joining the anus and mouth. Each statocyst originally consists of approximately six cells. Later this number increases to 13 and the cells sink below the surface to form a closed vesicle at the base of the foot and lateral to the esophagus. A large spherical statolith becomes conspicuous within each statocyst.

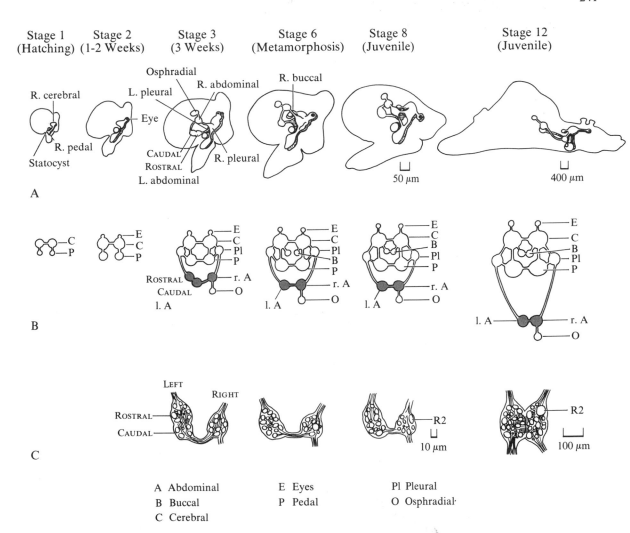

Stage 1 (Hatching) Stage 2 (1-2 Weeks) Stage 3 (3 Weeks) Stage 6 (Metamorphosis) Stage 8 (Juvenile) Stage 12 (Juvenile)

A

B

C

A Abdominal	E Eyes	Pl Pleural
B Buccal	P Pedal	O Osphradial
C Cerebral		

FIGURE 7-14.

The development of the nervous system of *Aplysia californica*. [After Kriegstein, 1977a.]

A. The major ganglia appear at the various developmental stages illustrated (see Figure 7-13 for the complete cycle of development). The cerebral and pedal ganglia are present at hatching (Stage 1); the eyes develop at one week (Stage 2); the osphradial, abdominal, and pleural ganglia appear at three weeks (Stage 3); and the buccal ganglia, which develop during the fifth week, are present at metamorphosis (Stage 6). During the juvenile stages (8 through 12) the pleuroabdominal connectives lengthen and the ganglia spread apart.

B. The formation of the nervous system is depicted schematically in stages corresponding to part A.

C. The abdominal ganglion develops from three cell clusters that appear at Stage 3, although precursor cells are present at Stage 1. The right hemiganglion develops as a single ganglion while the left arises as a fusion of two parts, one caudal and one rostral, which merge together prior to metamorphosis (Stage 6). The cholinergic neuron R2 becomes visible after metamorphosis (Stage 8) and grows rapidly during the juvenile stages.

The first ganglia to develop are the paired cerebrals and pedals, which form prior to hatching and within 10 days of fertilization (Figures 7-10B and 7-14). The ganglia first appear as slight thickenings on the ectoderm. The cerebral ganglia lie just above the mouth; the pedal ganglia lie beside the statocyst (Figures 7-10B4 and 7-14). The two cerebral ganglia are close together and are united by a broad commissure.

By Stage 1 the cerebral and pedal ganglia are linked ipsilaterally by short connectives that pass in front of the statocysts. The cerebral ganglia are joined by a commissure above the esophagus, but no commissure links the pedal ganglia at this stage. At Stage 2 the eyes develop from a proliferation of velar ectodermal cells adjacent to each cerebral ganglion. A small optic ganglion forms from a cluster of cells on the anterior dorsal surface of each cerebral ganglion that is in contact with the newly formed eyes. At Stage 3 the osphradial, pleural, and abdominal ganglia develop. The pleural ganglia are directly connected to the cerebral ganglia, upon whose lateral surfaces they lie, and have connectives passing in front of the statocysts to enter the ipsilateral pedal ganglion. The osphradial ganglion develops from a proliferation of epithelial mantle cells lining the superior border of the entrance to the mantle cavity. A short connective links the osphradial ganglion to the supraesophageal ganglion, which will become the *right* abdominal ganglion of the adult.

At Stage 5, approximately 32 days after hatching, the buccal ganglia develop on the posterior surface of the buccal muscle. From Stages 8 through 12 the pleuroabdominal connectives increase proportionately in length, and the right and left abdominal ganglia approach each other and begin to fuse. Between Stages 12 and 13 the bag-cell clusters form in the pleuroabdominal connectives near the abdominal ganglia. The genital ganglion is the last major ganglion to develop. It forms between Stages 12 and 13, arising from a small cell cluster within the genital nerve sheath at the junction of the nerve, the common hermaphroditic duct, and the nudimental gland.

Formation of commissures and connectives between ganglia has not yet been studied in *Aplysia.* In prosobranch gastropods, such as *Patella* and *Crepidula,* the formation of commissures requires cellular contact (Raven, 1966). The ganglia either approach each other until they touch or become united by a strand of migrating nerve cells (Smith, 1935; Moritz, 1939). Later, fibers from the ganglia grow out along the commissures. The cerebropedal connectives are usually the first to form and appear in *Crepidula,* even before the pedal ganglia are clearly delimited; these connectives are often entirely lined with ganglion cells (Conklin, 1897). The other connectives are usually formed by the free outgrowth of nerve fibers from the ganglia and do not require cellular contact. In the prosobranch *Haliotis* (Crofts, 1955), however, connective formation requires cell contact. Cell strands split off in the overlying ectoderm and precede the formation of connectives.

A specific example of the phased development of the nervous system of *Aplysia* is the formation of the abdominal ganglion. This ganglion begins to form at Stage 1 and consists of three groups of two to five cells each. By Stage 3 distinct clusters are evident; the adult ganglion arises as a fusion of these cell clusters (Figure 7-14C). One cell cluster, which appears to be homologous to the supraintestinal ganglion of more primitive opisthobranchs (Chapter 4), develops above the esophagus. This cluster becomes the *right* abdominal hemiganglion of the adult. The *left* hemiganglion develops *below* the esophagus and derives from two independent clusters of neurons, one rostral and the other caudal, each surrounding a central medulla of neuropil. The rostral cluster seems to correspond to the subintestinal ganglion of primitive opisthobranchs and becomes the left upper quadrant of the adult abdominal ganglion. The caudal cluster corresponds to the visceral ganglion of the primitive opisthobranchs and becomes the left lower quadrant of the adult ganglion.

Connectives link the right and left abdominal hemiganglia to the right and left pleural ganglia. Prior to metamorphosis, by Stage 6, the left abdominal hemiganglion consolidates into a single cell cluster slightly larger than the right hemiabdominal ganglion. After metamorphosis the abdominal commissure becomes shorter and the left abdominal ganglion moves from below to above the esophagus.

Approximately 35 days after metamorphosis a small cluster of cells, the bag cells (Frazier *et al.*, 1967), appears on the dorsal surface of each pleuroabdominal connective approximately 50 μm from the junction between the connectives and the abdominal ganglion. The bag cells increase in number until each cluster contains about 400 in the adult. These two clusters are the last components of the abdominal ganglion to develop.

Not only does the ganglion develop in stages, but different identified cells become morphologically recognizable at different times (Kriegstein, 1977a; Schacher and Kandel, 1977). The cholinergic cell R2, the largest cell of the abdominal ganglion, can be identified early. Initially (Stages 1 and 2) the right hemiganglion consists of only about 4 to 6 largely undifferentiated nerve cells (pre-ganglion cells), none of which have as yet sprouted an axon. These ellipsoid cells, about 4–6 μm in diameter, are embedded in a basket of nonneuronal support cells and surround a tract of incoming fibers from the pleuroabdominal connective. Some of the incoming fibers from the connective contact and form axosomatic synapses on the cell bodies of the pre-ganglion cells. This contact appears to be a signal for axon formation, for soon after its appearance the cell body sprouts a primary axon and secondary branches. Once the axon has formed, the soma synapse disappears. It is not as yet clear whether the soma contact is transient and later retracts or whether, with the growth of the axon, it moves down that process to its definite site (Schacher and Kandel, 1977). Although cell R2 does not differ in size from the other neurons in the right hemiganglion, it becomes recognizable early during Stage 3 once its major axon appears in the right connective. At metamorphosis (Stage 6), when the right hemiganglion already contains about 100 cells, R2 still resembles the other cells in size, being only 7–10 μm in diameter. But at Stage 8, in the days following metamorphosis, R2 suddenly becomes 20 μm in diameter, or twice as large as the other cells (Figure 7-14C). At the same time, a cell of similar size appears in the left pleural ganglion that can be identified as cell LP1. The increase in size of these two cells coincides with the development of the parapodia, the structures to which the two cells send their major axonal branches.

An interesting feature of the developing nervous system is that the support cells in which the neurons of the individual ganglia are embedded contain large secretory granules, 1–5 μm in diameter. Following metamorphosis the granules disappear; their disappearance coincides with a growth spurt on the part of the neurons. The granules can be prematurely released by exposing animals briefly to artificial seawater containing high K^+ and low Ca^{++}. Under these conditions, where other secretory cells as well as synapses appear to be unaffected (because low Ca^{++} blocks most types of release), the neurons undergo a premature growth spurt (Schacher and Kandel, 1977). Thus the granules may contain a factor, analogous to the nerve growth factor of vertebrates (Levi-Montalcini, 1975), that stimulates the late stages of neuronal differentiation. Since the incoming fibers that form axosomatic contacts on the relatively undifferentiated neurons seem to trigger an early stage of differentiation, one can envisage a program of differentiation whereby the synapse formed by the incoming fiber signals axon outgrowth and initial process formation, and the subsequent release of granules by the nonneural support cell signals the further and more extensive outgrowth necessary for synapse formation.

Since the *Aplysia* egg is determinate and the ganglia of the fully developed nervous system contain identified cells, it may be possible to study the lineage of various identified cells. Some preliminary information is already available about the origins of the nervous system of other molluscs (see Raven, 1966). The central ganglia come from ectoderm (therefore from one of the first three quartets of micromeres). The cerebrals come from the first micromere quartet, which forms the ectoderm of the head. The pedal ganglia arise from cell 2d, which gives rise to the ectoderm of the foot. This type of analysis could be advanced to the cellular level. For example, the cholinergic cells R2 and LP1 could be used to study the lineage of abdominal and pleural neurons, and the metacerebral cells, a pair of identified serotonergic cells in the cerebral ganglia of *Aplysia,* could be used to examine the lineage of a cerebral neuron.

If the origin of a specific identified cell were traced back to a particular neuroblast, it might be possible to examine the fate of the other daughter cells of that neuroblast and determine at what stage in the sequence of division all daughter cells might share common properties, such as transmitter biochemistry or behavioral functions.

In insects, where neuronal cell lineage has been studied, each neuroblast divides asymmetrically, yielding a daughter neuroblast and a small preganglion cell, the ganglion mother cell (Wheeler, 1893; Schrader, 1938; Nordlander and Edwards, 1968). The ganglion mother cell then divides symmetrically into daughter cells that differentiate into neurons. Each neuroblast tends to remain in an outer peripheral position and gives rise to a column of daughter cells below it. With continued proliferation of daughter cells, the columnar arrangements become distorted. Nevertheless, most neurons derived from a given neuroblast tend to remain together; only a few migrate to join groups of cells derived from other neuroblasts. In view of the pronounced tendency of the large identified cells to be at the periphery, it is possible that each large cell represents a differentiated form of the neuroblast that originally gave rise to the smaller cells below and surrounding it. (For recent studies in nematode worms, see Sulston, 1976.)

Torsion and Detorsion

As we have seen (pp. 232-235), a major event in the embryonic development of *Aplysia* and other gastropod molluscs is *torsion.* In newly hatched *Aplysia* veligers the visceral mass is twisted 120° counterclockwise in relation to the cephalopedal mass. During metamorphic Stages 7 and 8 *detorsion* begins, bringing the anus from its premetamorphic position alongside the mouth to its adult position on the posterior dorsal surface. The digestive system undergoes a 90° rotation clockwise (Figure 7-15).

The gill, the adult heart and nervous system all develop in their adult position. Postmetamorphic detorsion is therefore limited to two parts of the digestive system, the anus and the intestine; the anterior part of the mantle complex is unaltered. As a result, the gill, adult heart, and nervous system are essentially unaffected by the detorsional movements.

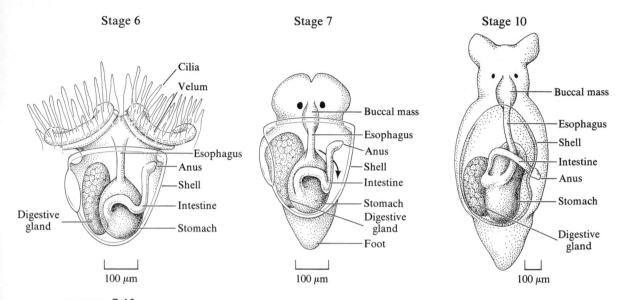

Stage 6   Stage 7   Stage 10

FIGURE 7-15.
Changes in the digestive system with detorsion. The lack of change in orientation of the digestive system during metamorphosis (Stage 7) contrasts with the subsequent change during detorsion (Stage 10). [After Kriegstein, 1977b.]

Detorsion occurs after the development of the abdominal ganglia and the pleuroabdominal connectives (Kriegstein, 1977b). However, because the pleuroabdominal connectives are short, even those parts of the nervous system that in some other gastropods are most affected by detorsion are little altered in *Aplysia* (Figure 7-16B).[8]

In prosobranchs the time of appearance of homologous ganglia varies in relation to time of torsion, as well as to whether the visceral ganglia are affected by torsion. In the mesogastropods *Viviparus* [*Paludina*] and *Bithynia* the parietal ganglia are at first symmetrical. During torsion the right ganglion moves dorsally and to the left to become the supraesophageal ganglion, while the left

---

[8]Complete streptoneury and euthyneury are actually two extremes of a continuum. As the visceral connectives get shorter and the visceral loop becomes more anterior, the condition of the nervous system goes from streptoneury to euthyneury even in gastropods that have undergone torsion. *Aplysia* exhibits neither extreme and therefore seems only partly torted in the larva and juvenile (Figure 7-16).

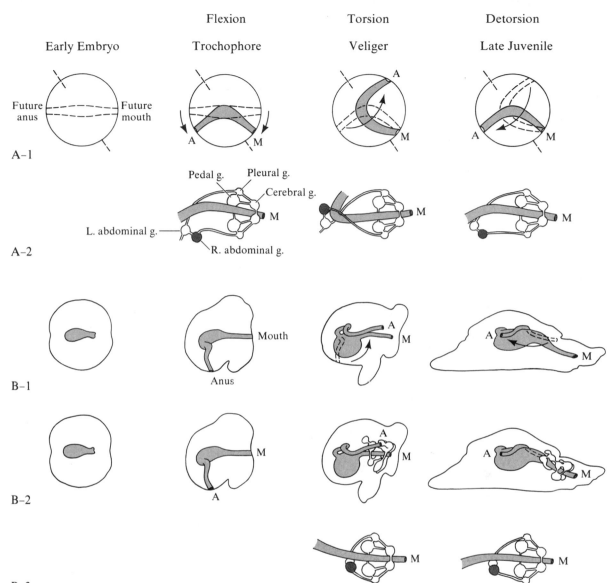

FIGURE 7-16.

Effects of anal-oral flexion, torsion, and detorsion on the nervous system.

A. Changes in the position of the abdominal ganglia that would occur if the ganglia were present during the period of torsion. **(1)** Illustration of the flexion and torsion sequence (see Figure 7-11B). **(2)** The theoretical consequences of that sequence on the abdominal ganglia were they present at the time of torsion. The darkened ganglion is the right abdominal (superintestinal) hemiganglion.

B. Illustration of actual changes in *A. californica* without the nervous system **(1)** and with the nervous system. **(2)**. The theoretical consequences of torsion and detorsion are illustrated in **3**.

parietal ganglion is displaced to the right and down to become the subesopha-geal ganglion (Erlanger, 1891a, b; 1892). Similarly, in the neogastropod *Busy-con* (= *Fulgur*) there is an actual twisting of the connectives into a figure eight during torsion and a crossing of the parietal ganglia from one side to another, a configuration known as streptoneury (Conklin, 1907). In the archaeogastropods *Haliotis, Patella, Calliostoma,* the subgenus *Ansates* of *Helcion* (previously classified as a separate genus, *Patina*), as in *Aplysia,* the supra- and infra-esophageal ganglia and the visceral ganglion are formed only after torsion is completed (for review see Hyman, 1967; Raven, 1966). Thus, they are situated from the beginning in their definitive places and connected immediately with the pleural ganglia of the *pretorsional* ipsilateral side. In *Haliotis,* for example, the right pleurovisceral connective grows leftward from the right pleural gan-glion toward the supraesophageal ganglion near the left gill. The left pleuro-visceral connective grows rightward toward the infraesophageal ganglion, which is formed later and simultaneously with the right gill. Both pleurovisceral connectives then continue beyond the parietal ganglia beneath the floor of the mantle cavity to unite and form part of the visceral loop (Crofts, 1937).

These comparative anatomical considerations illustrate that if the visceral ganglia arise in a part of the body that was torted, the connectives are twisted whether or not they develop before torsion. The tissue that gives rise to the ganglia moves along with the rest of the torted organs and determines where the ganglia will ultimately develop. In *Aplysia* the precursors of the cell regions of the abdominal ganglion escape full torsion because they mostly arise in the marginal zone of the head region, anterior to the zone of torsion and detor-sion. In some other opisthobranchs, and in most pulmonates, the visceral ner-vous system is untorted but the visceral organs are torted because the visceral ganglia and connectives do not arise in the caudal part of the body that under-goes torsion. Whatever the evolutionary advantage of torsion may be, it does not seem to derive from the twisting of the pleurovisceral connectives, which only follow the fate of other parts of the body.

Postmetamorphic Development

Coggeshall (1967) and Skyles (personal communication) have found a signifi-cant increase in the number of nerve cells and neuronal processes during the course of postembryonic development. Cell size also increases with age and neuronal cell membranes become more scalloped.

The best studied case of increasing cell number is the bag cells. These cells are absent in small juveniles prior to Stage 13 (weight 0.2 g) but slowly increase to about 800 cells in mature animals (weight >50 g). Since bag cells release a hormone that triggers egg-laying (Kupfermann and Kandel, 1970), an increase in their number may be necessary to supply adequate hormone for the animal's reproductive phase. Increases are clearly not restricted to the bag cells. Other small cells also increase, with the result that the total number of cells in the abdominal ganglion increases from about 1,000 in juveniles weighing 0.2 to 5.0 g, to about 2,500–3,000 in mature animals. The new cells seem to arise from the deep (neuropil) regions of the ganglion.

In addition to increases in cell number there are also increases in cell size. For example, in small ganglia of animals weighing less than 0.2 g the widest diameter of cell R2 is rarely bigger than 100 μm. In larger animals weighing 8 g the diameter increases to 300 μm, and in very large animals of more than 200 g it can reach 800–1,000 μm. As the ganglion increases in size, the connective tissue thickens and more compartmental dividers appear, separating cells into still smaller groupings (Figure 5-1, p. 151).

Neuronal Differentiation and Development of Peripheral Organs

If the central ganglia of cephalopods are separated by cutting the commissures, the ganglia still differentiate (Ranzi, 1932). Similarly, if the foot of a *Lymnaea* embryo is split in half, thereby cutting the pedal commissure, each pedal ganglion still differentiates normally (Raven and Beenakkers, 1955). Even if the cerebral ganglia are dislocated so that the ganglion on one side is far from and no longer connected to the pedal ganglion of the same side, both cerebral and pedal ganglia differentiate normally (Raven, de Roon, and Stadhouders, 1955).

Whereas the differentiation of central molluscan ganglia is surprisingly independent of interconnections with other central ganglia, their development is dependent on the presence of peripheral structures, as is the case with other invertebrates and vertebrates. For example, the optic ganglion of cephalopods is dependent on the eye. When the eye does not develop fully, the optic ganglion is reduced in size, and when the eye does not develop at all neither does the optic ganglion (Ranzi, 1928a, b). Similarly, growth of the statocyst seems to influence the pedal ganglia. When the statocyst is lacking on one side, the corresponding part of the pedal ganglion does not differentiate.

A similar situation occurs in developmental mutants of the fruit fly *Drosophila* (Power, 1943). In mutants with reduced (small and incomplete) eyes the volume of the optic glomerulus varies linearly with the number of ommatidia. In totally eyeless flies there is a complete failure to develop that part of the external glomerulus that receives the afferent optic fibers. Parts of the external glomerulus that receive fibers from neural structures other than the eyes are less affected. However, if the eye is removed at the time fiber formation is beginning in the ganglion, differentiation in the ganglion proceeds normally. (For review of vertebrate neurogenesis see Jacobson, 1978).

DEVELOPMENT OF BEHAVIOR

Aplysia veligers are complete organisms capable of carrying out all life functions except reproduction. Thus, the adult behaviors of locomotion, feeding, and defensive withdrawal have their parallel in the veliger. One can thus examine the transformation of the larval behavioral repertory into the adult one. Three larval behavioral responses, locomotion, feeding, and defensive withdrawal, have so far been studied (Kriegstein *et al.,* 1974).

Locomotion

As was discussed above, prior to metamorphosis *Aplysia* veligers swim by means of the coordinated beating of long velar cilia. Swimming is occasionally interrupted by a withdrawal response, by which the veliger retracts its velar lobes and foot into its shell. Deprived briefly of the action of locomotor cilia, the veliger sinks. The first sign of a transformation in the swimming pattern becomes evident about 30 days after hatching, when the propodium develops. At this point, a veliger has the ability to crawl but continues to spend most of its time swimming. As metamorphic competence is attained the veliger exhibits a greater tendency to crawl upon the substrate, and if the appropriate substrate is encountered the larva will settle upon it and partake in little or no further velar swimming. Within three hours after settlement, the velar lobes begin to shrink. Now the larvae can only crawl; they can no longer swim (Figure 7-17A).

The adult locomotor pattern develops gradually. At first crawling depends on the action of pedal cilia as well as on muscular contraction. The contribution of integrated pedal waves, characteristic of the adult, increases progressively during the early juvenile phase.

Feeding

Unlike locomotion, the adult feeding does not emerge until metamorphosis is complete. Prior to metamorphosis veligers feed by sweeping algal cells into their mouths using the water currents generated by velar and subvelar cilia.

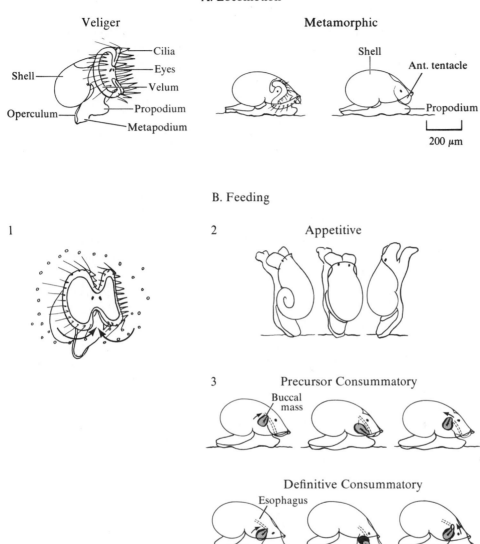

FIGURE 7-17.
Changes in locomotion and feeding at metamorphosis. [After Kriegstein *et al.,* 1974.]

A. Locomotion. The veliger (30 days) swims by means of velar cilia. The metamorphic animal (34 days) only crawls. The velar cilia are shed, the velar lobes shrink, and the larval propodium and metapodium develop into a crawling foot.

B. Feeding. **(1)** Veliger feeding on unicellular algae, which are swept into the mouth by velar cilia. **(2)** The appetitive component of feeding behavior. After the first day of metamorphosis (day 35) an animal exhibits the essential features of the appetitive response. It anchors its metapodium on a substrate and waves its head and body from side to side in an orienting fashion. **(3)** The consummatory component of feeding behavior. During the first and second days of metamorphosis (days 35–36) the animal only partially protracts and retracts the buccal mass but does not ingest food (precursor response). On day 37 the metamorphosed animal exhibits full consummatory behavior and ingests pieces of seaweed (definitive response). All animals are drawn to the same scale.

After settlement on the appropriate triggering substance, which is destined to serve as their diet after metamorphosis (*Laurencia pacifica* in the case of *A. californica*), veligers stop eating unicellular algae and do not feed until metamorphosis is complete. Three days after settlement (as early as 37 days posthatching), the metamorphosed *A. californica* resume eating by feeding on the growing tips of *Laurencia* (Figure 7-17B).

Adult feeding consists of two components: an appetitive response, in which the animal rears up its head and waves it from side to side orienting toward the seaweed, and a consummatory response, involving an all-or-none ingestive movement of the buccal mass and radula (see Chapter 8). These two components of the feeding behavior develop sequentially. The appetitive response emerges first. Even before metamorphosis, on day 32, a veliger will wave its propodium from side to side, apparently searching while crawling. Once the anterior tentacles take shape, halfway through the metamorphosis, an animal will exhibit all the essential elements of the appetitive response. If removed from the seaweed it will anchor the posterior segment of its foot and rear up its propodium and head, swaying its body from side to side (Figure 7-17B).[9] At this same stage an animal exhibits only incomplete aspects of the consummatory response. When anchored to the seaweed it will rotate its radula and buccal mass in excursions that resemble the consummatory response, but the behavior is only partial, and the animal will never bite or swallow seaweed.

The full consummatory response—biting and swallowing—does not appear until one day later (day 37), when metamorphosis is complete (Figure 7-17B). A metamorphosed animal exhibits fully integrated adult feeding behavior: the appetitive component followed by the full consummatory response.

Defensive Withdrawal

Veligers have a defensive response consisting of withdrawal of the head and foot into the shell. This response is analogous to the adult defensive withdrawal of the head into the body mass. During metamorphosis the withdrawal reflex

---

[9]The appetitive component of feeding may appear first because it is a more generalized searching behavior that can occur independent of feeding; for example, it could be important in the searching for a new substrate on which to crawl. In contrast to *Aplysia*, in the predatory snail *Natica* the consummatory feeding response appears full-blown upon the first feeding attempt (Berg, 1976).

of the head–foot is still present. Larvae disturbed during the first day of metamorphosis will withdraw into the shell and close the shell orifice tightly against the operculum, maintaining the metapodial point of attachment. Animals disturbed at later stages of metamorphosis will withdraw into their shells upon stimulation but will begin to crawl when they recover.

Bodily changes at metamorphosis also lead to changes in the defensive behavior. The growth of the head and foot and the change from a coiled to a flattened shell make it impossible for the animal to withdraw its head–foot into the shell. Instead, tactile stimulation of the head results only in generalized contraction. In addition, a new, more restricted mantle defensive reflex develops: withdrawal of the newly formed gill, siphon, and mantle shelf. The withdrawal is very similar to the response in the adult.

Thus, the adult behavioral repertory appears in a specified sequence (Figure 7-18). The adult feeding pattern appears first. The adult locomotor pattern begins to develop next. Other adult behavioral responses, such as the reflex responses and fixed-action patterns of the mantle organs, siphon, gill, and mantle shelf, appear only later. This order of development seems to parallel the sequence with which the critical ganglionic controls develop (see Figure 7-15).

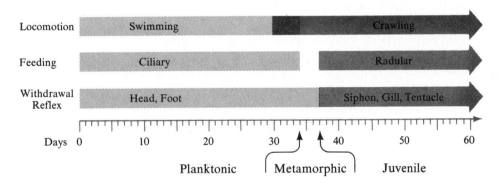

FIGURE 7-18.
Time table for the transformation of locomotor and feeding behaviors from larval to adult form in *A. californica*. This table describes the minimal time necessary for the behavioral changes at 22°C. Whereas larval and adult locomotor behaviors overlap for four days at the onset of metamorphosis, the termination of ciliary feeding is separated from the onset of radular (adult) feeding by three days. During these three days of metamorphosis the animal does not feed. Generalized defensive withdrawal of the head–foot is replaced by localized siphon, gill, and tentacle responses during metamorphosis. The time scale begins at hatching and extends to reproductive maturity. [From Kriegstein *et al.,* 1974.]

In feeding, the appetitive behavior appears first and is complete before the consummatory behavior develops. The consummatory response in turn undergoes a precursor stage before developing into the stereotypic adult response. It would be interesting to see whether the precursor movements are a necessary prerequisite for proper development of the adult response, and whether they are due to a lack of fully matured neuronal control or of necessary maturation of the feeding structures, or to both.

INTERACTION BETWEEN ENVIRONMENT AND DEVELOPMENT

Perhaps the most remarkable feature of the development of *Aplysia* and other invertebrates having an obligatory planktonic larval phase is that metamorphosis, and therefore subsequent development, is dependent on self-initiated behavior by the animal. Whereas in lower vertebrates and in some invertebrates metamorphosis involves the response of the organism to internal hormonal events that do not require an environmental triggering factor, in *Aplysia* a behavioral step is interposed in the developmental process. Before metamorphosis can begin *Aplysia* larvae must find and settle upon an external triggering substance, a specific species of seaweed. This external triggering factor in turn sets off a series of internal metamorphic events that permit the animal to assume adult behaviors. Thus, the life cycles of *Aplysia* and related opisthobranchs provide an opportunity for studying how external triggering factors and internal processes interact in determining the development of the nervous system and of behavior. This could be accomplished by identifying the critical ingredient in *Laurencia pacifica* that triggers metamorphosis and to learn how it is recognized by the animal and how it initiates the changes characteristic of metamorphosis.[10]

DEVELOPMENTAL MALFORMATIONS

It is possible to experimentally induce certain abnormalities in *Aplysia* that may be useful in studying the mechanisms of development. For example, Peltrera (1940) created double *Aplysia* larvae (double monsters) by centrifuging eggs at the two- or four-cell stage. When the two halves were separated

[10]See Hadfield and Karlson (1977) for a discussion of the triggering substance in the coral *Porites* that stimulates the nudibranch *Phestilla sibogae* to metamorphose.

each one formed a *dwarf larva*. Peltrera interpreted these experiments to indicate that the fate of cells had not yet been irrevocably determined at early cleavage stages. This contrasts with the general view, discussed above, that eggs undergoing spiral cleavage are a mosaic and that determining substances are already distributed among the blastomeres. Peltrera (1940) assumed that centrifugation leads to an abnormal distribution of cytoplasmic substance within the egg. This need not occur, however. As Morgan (1927) pointed out, although subcellular organelles can be moved around by centrifugation, the determining substances, whose location is unknown, may not be affected. Clement (1968) found that the morphogenetic substances present in the vegetal pole of the eggs of the snail *Ilyanassa* are not moved by centrifugal forces that displace yolk, nuclei, and yolk droplets because some of the substances are in the cell membrane. In addition, Peltrera did not show enough detailed histology to prove that each of the two halves forms a complete larva.

Although double monsters are the only malformation so far described in *Aplysia,* Ranzi (1932) reported a number of interesting deformities of the head and eye of the cephalopod *Loligo* caused by treating the fertilized egg with lithium. They include (1) *synophthalmia,* where the eyes approach one another and the cerebral ganglia are slightly reduced but the other ganglia are normal; (2) *cyclopia,* in which the eyes are fused in the midline, the originally paired ganglia become fused during later development into a single ganglion, and the cerebral ganglion is small and unpaired or is lacking altogether; (3) *anophthalmia,* in which the eyes, optic ganglia, and cerebral ganglia are lacking altogether; and (4) *acephaly,* in which the only organs formed are the shell gland, pedal ganglia, and gut.

Similar malformations of the head region have been obtained by treating the eggs of the basommatophoran pulmonate *Lymnaea stagnalis* with lithium. Raven (1942, 1949, 1952) found that the effects of lithium are primarily limited to the posterior part of the apical plate. No changes in mesodermal or endodermal organs were observed. Raven suggests that lithium treatment suppresses differentiation at the animal pole because the ion's effects decrease with the distance from the pole (Raven, 1942, 1949). Recent studies (reviewed in Davidson, 1976) indicate that lithium induces cells otherwise destined to be ectodermal to behave as endoderm, with the result that treated embryos also develop an enormous gut. Moreover, non-mesodermal cells seem to participate in the production of primary mesenchyme (Davidson, 1976). Runnström and Markman (1966) found that lithium's action is blocked by actinomycin D. which interferes with RNA synthesis. These results suggest that lithium interferes with the expression of the genetic program in the affected cells.

TYPES OF DEVELOPMENT IN OPISTHOBRANCHS

In some opisthobranch molluscs development proceeds smoothly from the embryonic stage to reproductive maturity without an intervening free-swimming larval stage. These forms are said to have *direct* development. Most opisthobranchs resemble *Aplysia* in having a well-defined *veliger* larval stage interposed between hatching and metamorphosis. These forms are said to have *indirect* development. Thompson (1967) has subdivided opisthobranch development into four types. Types 1 and 2 have indirect development; types 3 and 4 have direct development (for discussion of types of development see Tardy, 1970).

Type 1 development.   The larvae pass through a long and obligatory intervening plankton feeding stage (planktotrophic larvae). This type includes *Aplysia* as well as genera from all the major orders of opisthobranchs (Table 7-3) and is characterized by small ova ranging in diameter from 40 to 170 μm that are produced in large numbers (about 10^6 per spawn mass) with multiple embryos (up to 100) per egg capsule.[11] Veligers hatch after a relatively short embryonic period ranging from 2 to 28 days. The newly hatched veliger usually lacks eyes and a visible propodium rudiment, but possesses a velum, larval retractor muscles, larval kidney, nephrocysts, metapodial mucous glands and operculum (Figure 7-19). The mantle fold is still undifferentiated and the larva feeds on microorganisms. The factors necessary for triggering metamorphosis are known for only a very few type 1 opisthobranchs, such as *Aplysia californica*, *Aldera modesta*, and *Onchidoris fusca*, partly because of difficulties in rearing veligers to this stage.

The evolutionary advantages of this type of development include the large number of larvae and their wide distribution (Thorson, 1950). Disadvantages include the great possibility of death, especially by predation, the scarcity of food in certain regions and in certain seasons, and possible failure to locate the specific factor that triggers metamorphosis.

Type 2 development.   The larvae undergo complete larval development while subsisting entirely on the original yolk supply of the egg (lecithotrophic larva). This type includes representatives of the notaspids, sacoglossans, and nudibranchs but not any anaspids or cephalaspids (Table 7-3).

---

[11]Although the various *Aplysia* species so far examined have been found to be type 1, there are variations in types of development even within genera. Thus the nudibranch *Phestilla sibogae* is lecithotrophic while *Phestilla melanobranchia* is planktotrophic (Harris, 1975).

TABLE 7-2.
Characteristics of three types of embryonic development of opisthobranchs.

	Type 1		Type 2		Type 3	
	Alderia modesta (15°C)	*Onchidoris fusca* (10°C)	*Tritonia hombergi* (10°C)	*Adalaria proxima* (9–10°C)	*Retusa obtusa*	*Cadlina laevis* (10°C)
	Day	Day	Day	Day	Day	Day
Four-cell stage	0.5	1	1	1	0.4	
Beginning of gastrulation	1	6	5	5	4	15
Greatest size of blastopore	1.5	7	5.5	6	5	
Shell-gland	2	9	8	7	6	21
Mouth	2	10	9	9		
Foot	2	10	9	9		
Cilia	2	10	8	8	6	
Anal cells	2.5	12	9	10	7	
Operculum	2	12	20	20	15	
Metachronism of velar celia	3	11	11	14	8	
Velar cilia can be halted*	4		14			
Hatching	5–5.5	17–19	36–38	36–39	28	50

*Presumed first behavioral sign of cerebral ganglia function.
SOURCES: Types 1 and 2: Thompson, 1967, Type 3: Thompson, 1967 (*Retusa*); Smith, 1967 (*Cadlina*).

Mature animals in this group lay a smaller number of eggs than those of type 1, with a maximum of only 51,000 per spawn mass. The eggs are larger, ranging in diameter from 110 to 250 μm. Veligers hatch after a relatively long embryonic period ranging from 4 to 42 days. At the time of hatching the free-swimming veliger has eyes, a radula, and a propodium rudiment in addition to a velum, larval retractor muscles, larval and adult kidneys, metapodial mucous glands, and operculum (Figure 7-19). These veligers may feed, but food is not necessary for development. The veliger larval phase is brief, usually lasting two days or less. Metamorphosis has been studied in several species and, as is the case in *Aplysia,* is triggered by a specific (perhaps unique) nutrient.

Type 3 development. This type includes species with direct development without metamorphosis, including representatives of cephalaspids, sacoglossans, and nudibranchs but not any anaspids or notaspids. Type 3 is characterized by large eggs (210–400 μm) produced in small batches (up to 5,850 per

Type 1

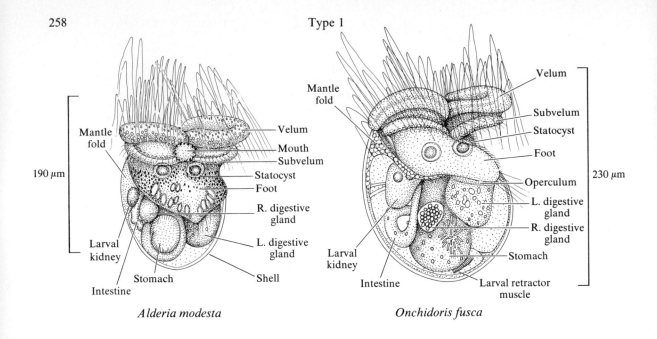

Alderia modesta

Onchidoris fusca

Type 2

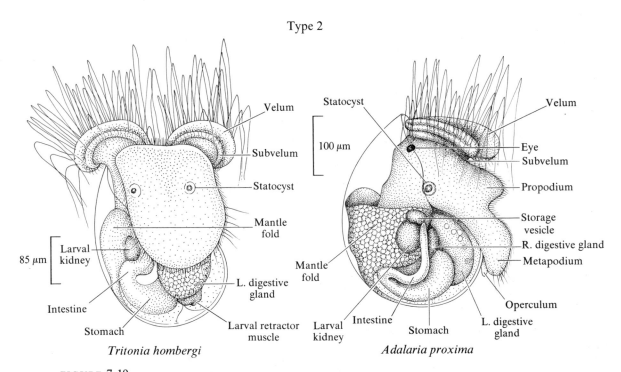

Tritonia hombergi

Adalaria proxima

FIGURE 7-19.
Types 1 and 2 of opisthobranch development. Camera lucida drawings from life of newly hatched veliger larvae. [After Thompson, 1967.]

spawn mass). More than one embryo per egg capsule is rare. The embryo hatches as a juvenile after a long embryonic period lasting 13 to 50 days. The juvenile has eyes and a radula but lacks larval retractor muscles, velum, metapodial mucous glands, external shell, and operculum. The mantle fold forms part of the dorsal body surface. Some species pass through a typical veliger stage before hatching; others develop certain veliger structures (for details see Thompson, 1967).

Type 4 development.   This type includes viviparous forms. It has been found only in the cephalaspid *Pluscula cuica,* in which Marcus (1953) found veligers in the reproductive tract.

METAMORPHOSIS AND POSTMETAMORPHIC DEVELOPMENT IN OPISTHOBRANCHS

As with *Aplysia,* other opisthobranch veligers also need and search for a suitable substrate on which to settle and metamorphose (for general discussion on settlement in gastropods see Scheltema, 1974). For example, veligers of the nudibranch *Amphorinae doriae* die without metamorphosing unless they come into contact with the hydroid *Kirchenpaueria,* on which the adult feeds. An hour's contact with this hydroid is sufficient to initiate metamorphosis (Tardy, 1962, 1970). The veligers of the nudibranch *Adalaria proxima* need the bryozoan ectoproct *Electra pilosa,* and those of *Tritonia hombergi* need the colonial coelenterate *Alcyonium digitatum* (Thompson, 1958, 1962).

The end of the planktonic phase is characterized by searching for the appropriate substrate. In *Adalaria* this behavior can continue for up to two weeks until either the substrate is found or the veliger dies. During searching several premetamorphic changes occur: the pleural and optic ganglia and the rudiments of the rhinophores develop. Later the shell is cast off and the operculum and the velar lobes are reduced in size (Figure 7-20). If the *Adalaria* larva encounters an obstacle while searching, it retracts its velar lobes and the upper part of the propodium and creeps over the surface of the substrate. If the substratum is not a live colony of *Electra pilosa,* the larva again protrudes its velar lobes and swims away. When the larva encounters the correct substrate, creeping continues and the larva never again uses velar cilia for locomotion. At this point metamorphosis begins rapidly and in one day the shell and the operculum are cast off. Metamorphosis involves the numerous structural changes summarized in Figure 7-20. Postmetamorphic growth is also very rapid as *Adalaria* continues feeding on the ectoproct on which it settles.

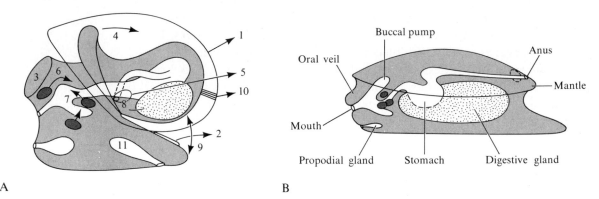

FIGURE 7-20.
Larval metamorphosis in *Adalaria proxima* showing **(A)** a late veliger larva and **(B)** a young post-larval benthic stage. Arrows indicate direction of repositioning of anatomical features. The developmental changes illustrated are: **(1)** loss of shell; **(2)** loss of operculum; **(3)** transformation of velar lobes into oral veil; **(4)** reflexion of the mantle fold; **(5)** movement of the anal and renal opening to the rear; **(6)** rearward movement of cerebral ganglia and commissure; **(7)** closer approximation of all the ganglia; **(8)** extension of the digestive gland anteriorly; **(9)** fusion of the visceral sac with the metapodium; **(10)** loss of larval retractor muscle; and **(11)** loss of metapodial mucous gland. [After Thompson, 1958.]

The *Tritonia hombergi* larvae settle and metamorphose when they encounter a live colony of *Alcyonium*. The larvae retract their velar lobes and the upper part of their propodia creep over the surface of the *Alcyonium* and never use the velar cilia again for locomotion. Within a day or two of settling, the shell and operculum are lost (Figure 7-21). During the next 24 hours metamorphosis is completed: the animal transforms its life style from a free-swimming form to a crawling one. The velar locomotor cells are resorbed and the area where they were becomes the oral veil. In the weeks following settling and metamorphosis the juvenile increases rapidly in size. (See also Bonar and Hadfield, 1974, on metamorphosis in the nudibranch *Phestilla sibogae*, and Bridges, 1975, on the anaspid *Phyllaplysia taylori*.)

SUMMARY AND PERSPECTIVE

The molluscan eggs undergo spiral cleavage. The first and second cell divisions are at right angles to each other (meridional), forming four cells of somewhat unequal size. But subsequent divisions are oblique, resulting in alternating right and left displacement of the cleaving cells.

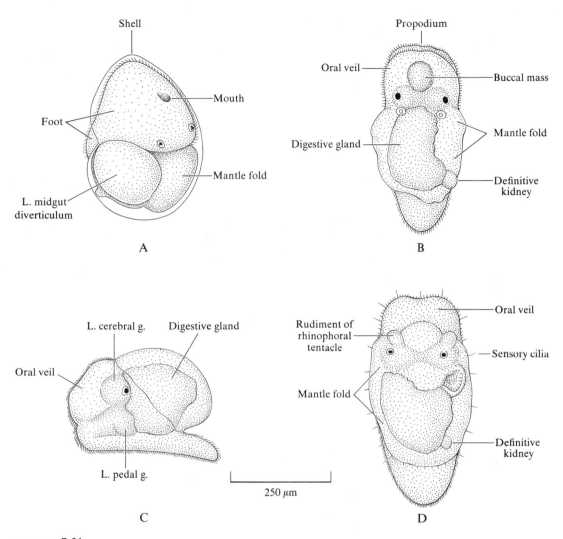

FIGURE 7-21.
Metamorphosis of *Tritonia hombergi*. **A.** Late shelled stage. **B.** Shell-less stage, dorsal aspect. **C.** Shell-less stage, left lateral aspect. **D.** Later shell-less stage, dorsal aspect. [After Thompson, 1962.]

The eggs undergo determinate development. At the very earliest divisions each blastomere receives a recognizable part of the original egg cytoplasm and surface membrane that endows it and its cell lineage with a specific ability to give rise to parts of the veliger larva. These morphogenetic or determining substances are present in the egg before cleavage begins. As a result of the selective distribution of these determinants, the fate of each blastomere is to some degree independent of the presence of the other blastomeres. The determinants are synthesized during oogenesis and stored in the egg cytoplasm and membrane until just before or just after fertilization. The distinct patterns of gene activity necessary for cellular differentiation are thought to result from the interaction of the totipotential genome of the blastomeres and the specific determinants for each cell.

The molluscan embryo has a small number of cells that split off in a predetermined manner. As a result, cells in the *Aplysia* blastula assume recognizable shapes and positions. In early stages the embryo is almost identical, cell for cell, to that of annelids. In each group the mesoderm derives from a primitive cell that arises from primitive endoderm.

The development of *Aplysia californica* involves four major phases: (1) embryonic development, from fertilized egg to hatching; (2) larval development, from hatching to metamorphosis; (3) metamorphic development; and (4) postmetamorphic or juvenile development, from metamorphosis to reproductive maturity.

Embryonic development takes approximately 9–14 days at room temperature. Four to five days after the eggs are laid movements of the embryo are first seen inside the egg capsule and spontaneous hatching occurs six to seven days later. When the egg hatches, a free-swimming larval form (the veliger) emerges.

The second phase, larval (veliger) development, takes about 34 days at 22–25°C. The veliger larvae swim as plankton in the oceans and feed on unicellular algae. The veliger is a complete and independent organism that performs all the necessary life functions except reproduction. To achieve reproductive maturity the veligers must metamorphose and undergo a radical change in form, diet, behavior patterns, and life style.

The third phase, metamorphic development, takes three days (days 34 to 37 posthatching). During this period the animals do not eat and do not grow in size; rather, they undergo preparatory changes that allow them to switch their diet from unicellular alga to seaweed, and their feeding habits from ciliary to radular feeding. They also stop ciliary swimming and begin to crawl.

The final stage, postmetamorphic (juvenile) development, begins after metamorphosis and extends for about 85 days, when reproductive maturity is reached.

The nervous system of *Aplysia* develops in a specific sequence over a period of 140 days. The statocyst and the cerebral and pedal ganglia develop at hatching; the abdominal, pleural, and osphradial ganglia three weeks after hatching; and the buccal ganglia at four weeks. Certain peripheral ganglia, such as the genital ganglion, do not develop until the end of the juvenile stage.

The origin of the abdominal ganglion is complex; its anlage forms during the first three weeks after hatching from three larval ganglia that fuse to form the abdominal ganglion. Individual cells cannot be distinguished from one another by their location within the ganglion or by their appearance alone until metamorphosis at five weeks. After metamorphosis, the identified neuron R2 becomes recognizable because of a significant increase in size.

The generation time of *Aplysia,* formerly thought to be about a year, is 19 weeks. This time is still far longer than that of genetically more useful preparations, such as the nematodes (two days) and *Drosophila* (14 days), but is comparable to the generation time of crickets (six weeks). Since these genetically useful preparations are not as favorable as *Aplysia* for cellular biological studies, this animal could prove a useful compromise for selected genetic problems. Moreover, *Aplysia* is hermaphroditic, and it may be possible to experimentally induce self-fertilization. One might be able to generate isogenetic lines within five generations (16–20 months). Since each animal lays egg strands containing over a million fertilized eggs, there is ample opportunity to isolate structural and behavioral mutants using screening techniques of the type used successfully with nematodes, *Drosophila,* and crickets.

SELECTED READING

Cather, J. N. 1971. Cellular interactions in the regulation of development in annelids and molluscs. *Advances in Morphogenesis,* **9:**67–127. An up to date review of spiral cleavage and development.

Davidson, E. H. 1976. *Gene Activity in Early Development,* 2d ed. New York: Academic Press. A clearly written review of the biochemical basis of differentiation. Chapter 7 contains an excellent discussion of regulative and mosaic development and of the universality of morphogenetic determinants in egg cytoplasm.

Fretter, V., and A. Graham. 1962. *British Prosobranch Molluscs: Their Functional Anatomy and Ecology.* London: The Ray Society. Excellent, detailed discussion of prosobranch gastropod biology, including a chapter on the nervous system and development.

Hyman, L. H. 1967. *The Invertebrates, Vol. VI, Mollusca 1.* New York: McGraw-Hill. Includes a comprehensive review of the biology of the whole class Gastropoda, with a chapter on opisthobranchs.

Jacobson, M. 1978. *Developmental Neurobiology,* 2nd ed. New York: Plenum Press. A description of the development of the vertebrate central nervous system.

Kume, M., and K. Dan, eds. 1957. *Invertebrate Embryology,* translated by J. C. Dan. National Technical Information Service. U.S. Department of Commerce. A detailed review of molluscan development.

Raven, C. P. 1966. *Morphogenesis: The Analysis of Molluscan Development,* 2nd ed. New York: Pergamon. Another good review of molluscan development.

Saunders, J. W., Jr. 1970. *Patterns and Principles of Animal Development.* London: Macmillan. A good introduction to developmental biology.

Watson, J. D. 1976. *Molecular Biology of the Gene,* 2nd ed. New York: Benjamin. Chapter 17, "Embryology at the Molecular Level," is a concise discussion of a molecular biological approach to development emphasizing the usefulness of simple systems. such as slime molds.

CHAPTER 8

Species-Specific Behavior of Opisthobranchs

At the beginning of the twentieth century students of behavior relied heavily on comparative studies, especially on studies of various invertebrate animals, in their effort to develop an objective experimental basis for the study of behavior (see for example Verworn, 1889; Jennings, 1906; Loeb, 1918). Following the discovery of associative conditioning by Thorndike and Pavlov, however, interest in comparative studies declined, and with it the interest in invertebrates. Because associative learning was initially difficult to demonstrate in many invertebrates, psychologists concentrated on vertebrates and in particular on a few types of animals (pigeons, rats, cats, and dogs). The dramatic decline of the work on comparative aspects of behavior was documented in 1950 by Frank Beach in his presidential address to the American Psychological Association. Using the *Journal of Animal Behavior* and its successor, the *Journal of Comparative and Physiological Psychology,* as the basis of his analysis, Beach found that between 1911 and 1920 about 35 percent of published articles concerned invertebrates. In 1923 the number of articles on invertebrates suddenly dropped to zero, and remained between zero and five percent from 1923 to 1948. During the same period the number of published articles on the rat rose from 12 percent to 70 percent! A reversal also occurred in the types of behavior studied. In 1911, 35 percent of the articles were devoted to conditioning and learning and 25 percent were devoted to sensory capacities, reflexes, and simple behavior patterns. By 1939, 78 percent of the articles were devoted to learning and only five percent to sensory capacities, reflexes, and simple behavior.

The extensive study, since 1923, of only a few animal groups turned the field away from the comparative study of behavior (Beach, 1960; Lockard, 1971). Only now, half a century later, is this trend being reversed (Hodos, 1974; Tobach *et al.,* 1973). The impact of ethological thinking, the discovery of the biological constraints on learning, and the ability to analyze the behavior of invertebrates on the cellular level have all been factors in the renewed interest in invertebrates and in comparative psychology (for discussion see *Cellular Basis of Behavior,* Chapters 1 through 3).

In addition to being supplanted by studies of associative learning in the pigeon and the white rat, interest in the comparative study of behavior declined during the period 1920–1950 because it did not prove fruitful. Most of the early comparative studies selected both animals and behavior arbitrarily, without concern for phylogenetic relationships, on the one hand, or behavioral homologies on the other. These studies then tried to explore how the difference in their brains might explain the differences in behavior. This proved a crude, relatively unsuccessful approach. It is now appreciated that to be successful comparative studies must fulfill certain criteria (Beach, 1960; Hodos, 1974). First, the studies must be carried out in closely related animals that are specifically selected for their phylogenetic interrelationships. Second, the studies should focus on naturally occurring (species-specific) responses because only these behaviors are present in similar form in all members of a species of the same sex. Third, the behaviors selected should be at least crudely homologous. Fourth, an attempt should be made to compare homologous neural structures with these homologous behaviors. Finally, the studies should be based on classes of behavior that are widely distributed phylogenetically so as to maximize the opportunity for interspecific comparisons and increase the likelihood of arriving at broad insights into behavioral responses.

As I have indicated in the Introduction, opisthobranchs, and in particular the various species of *Aplysia,* are useful animals for comparative studies because of the existence of many closely related forms with interesting differences in their behavior. Despite these advantages, little was known about opisthobranch behavior prior to 1965 and the onset of the current interest in the neurobiology of opisthobranchs. Since then, however, much effort has gone into studying their behavioral capabilities. This effort has been surprisingly successful. A significant body of knowledge has been gained about five major classes of behavior: (1) defensive responses; (2) autonomic concomitants; (3) locomotion; (4) feeding; and (5) mating and egg-laying.

Since the behavioral repertoire of opisthobranchs is quite limited, these five classes represent a significant aspect of an individual animal's *total* behavioral capability. As a result, a significant fraction of the total behavioral capability of individual opisthobranchs has been analyzed at the cellular level. In these animals one can begin to explore a number of higher-order questions in the neurobiology of behavior. What are the rules that relate different types of behavior? How do carnivores and herbivores differ in the neural circuitry that controls feeding behavior? How are the neuronal circuits of related behavioral responses altered by a common learning situation? How do the cellular mechanisms of similar learning processes vary between close relatives? How general are the mechanisms whereby a common arousal stimulus affects different classes of behavioral responses? In a comparative context each of these questions takes on a broader significance and allows us to examine the generality of the answers.

So far only a very few opisthobranch species have been studied, and these were not selected specifically for their phylogenetic interrelationships. Consequently, the results do not as yet lead to a coherent view of the comparative behavioral biology of the key opisthobranch orders—or of the relationship of these to pulmonates. Nonetheless, some preliminary ideas are emerging. The animals studied vary in their habitat (land and marine, intertidal, and sublittoral), body structure (shelled and unshelled; location or absence of gills), feeding behavior (herbivorous and carnivorous), sensory capabilities (distance- versus contact-receptors), and brain organization (distributed and concentrated; simple and complex). By comparing common classes of behavior in different groups of animals one can begin to see how these biological constraints restrict the behavior and learning of opisthobranch and pulmonate molluscs.

In this chapter I shall review the behavioral capabilities of several well-studied species of *Aplysia*. I shall also compare studies of *Aplysia* with parallel studies of representatives of several other opisthobranch orders: the cephalaspids, the notaspids, and the nudibranchs. Finally, I will describe behavioral studies of representatives of several orders of pulmonates, in particular the basommatophoran *Helisoma* and the stylommatophorans *Helix* and *Limax*.

I shall focus primarily on the behavioral studies because the related cellular neurobiological studies have been described in the companion volume, *Cellular Basis of Behavior*. I shall, however, briefly summarize the relevant cellular data to indicate the extent to which each behavior has been analyzed on the cellular level.

CLASSIFICATION OF BEHAVIOR

For behavioral comparisons between opisthobranchs, as between any animals, it is crucial to have a general scheme for classifying behavior. Of the many schemes that have been devised, I will use a particularly simple one that distinguishes between two broad categories at the extremes of a behavioral spectrum: *graded* behaviors and *all-or-none* behaviors. This distinction is based on the stimulus–response characteristics of a given behavioral response, that is, the manner in which the intensity and pattern of the sensory stimulus affects the amplitude and pattern of the motor response.

In graded responses, such as the flexion withdrawal or pupillary response of vertebrates, the response amplitude is a function of stimulus intensity and the response pattern is largely determined by the temporal characteristics of the stimulus (see for example Sherrington, 1906; Lloyd, 1957). By contrast, in all-or-none (stereotypic) behaviors, such as swallowing, vomiting, sneezing, and orgasm (see for example Morris, 1958; Doty, 1968), the response amplitude and pattern are relatively independent of the strength and pattern of the stimulus. A stimulus elicits either a full response or none at all.

The distinction between graded and all-or-none responses is perhaps the most objective criterion for distinguishing between *reflex acts* and *fixed acts*. Reflex acts and fixed acts are instances of *elementary behavioral responses,* which are defined as a single episode of behavior in a single motor pathway. The stimulus–response curve for a reflex act has the gradual slope characteristic of a graded response. The precise form of a reflex act is thus closely tied to the nature and duration of the stimulus. On the other hand, the stimulus–response curve for a fixed act has a steep slope resembling a step function, which is typical of an all-or-none response. The expression of the fixed act is thus stereotypic and largely independent of the eliciting stimulus.

Elementary behavioral acts can be combined to form *complex behavioral responses,* which are defined as recurring episodes in a single motor pathway or motor patterns involving more than one motor pathway. Complex behavior is of two types: *reflex patterns* (such as orientation to food) and *fixed-action patterns* (such as escape or courtship maneuvers). To some degree, the distinction between graded and all-or-none behaviors is also useful in distinguishing reflex and fixed-action patterns. In complex behaviors, however, another dimension is added: the amplitude and sequence of the individual behavioral components may vary. In a reflex pattern the sequence as well as the amplitude of each component can vary, whereas in a fixed-action pattern both amplitude and sequence tend to be stereotypic. Finally, certain types of complex behavior, such as feeding or copulation, involve combinations of reflex and fixed-action patterns.

An alternative classification is based on the degree to which behavior is primarily inherited or acquired. Since "inherited" refers to the genotypic aspects of behavior and "acquired" to the phenotypic, the two terms are not mutually exclusive. Ernst Mayr (1974) has therefore proposed that behavior might be best classified in relation to its genetic program. A *closed program* is a genetic program for a behavior that does not allow significant modifications during the translation into the phenotype. In a closed program nothing can be added by experience. An *open program* is a genetic program that allows for experiential adjustments on the part of the animal.

The two classifications thus have features in common, particularly as applied to the behaviors of invertebrates. Many all-or-none responses (fixed acts and fixed-action patterns) result from closed programs, whereas graded responses (reflex acts and reflex patterns) tend to result from open programs.

REFLEX AND FIXED ACTS

Perhaps surprisingly, only a few simple reflex and fixed acts have been studied in detail. Among the best examples are the gill- and siphon-withdrawal reflex acts and the fixed act of inking in *A. californica*.

Defensive Reflexes

Opisthobranchs have a soft unprotected head and several exposed appendages: anterior tentacles, rhinophores, siphon, and gill. Each of these organs shows a graded defensive-withdrawal reflex. For example, the head, anterior tentacles, and rhinophores of *Aplysia* withdraw to touch. In addition, the siphon of *Aplysia* typically projects out from between the parapodia, and the gill often protrudes slightly from the mantle cavity; both are reflexively withdrawn when the siphon is touched. The intensity of the withdrawal is a function of the intensity of the stimulus (Figure 8-1A).

A large part of the neuronal circuit of the siphon- and gill-withdrawal reflexes has been worked out in cellular detail (Kupfermann and Kandel, 1969; Peretz, 1969; Kupfermann *et al.*, 1974; Castellucci *et al.*, 1970; Byrne *et al.*, 1974). There are five central motor cells located in the abdominal ganglion that move only the gill, seven motor cells that move only the siphon, and one cell that moves the siphon, gill, and mantle shelf. In addition, there is a group of about 30 peripherally located siphon motor cells (see p. 203). The central as well as the peripheral motor cells are innervated by several interneurons and by two groups of mechanoreceptor sensory neurons each containing about 24 neurons. One of these groups innervates the siphon skin and posterior mantle shelf; the other innervates the anterior mantle shelf (Figure 8-2).

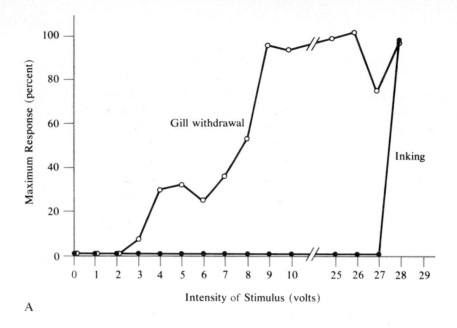

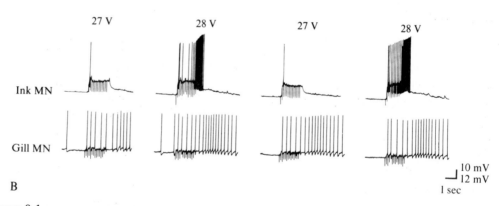

FIGURE 8-1.

Comparison of a graded, reflex act and an all-or-none, fixed act in *Aplysia*. [From Carew and Kandel, 1977a.]

A. Gill withdrawal and inking produced in intact, restrained animals by electrical stimulation of the siphon. Gill withdrawal was measured by means of a photocell and inking by means of a spectrophotometer (for details of procedure see Carew and Kandel, 1977a). The data are normalized by expressing each response as a percentage of the maximum response exhibited by the animal. Electrical stimuli (1.5-msec biphasic pulses, 6/sec for 800 msec) of increasing intensity were delivered to the siphon skin (once every 20 min) by means of platinum wires sewn into the skin. These curves are representative data from one of five such experiments. In all experiments the threshold for inking was higher than for gill withdrawal by at least an order of magnitude and the stimulus-response characteristic for inking was very steep compared to that for gill withdrawal.

B. Comparison of input-output characteristics of ink-gland motor cells and gill motor cells. Simultaneous recordings from an ink-gland motor cell (L14A) and a gill motor cell (L7) during trains of brief (2 msec) electrical pulses (6/sec for 2 sec). Slight increases in stimulus intensity, from 27 to 28 V, trigger an accelerating burst in the ink-gland motor cell, whereas the gill motor cell's output is relatively unchanged.

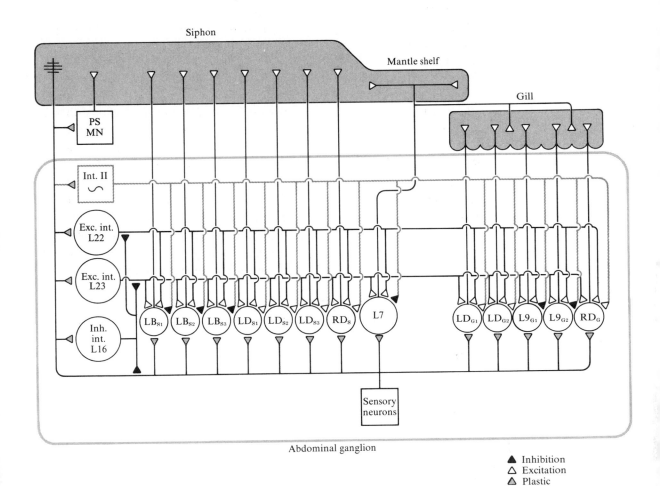

Siphon

Mantle shelf

Gill

PS MN

Int. II

Exc. int. L22

Exc. int. L23

Inh. int. L16

LB$_{S1}$ LB$_{S2}$ LB$_{S3}$ LD$_{S1}$ LD$_{S2}$ LD$_{S3}$ RD$_{S}$ L7 LD$_{G1}$ LD$_{G2}$ L9$_{G1}$ L9$_{G2}$ RD$_{G}$

Sensory neurons

Abdominal ganglion

▲ Inhibition
△ Excitation
△ Plastic

FIGURE 8-2.
Neural circuit controlling gill- and siphon-withdrawal reflexes in *Aplysia* in response to a tactile stimulus applied to the siphon skin. A weak stimulus activates a number of mechanoreceptor cells that excite all the gill and siphon motor cells as well as two excitatory interneurons (L22 and L23) and one inhibitory interneuron (L16). A stronger stimulus also excites a population of command elements for respiratory pumping (Interneuron II). These cells produce a motor sequence by exciting some cells first and then others only as a result of rebound excitation following an initial inhibition.

Siphon and gill withdrawal are each elementary *reflex acts*. They are part of a larger, more complex *reflex pattern* that includes contraction of the parapodia and mantle shelf (Kupfermann and Kandel, 1969; Hening *et al.*, 1976). If the stimulus is sufficiently strong the sensory input will also trigger a fixed act: respiratory pumping. The pumping movements are mediated by a command population that excites some of the gill and siphon motor cells and inhibits others (which then fire later by postinhibitory rebound excitation), thereby producing a sequential pumping action that circulates seawater through the mantle cavity (Figure 8-2; for details of the respiratory command cells see Figure 8-5).[1] The pumping movements can also occur in the absence of tactile stimuli (see discussion in Chapter 3 and below). Thus, as a result of having central motor cells for siphon and gill that can be independently and selectively activated, the same effector system can be utilized in three different ways: (1) as a reflex act; (2) as a fixed act triggered in response to a strong mechanical stimulus to the siphon; and (3) as a spontaneously occurring fixed act without obvious external stimulus (for discussion see Kupfermann and Kandel, 1969; Peretz, 1969).

If the stimulus strength to the siphon is further increased the animal will secrete ink and exude a viscous white substance (see below). Similar defensive reflexes occur in most molluscs.[2]

As we have seen (Chapter 6) individual pinnules of the gill of *Aplysia* also contract in a graded manner in response to direct stimulation (Peretz, 1970; Carew *et al.*, 1974).

Defensive Fixed Act: Inking

One of the most dramatic behavioral responses in *Aplysia californica* (also found in certain other *Aplysia* species) is the sudden massive release of purple ink when the animal is disturbed (Tobach, Gold, and Ziegler, 1965; Carew and Kandel, 1977a). The ink is released from Blochmann's gland (the ink or purple

---

[1]Until recently it was not known whether the command for respiratory pumping was mediated by one or by a population of cells. For convenience, this unidentified element was called Interneuron II, since L10, the command cell for heart rate increases, was called Interneuron I (see for example Koester *et al.,* 1974; Kandel, 1976). Recently, Byrne, Dieringer, and Koester (1977) identified part of the command system and found it consisted of several cells. They have therefore named the group (formerly called Interneuron II) the *command population for respiratory pumping.*

[2]For some recent studies on siphon withdrawal in the clam *Spisula* see Prior (1972a). For earlier studies of withdrawal reflexes in molluscs see Arey and Crozier (1919) and Bethe (1903) and the review by Willows (1973).

gland) located at the edge of the mantle shelf (see Figure 3-11B, p. 71). When elicited by mechanical stimulation, inking is a high-threshold, all-or-none behavior. A weak tactile stimulus that produces a brisk reflex response of the siphon and gill does not cause inking. Ink is only released, and then massively, when the stimulus intensity surpasses a high and fairly sharp threshold (Carew and Kandel, 1977a).

Cellular studies have shown that the differences between the low-threshold, graded reflex acts of gill and siphon withdrawal and the high-threshold, all-or-none fixed act features of inking reside in the properties of the motor cells (Figure 8-1B; Carew and Kandel, 1977a). Unlike the central siphon and gill motor cells, which are independent elements, the three ink motor cells (L14A, L14B, and L14C) are electrically interconnected. Whereas the gill and siphon motor neurons are spontaneously active and show a graded increase in firing to a graded increase in input, the ink motor cells are silent, have a high resting potential (-65 mV), and a high threshold for spike generation. But once triggered by a strong and prolonged input, the ink motor cells suddenly fire a high-frequency burst of action potentials that causes inking (Figure 8-1B; Carew and Kandel, 1977a).

It is not known whether stimuli other than mechanical ones can trigger inking. Although the secretion of the purple gland may mask an animal confined to a small tidal pool, the ink is not effective as a screen for sublittoral animals. Its function in sublittoral animals is therefore not known. Conceivably, the ink is noxious to some animals that compete with *Aplysia* for food. Alternatively, or perhaps in addition, the ink could be a defensive secretion. While alive, opisthobranchs, including *Aplysia,* are unpalatable to fish. But many species of fish will eagerly eat dead or damaged opisthobranchs (Thompson, 1960a, b). The unpalatability of living opisthobranchs is thought to be due to a defensive (acid) substance secreted by the animals when perturbed (Thompson, 1960a, b). Perhaps in *Aplysia* ink serves this purpose. However, the few animals that prey on *Aplysia* in nature (such as the sea anemone *Anthopleura xanthogrammica* and the opisthobranch *Navanax*) do elicit inking (Winkler and Tilton, 1962; Dieringer, Koester, and Weiss, 1978).

Of the five subgenera of *Aplysia,* all but the subgenus *Aplysia* (species *A. depilans, A. juliana, A. vaccaria, A. cedrosensis, A. dura,* and *A. nigra*) have a well-developed purple gland. Comparative studies might therefore provide insights into the function that ink serves in the behavioral repertory of the animal (see Chapter 10 for further details). The primitive cephalaspid *Acteon* and all cephalopods (except *Nautilus*) also release a dark secretion in response to noxious stimuli.

In contrast to inking, which does not occur in all species, the secretion of a thick, white, mucus-like substance occurs in all species when the animals are seriously disturbed. The secretion, called opaline, is released by the opaline (Bohadsch's) gland, which lies beneath the floor of the anterior part of the mantle cavity and discharges into the left anterior portion of the mantle cavity. Opaline produces convulsions and paralysis when injected into coelenterates, jellyfish, worms, echinoderms, molluscs, arthropods, and fish (Flury, 1915). Opaline also inhibits the feeding response of crabs (Kittredge, unpubl. observ.). Neither the neural control of opaline secretion nor its input-output characteristics have yet been worked out.

Effects of Environment on Elementary Behavior

As we have seen (Chapter 2), a given species of *Aplysia* can live in two radically different shore environments. Animals living in the intertidal zone are buffeted by the tides, whereas those in the sublittoral zone live in a relatively calm and stable environment.

What consequences do these radically different environmental conditions have on defensive behavior? This question has been explored by Carew and Kupfermann (1974) in *A. californica*. They found that the protected sublittoral animals have a lower reflex threshold for siphon withdrawal and inking than do intertidal animals (Figure 8-3).

In some species the threshold for inking may be altered by the social environment. Thus, Tobach, Gold, and Ziegler (1965) found that the threshold for inking in *A. dactylomela* is lower in animals that are grouped than in isolated animals. Whether this is due to contact with other animals is not known.

COMPLEX BEHAVIOR

In contrast to the paucity of data on elementary behavior in opisthobranchs, there is now a great deal of information on a variety of complex behaviors, particularly locomotion and feeding.

Autonomic Regulation

RESPIRATORY PUMPING MOVEMENTS IN APLYSIA

The elementary, fixed-act movements of the siphon and gill are linked to movements of the parapodia and to inhibition of heart rate to produce a complex fixed-action pattern: a respiratory pumping movement sequence that

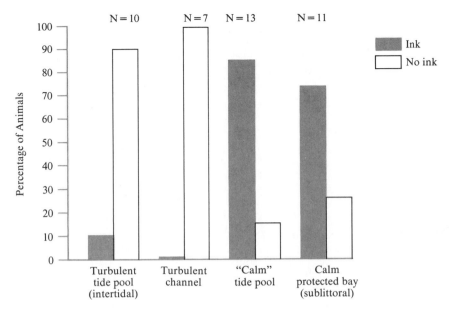

FIGURE 8-3.
Behavioral determinants of environment. Comparison of inking in animals living in calm and turbulent environments in response to a pin inserted through a parapodium. Data are expressed in terms of the percentage of animals of each group that inked when pinned. Animals living in turbulent water have a higher threshold than those living in calm water. [After Carew and Kupfermann, 1974.]

circulates fresh seawater through the mantle cavity. This aerates the gill and flushes debris out of the mantle cavity (Figure 8-4). The neural circuit for this behavior has been analyzed in part (Figure 8-5; see Kupfermann and Kandel, 1969; Peretz, 1969; Kupfermann *et al.,* 1974; Hening, Carew, and Kandel, 1976; Byrne and Koester, 1978). The program for this fixed-action pattern is centrally located and does not require proprioceptive feedback from peripheral structures. The program is triggered by a command population (formerly called Interneuron II) located in the abdominal ganglion. An interesting feature of the command population is that it makes reciprocally inhibitory connections with command cells for related functions (such as heart rate increases), so that only one of a family of commands is carried out at a time. Although some components of the command population for respiratory pumping now have been identified, details of the entire population have not as yet been worked out (see Byrne and Koester, 1978).

Respiratory pumping and fecal extrusion are found in other gastropods as well as in certain cephalopods and bivalves. In some opisthobranchs, e.g., the

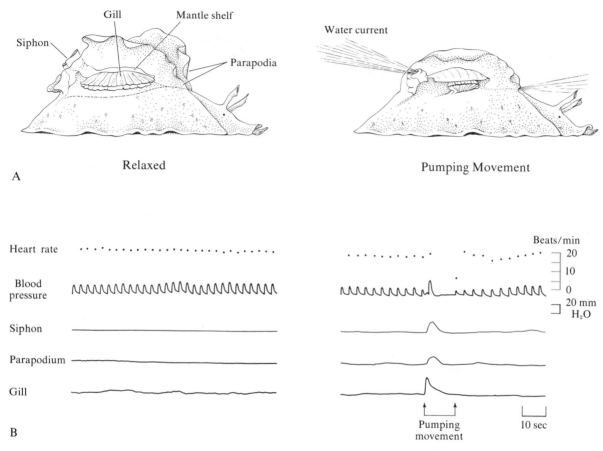

FIGURE 8-4.
Respiratory pumping in *Aplysia*. **A.** Dorsal view of relaxed and contracted animals. In the latter pumping movements that circulate seawater through the mantle cavity are illustrated. **B.** Synchronous recruitment of different behavioral components in the respiratory pumping fixed-action pattern. Simultaneous strain-gauge recordings were used to obtain records from the siphon, parapodia, and gill. The aorta was cannulated to obtain records of blood pressure and heart rate. [From Byrne and Koester, 1978.]

anaspid *Notarchus punctata,* respiratory pumping is linked to swimming, thereby adjusting the rate of gas exchange to motor activity (Purchon, 1968). A similar linkage should be looked for in swimming species of *Aplysia*.

CIRCULATORY ADJUSTMENTS IN APLYSIA

Other behavioral responses are also accompanied by homeostatic adjustments. For example, a large increase in heart rate accompanies escape behavior (Dieringer, Koester, and Weiss, 1978) and, as we shall see in Chapter 9, a moderate increase accompanies the arousal to food. *Aplysia* also show an emersion response similar to the diving reflex of vertebrates. An increase in heart rate accompanies exposure to air and a decrease occurs with immersion in seawater (Feinstein *et al.,* 1977; for details see Chapter 3).

The motor neurons controlling circulation have been delineated (Mayeri *et al.,* 1974; Koester *et al.,* 1974; Liebeswar *et al.,* 1975). There are two excitatory motor cells to the heart, one of which, the serotonergic cell RB_{HE}, is particularly important because a brief burst of activity in this cell produces a prolonged increase in heart rate. There are also two (cholinergic) inhibitory neurons and three (cholinergic) vasoconstrictor cells. The motor cells are in turn innervated by several command cells. One of these is a cholinergic dual-action cell (L10), which increases heart rate and cardiac output by exciting RB_{HE} while inhibiting the inhibitors and the vasoconstrictors (Figure 8-5). An interesting feature of cell L10 is that it is capable of endogenous burst activity. In this mode of firing in L10 each burst serves as the central program for a phasic increase in heart rate. Thus, L10 also serves as the pattern generator providing the central program for a homeostatic fixed-action pattern (Figure 8-6; Koester *et al.,* 1974; for review see Kandel, 1976).

Locomotion

Opisthobranchs move by crawling (or creeping) and by burrowing. In most orders some species can also swim. Each of these locomotor patterns is a complex fixed-action pattern involving a sequence of rather stereotypic movements. When an opisthobranch creeps over a surface, part of the ventral surface of the foot is fixed to the substratum while the remainder of the foot moves forward (Gray, 1968). These coordinated movements produce waves of muscular contractions (followed by relaxation) that pass along the length of the foot and propel the body forward. The waves may be *direct,* moving from posterior to anterior in the direction of travel, but more commonly they are *retrograde,* passing from anterior to posterior, opposite from the direction of travel (Figure 8-7; Vles, 1907). Each wave may be *monotaxic,* occupying the whole breadth of the foot, or *ditaxic,* occupying only one half of the foot.

CRAWLING IN APLYSIA

In *Aplysia* the pedal waves are retrograde and monotaxic (Figure 8-8; and Parker, 1917). Thus, *A. californica* crawls by releasing suction, lifting its head, and raising the anterior part of its foot. After extending the anterior part of its foot half the length of the animal, it reattaches the anterior part of its foot.

278

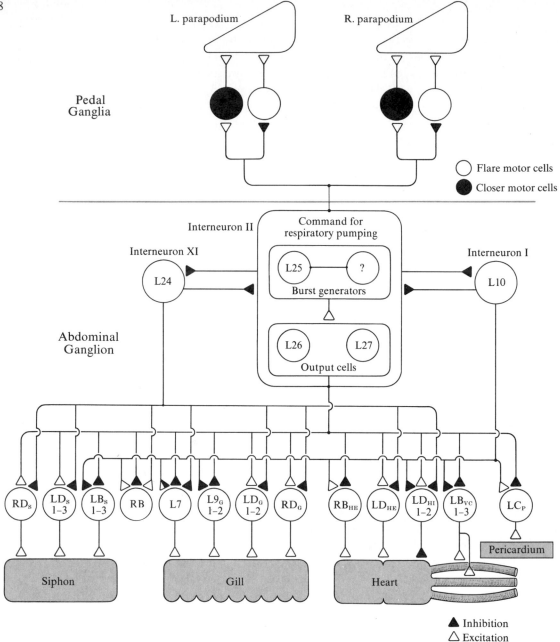

FIGURE 8-5.
Neural circuit of respiratory pumping and associated behavioral responses. The independent motor cells for the siphon, gill, and heart are located in the abdominal ganglion; those for the parapodia are in the pedal ganglion. The motor cells are controlled by several command elements. One of these is the command population (formerly called Interneuron II) that drives respiratory pumping. This command population consists of burst-generating cells (thought to be electrically coupled to one another) and output cells that connect to the motor neurons. Some members of this command population (perhaps the output cells) and the two other command cells L24 and L10, make reciprocal inhibitory connections with one another. [After Byrne and Koester, 1978; Hening *et al.*, unpubl.]

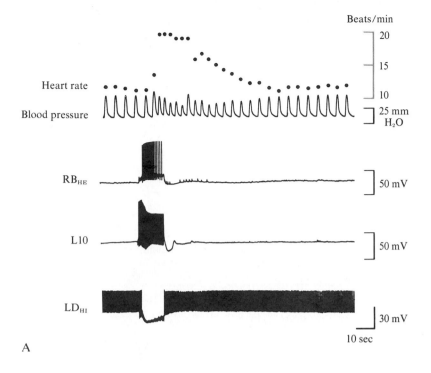

A

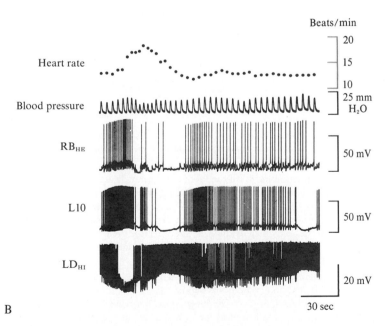

B

FIGURE 8-6.

Control of the circulation by the dual-action command cell L10. **A.** The effect of intracellular stimulation of cell L10 on cardiovascular motor neurons. A 12-sec burst of spikes in L10 excites RB_{HE} and inhibits LD_{HI}. This is associated with a long-lasting increase in heart rate and a short increase, followed by a decrease, in blood pressure. **B.** Spontaneous bursting. Each burst of activity in L10 results in excitation of RB_{HE} and inhibition of LD_{HI}. These alterations in firing pattern are associated with a phasic increase in heart rate plus an initial slight increase in pulse pressure followed by a more prolonged decrease. [From Kandel, 1976.]

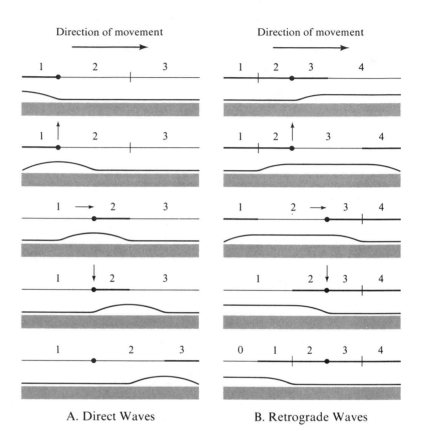

Direction of movement   Direction of movement

A. Direct Waves   B. Retrograde Waves

FIGURE 8-7.

The passage of a wave of muscular contraction along the foot of a gastropod. [After Smith *et al.*, 1971.]

A. Direct waves of contraction move over the foot in the same direction as the animal is traveling (left to right). In the figure successive regions of the foot are numbered 1, 2, and 3; the region contracting is indicated by the thick line. Forward movement is noted by the movement of the dot (indicating the junction between regions 1 and 2). The arrow indicates the elevation of the foot region under consideration in relation to the substrate. When the arrow points up, that region is lifted off the substratum; when it points to the right the region is moved forward; when it points down the region has returned to the substratum.

B. Retrograde waves move in a direction (right to left) opposite to that in which the animal is traveling (left to right). The foot regions are numbered 1, 2, 3, and 4 as in part A, but their contraction occurs in the order 4, 3, 2, 1. The junction of foot regions 2 and 3 is indicated by the dot. Note that the region at which contraction occurs is, in general, attached to the ground and that detachment occurs at the point of relaxation. As in part A, forward movement (to the right) is noted by the movement of the dot (the junction between regions 2 and 3) to the right.

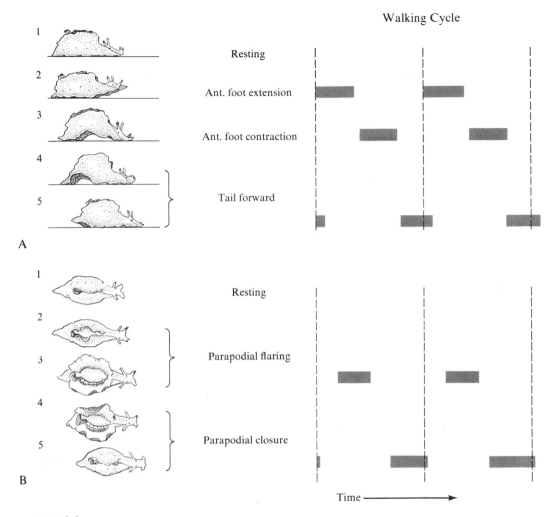

FIGURE 8-8.
The gait of *Aplysia*. [After Kandel, 1976; Hening, Carew, and Kandel, unpubl.]

A. The crawling gait of *Aplysia* showing one full step. The animal moves by lifting its head and releasing suction in order to raise the anterior margin of its foot **(2)**, which is then extended for a distance equal to half the length of the animal **(3)**. As the anterior edge of the foot is brought down and attached to produce an arch, the site of release and extension moves to the posterior portion of the foot **(4)**. The animal then contracts its anterior end to take up the slack produced by the continued extension at the posterior end of the body **(5)**. The anterior and then the middle part of the foot attach sequentially and the posterior portion of the foot is released and brought forward by a distance equal to that moved by the head **(3-5)**.

B. Parapodial sequences during crawling. Dorsal view of the animal showing parapodial movements that accompany the step phases illustrated in part A. The step is accompanied by a characteristic flaring **(2-3)** followed by closing of the parapodia **(4-5)**. Videotape analysis of key aspects of the crawling cycle (on the right) illustrates stereotypic sequence of the locomotor pattern.

The animal then contracts its anterior end to take up the slack while keeping the posterior end of its body extended. Once the anterior and then the middle part of the foot have been attached sequentially, the posterior portion of the foot is released and is brought forward by a distance equal to that moved by the head (Parker, 1917; Hening, Carew, and Kandel, 1977).

The phase relationships of this stereotypic locomotor pattern are preserved over a fourfold range of locomotor velocities extending from two to eight cycles per minute (Figure 8-8A). In addition to the sequential movement of the pedal wave, the locomotor sequence also involves movements of the parapodia and respiratory pumping (Figure 8-8B).

Hening and co-workers (1977) and Jahan-Parwar and Fredman (1978a, b) have identified several populations of neurons in the pedal and cerebral ganglia that produce movements of the parapodia, body wall, foot, and tail. Some of these produce unilateral movements while others produce bilateral movements. Among cells producing unilateral movements are presumed motor neurons in the pedal ganglion that send axons to the parapodia, foot, and tail. These neurons are rhythmically active during that phase of walking in which their target organs exhibit characteristic activity (Figure 8-9).

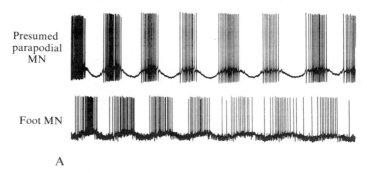

A

FIGURE 8-9.
Sequence of firing in presumed motor neurons controlling locomotion. **A.** The rhythmically recurring bursts in the two motor neurons are consistently 180° out of phase. **B.** By hyperpolarizing the two cells the phase-locked spike bursts followed by periods of silence are seen to be produced by bursts of excitatory and inhibitory postsynaptic potentials, respectively. [From Hening, Carew, and Kandel, unpubl.]

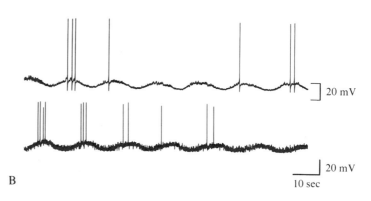

B

To determine whether the locus for generation of the walking pattern is central or peripheral, Hening and co-workers (1977) isolated the ring ganglia (the cerebral, pedal, and pleural ganglia) from the periphery and found that the typical cyclical activity of the presumed motor neurons persisted. Moreover, this cyclical activity could be evoked in the completely isolated ganglia by electrical stimulation of peripheral nerves. These data indicate that, as in many other locomotor sequences (in both invertebrates and vertebrates), the pattern generator for crawling is located centrally (within the ring ganglia) and that its activity does not require patterned afferent input (Figure 8-10; for a review of the role of central pattern generation in locomotion, see Pearson, 1972; Grillner, 1975).

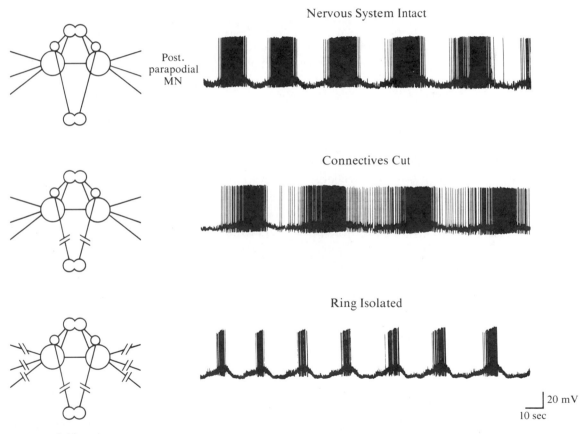

FIGURE 8-10.
Central pattern generation for walking. Comparison of the typical recurrent locomotor rhythm as seen in one presumed motor cell in the intact animal, in the animal with the connectives cut, and in the isolated ring ganglia. These experiments illustrate that the basic rhythm is preserved in the isolated ring ganglia, although the robustness of the bursting is reduced. [From Hening, Carew, and Kandel, unpubl.]

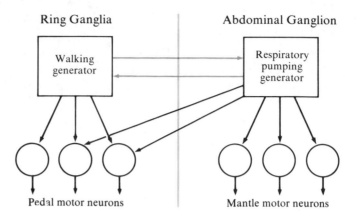

FIGURE 8-11.
Relationships between the pattern generators for walking and respiratory pumping and between the two pattern generators and the pedal and mantle motor neurons. Because the motor cells are not interconnected, there is independent central control over various aspects of motor output.

As is the case for the central gill, siphon, and heart motor neurons (Figure 8-5), the presumed motor neurons for crawling are independent cells organized in parallel; they receive both common and unique synaptic input, but form no synaptic interconnections. As a result, they can be independently activated so that different subsets of motor neurons can be selected in order to produce the patterned actions of different behaviors. Thus the same motor neurons are active in walking, a behavior commanded by cells in the head ganglia, and in respiratory pumping, a behavior commanded by cells in the abdominal ganglion (Figure 8-11).

CIRCADIAN PERIODICITY IN CRAWLING

Using time-lapse cinematography, Kupfermann (1965, 1968) and Li and Strumwasser (cited in Strumwasser, 1965, 1971) independently discovered a circadian rhythm in the locomotor activity of *A. californica* (Figure 8-12). A similar circadian rhythm has been found by Jacklet (1972) using an activity wheel (for review see Lickey *et al.*, 1976). Animals placed on an entrainment schedule of 12 hours of light and 12 hours of darkness are active primarily during light. On the activity wheel they show no preference for moving toward or away from the light source. The locomotor circadian activity persists in reduced form during the next 48 hours of constant darkness (Figure 8-12;

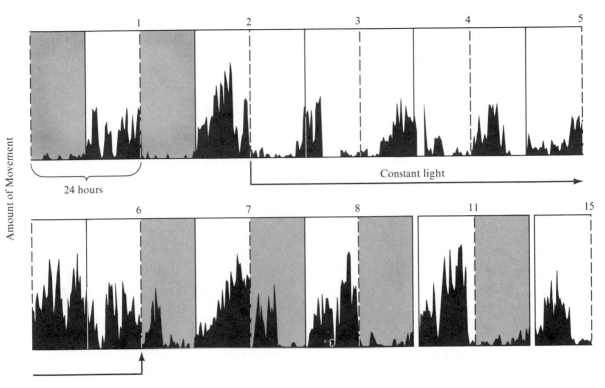

FIGURE 8-12.
Circadian rhythm of the locomotor pattern of *Aplysia californica.* Graphs of locomotor movement as a function of time of day (all data from the same animal). The frame numbers indicate the number of calendar days; periods of darkness are shaded. The first eight frames (calendar days 1 to 8) are continuous days; data have been omitted between days 8 and 11 and between days 11 and 15. [After Strumwasser, 1971.]

Kupfermann, 1967). Individual animals may free run for up to two weeks (Strumwasser, 1965; Lickey *et al.,* unpubl. observ.).

In the light, animals crawl 24–40 meters per day at medium intensities of light and 60 meters per day in high intensities of light (Kupfermann, 1967; Strumwasser, 1965; Jacklet, 1972). Tactile stimulation or agitating the seawater inhibits locomotor activity.

Because the isolated eye of *Aplysia* contains photoreceptors and an oscillator or clock and itself shows a circadian rhythm of neuronal activity (Chapter 3), it is possible that the rhythm in the eye is responsible for driving the locomotor rhythm. Indeed, Strumwasser and his colleagues (1974, and personal communication) have found that under weak or moderate intensity of illumination the eyes are necessary for entrainment. However, in addition to its modulation by the ocular photoreceptors and circadian clock, the locomotor rhythm can also be modulated by extraocular receptors and clocks whose location is as yet unknown (Lickey *et al.*, 1976). Thus, Block and Lickey (1973) and Jacklet (1973a) found that some blinded animals (5–20 percent) could be driven to follow a cyclic locomotor rhythm, even a circadian one, albeit slightly different from normal, by a given light–darkness schedule. Unlike the eyes, the extraocular photoreceptors are sensitive to long wavelengths of light (see Chapter 3). The extraocular photoreceptors therefore seem to come into play most effectively when the eyes are removed or when red light is used as a Zeitgeber. Figure 8-13 illustrates the two photoreceptors and clocks converging on a common locomotor system.

The existence of several independent oscillators raises the question of how they become synchronized in relation to one another. The available evidence suggests that the ocular and extraocular oscillators are not coupled to one another but are synchronized by a common timing source (Lickey *et al.*, 1976). How timing signals are transmitted from the ocular and extraocular oscillators to the pedal motor cells for activation of the circadian crawling pattern is not known. There is a suggestion that the timing signals are neurally rather than

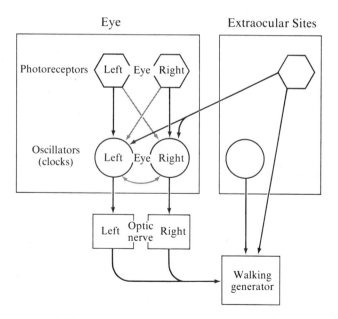

FIGURE 8-13.
Independent ocular and extraocular photoreceptors and clocks converging upon the walking generator. [After Lickey *et al.*, 1976.]

hormonally mediated. The clock in the eye seems to influence the circadian lo-comotor rhythms by means of signals conducted down the optic nerve. Cutting the nerve eliminates the influence of the eye on locomotion (Lickey *et al.,* 1976).

An interesting contrast to these studies on *A. californica* are Carefoot's (1970) studies of two tropical species, *A. dactylomela* and *A. juliana* off the coast of Barbados in the Caribbean Sea. These animals were most active at night, and hid in crevices during the day. This locomotor behavior is clearly different from that of *A. californica* observed in the laboratory. It would be interesting to determine whether this difference is due to species, temperature, or environ-ment. Carefoot's preliminary observations point toward environment because his animals gradually lost their habit of nocturnal activity in captivity, becom-ing active in the daytime. On the other hand, Carefoot (1970) states that in the field the European species *A. punctata, A. depilans,* and *A. fasciata* are active both day and night.

CRAWLING IN PULMONATES

Agriolimax and *Helix pomatia* have direct waves of longitudinal contraction and relaxation that pass along the foot (Figure 8-14). They appear as dark and light bands. Dark bands are regions of horizontal displacement of the foot caused by longitudinally contracted furrows that lift parts of the foot off the ground and move it forward. Light bands are the adjacent elongated, resting areas of the foot (Lissmann, 1945a, b). When one segment of transverse pedal surface undergoes contraction, its posterior edge is lifted up first and this edge is also placed on the ground first in a new position of rest. Thus, each of the many segments acts somewhat like a vertebrate foot, in which the heel is both lifted off and placed on the ground before the toes. Because there are many segments in the foot of *Helix,* each being lifted and advanced in turn, the total locomotor pattern resembles that of a millipede or caterpillar (Lissmann, 1945a). When *Helix* creeps slowly, most of its foot touches the ground, but when moving swiftly some species adopt a loping or galloping gait that involves retrograde waves and resembles the gait of *Aplysia.*[3]

Like *Aplysia californica,* certain pulmonates, such as the garden slug *Limax maximus,* have a clear circadian locomotor rhythm (Sokolove *et al.,* 1977). Following entrainment the rhythm is considerably more robust in constant darkness or constant light than that of *Aplysia californica.* Suggestive but less complete evidence for circadian rhythm of locomotion is available for *Agrio-limax reticularis* (Dainton, 1954) and *Arion ater* (Lewis, 1969).

[3]For a review of locomotion in prosobranchs see Miller (1967).

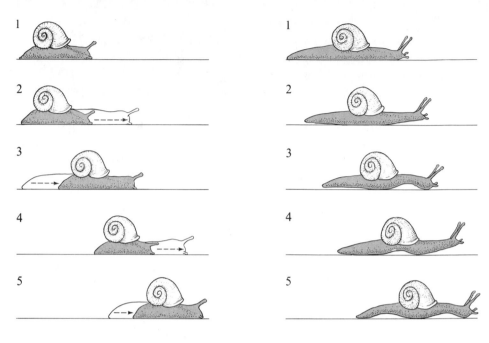

A. Creeping Gait   B. Galloping Gait

FIGURE 8-14.

Two types of gait in the land snail *Helix*.

A. The creeping gait of *Helix* involves five stages. **(1)** Rest; **(2)** elongation of the anterior half of the animal and longitudinal contraction while the posterior half acts as a holdfast; **(3)** sliding of the posterior part of the animal forward while the anterior part acts as a holdfast; **(4)** same as Stage 2; and **(5)** same as Stage 3. [After Lissmann, 1945b.]

B. In the galloping gait contact with the ground is restricted to points lying at the end of the arches. **(1)** No retrograde waves; **(2)** the head is lifted off the ground; **(3)** the head is placed on the ground forming an arch; **(4)** the arch moves backward; **(5)** a second arch is started at the anterior end. [After Gray, 1968; Carlson, 1905.]

BURROWING IN APLYSIA

Some species of cephalaspid and anaspid opisthobranchs have head modifications that allow them to burrow, e.g., the cephalic shield of *Acteon*. Several species of *Aplysia* also burrow, including *A. brasiliana, A. dactylomela, A. juliana,* and to some degree *A. californica.* In *A. brasiliana* (Figure 8-15) Aspey and Blankenship (1976a, b) found that burrowing is primarily engaged in by small (<250 g), presumably youthful animals prior to copulation. For them, burrowing may be a form of estivation or metabolic reorganization necessary

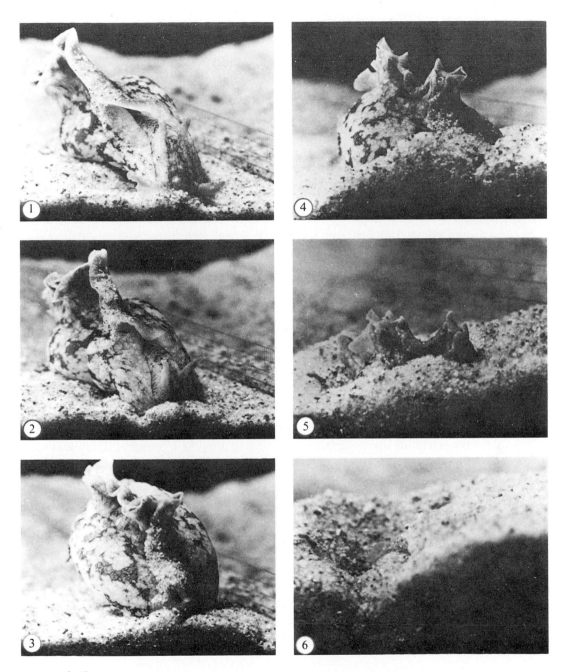

FIGURE 8-15.
Aplysia brasiliana burrowing into the sand. [From Aspey and Blankenship, 1976a.]

for sexual activity. In this species burrowing appears to be triggered by a pheromone. Even large (>250 g), presumably older, swimming animals that normally burrow only infrequently can be induced to burrow when placed in an aquarium containing an animal that has burrowed. The burrowing pheromone may act to regulate the size of the population by inducing burrowing in conspecifics (Aspey and Blankenship, 1976b).

SWIMMING IN APLYSIA

In several species of *Aplysia* (*A. brasiliana, A. fasciata,* and *A. juliana*) the parapodia are well developed and used for forward swimming. Thus, when an animal like *A. fasciata* is on a firm substratum it clings to the substratum and crawls, but when the substratum is removed the animal moves its parapodia and swims. The parapodial movements appear to be controlled by the pedal ganglia and may involve a central program (Chapter 4). The outstretched parapodia are large. In one animal 220 mm long the parapodia measured 190 mm across from edge to edge. Because living animals are only slightly more dense than seawater and the parapodia have a large surface area, only a minimal pressure on the water is necessary to stay afloat (Neu, 1932). The details of swimming in *A. fasciata* have been studied by Neu (1932)[4] and in *A. brasiliana* by Jahan-Parwar and Fredman (unpubl.) and by Hamilton and Ambrose (1975).

During swimming in *A. brasiliana* there are no pedal waves; rather, the edges of the foot are rolled into a close approximation forming a narrow groove along the length of the foot (Jahan-Parwar and Fredman, unpubl.). The parapodia are initially turned inward with the left parapodium overlapping the right, giving the animal the appearance of being enveloped by its parapodia. With the caudal part of the parapodia still overlapped, the rostral part starts to unroll, forming a funnel whose mouth faces rostrally and takes up water. At this point the caudal part of the parapodia begin to unroll. Just before this wave spreads to the most caudal segments, the rostral edges of the parapodia start to roll inward, forming a funnel with a mouth that now faces caudally and moves slowly backward. By these movements the water mass taken into the parapodial funnel, when it faces rostrally, is slowly moved backward and expelled caudally at the posterior edge of the parapodia. The entire sequence lasts about one second (Figure 8-16).

---

[4]Neu studied the details of swimming on film. He stated that he studied *A. depilans,* but Weevers (personal communication) states that *A. depilans* does not swim and that Neu must have been studying *A. fasciata.*

FIGURE 8-16.
Dorsal view of jet-propulsion swimming in *A. fasciata* (drawings based on movie frames). Jet-propulsive movement is achieved by taking water into the rostral part of a funnel created by the parapodia, moving the water backward, and expelling it posteriorly. The sequence takes 1.26 sec. [After Neu, 1932.]

Thus *Aplysia* does not swim simply by beating its parapodia up and down and moving two masses of water backward. Rather, it also uses a type of jet propulsion by moving one mass of water backward over the median line of its body. Only a small component of the extended parapodia—the swimming surface that keeps the *Aplysia* afloat—presses on the water surface. Most of the forward motion comes from the drawing in and pushing back of a seawater mass.

The transition from crawling to swimming has been studied by Weevers (1971) in a suspended preparation of *A. fasciata* in which neuronal responses can be recorded intracellularly while the animal makes swimming motions. As we have seen, in addition to swimming, parapodial movements facilitate respiration and the excretion of fecal material from the mantle cavity (Guiart, 1901; Eales, 1921). This may explain why brisk powerful closing movements of the parapodial lobes and spontaneous contraction of the gill and siphon also occur in *A. californica* and other species that do not swim (see Figure 8-4). Parapodial movements may also be used in the righting reflexes of both swimming and nonswimming species (Hamilton and Ambrose, 1975).

SWIMMING IN OTHER OPISTHOBRANCHS

Opisthobranchs have evolved a remarkable variety of swimming patterns (Farmer, 1970; Thompson, 1976). The simplest is *dorsal-ventral undulation,* a series of ventral and dorsal flexions involving the whole body-wall musculature. This type of undulation is used by the notaspid *Pleurobranchaea californica* and the nudibranchs *Tritonia* and *Hexabranchus* (Figure 8-17), all of whom have a dorsoventral flattening of the tail or metapodium. The cellular mechanisms of undulation have been studied by Willows (1971) and Getting (1977) and, as we shall see below, a partial neural circuit of this behavior has been delineated. Dorsal-ventral undulation typically takes the form of thrashing movements and is not an efficient means of locomotion. It is useful primarily for escape from predators. A modified form of swimming by undulation, called lateral bending, involves side-to-side movements that bend the head in relation to the body. Lateral bending is used by the nudibranchs *Melibe leonina* (Figure 8-17; for cellular studies see Thompson, 1974) and *Dendronotus iris.* The enlarged rhinophore sheaths are thought to be important in translating the lateral undulations of the body into forward movement.

Whereas undulation and lateral bending involve the whole body and are primarily designed for escape, other opisthobranchs use more subtle swimming movements which are effective for travel over considerable distances. For example, the aeolids *Glaucus atlanticus* and *G. marginata* float along the sur-

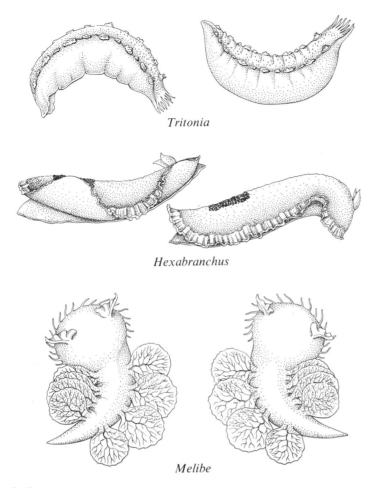

Tritonia

Hexabranchus

Melibe

FIGURE 8-17.
Swimming by undulation and lateral bending. *Tritonia* and *Hexabranchus* swim by undulation; *Melibe* swims by a modified form of undulation called lateral bending. In *Tritonia* and *Melibe* swimming is normally elicited in response to skin contact with certain starfish, but *Melibe* also swims spontaneously. In each case propulsion is achieved by a short sequence (three to ten cycles) of dorsal and ventral flexions. The lateral margins of the body wall are flapped up and down in a thrashing motion with no apparent control over the direction of swimming. [After Willows, 1973.]

face of the water by swallowing air and holding the air bubble in their stomach (Thompson, 1976). The more powerful swimmers among the opisthobranchs use the parapodia to propel their movements. Thus the notaspid *Pleurobranchus membranaceus* swims by flapping its parapodia asynchronously (Thompson and Slinn, 1959). As one parapodium starts to beat, the other goes through its recovery stroke. The mantle lies underneath the body and functions like the keel of a ship (Figure 8-18A).

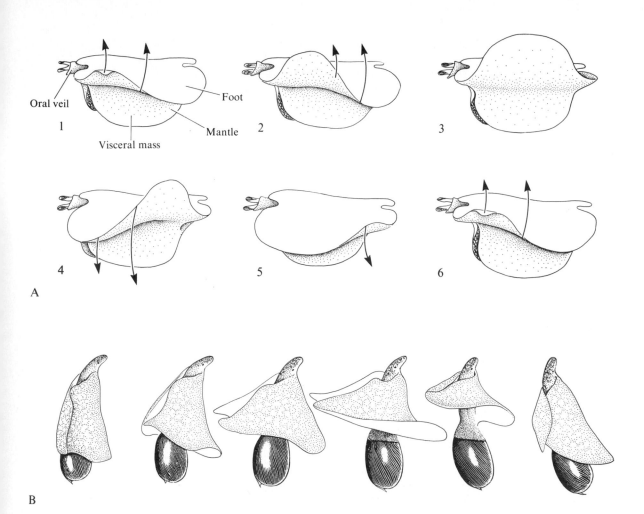

A

B

FIGURE 8-18.
Swimming by parapodial flapping.

A. The swimming pattern of *Pleurobranchus* as viewed from the right side. As one parapodium goes through its upstroke (recovery stroke) the other begins its upstroke; the animal thus rolls 45° to each side with each parapodial upstroke, causing considerable instability in forward motion. The recovery stroke starts anteriorly **(1)** and passes caudally until the whole parapodium is poised above the animal **(3).** The downstroke, which actually propels the animal, also starts anteriorly **(4)** and passes caudally **(5, 6)** with a powerful beat. In the illustration the left parapodium is kept stationary so as to better illustrate the movement of the right parapodium. [After Thompson and Slinn, 1959.]

B. The swimming pattern of *Akera bullata*. The animal is illustrated swimming from left to right. [After Morton and Yonge, 1964.]

Jet propulsion, the pattern used by the swimming species of *Aplysia,* is also used by other anaspid species, e.g., *Dolabella* and *Notarchus punctatus.* Among the anaspids, a particularly graceful swimmer is *Akera,* whose large parapodia form a bell-like cloak. The parapodia open and close in a series of delicate movements that carry the animal along in short spurts (Figure 8-18B). The most effective opisthobranch swimmers are found among the gymnosomes and thecosomes. These animals spend much of their lives swimming near the surface of the ocean and use their delicate parapodia for sculling and rowing movements (Morton, 1958a, 1964).

Escape and Predator-Prey Recognition

A fascinating variant of locomotion is escape: a stereotypic fixed-action pattern in response to a specific class of noxious stimuli. An animal's response to any stimulus implies recognition of the eliciting stimulus. The degree of recognition, i.e., the specificity of the stimulus required to elicit a response, varies considerably, however. The range of stimuli for an elementary defensive withdrawal is typically quite large. Thus, the siphon of *Aplysia* withdraws to a variety of mechanical stimuli (touch, pressure, vibration, noxious stimuli) as well as to light. More complex response patterns require more specific stimuli with more complex features. These specific stimuli are called *sign stimuli* by ethologists (Hinde, 1970; Marler and Hamilton, 1966). An interesting question in the comparative study of behavior is how the sign stimuli that elicit homologous behavioral responses vary in closely related species.

Various starfish predators are good examples of a class of sign stimuli effective in a variety of prosobranchs and opisthobranchs, eliciting homologous escape behavior (for review see Feder and Christensen, 1966). This behavior is usually triggered by chemical stimulation that occurs upon contact but the chemical can sometimes also be sensed at a distance (Phillips, 1975; Kohn, 1961; Feder and Christensen, 1966). For example, on contact with a certain species of starfish, the mud snail *Nassarius,* the whelk *Buccinum,* and the archeogastropods *Haliotis* and *Acmaea* make violent escape movements. They raise their shell above the substratum while rocking and turning from side to side, and rapidly glide or gallop away from the starfish. If the starfish's feet grip the animal, so that a straight pull is not sufficient for escape, the upward raising of the shell is followed by violent twisting movements of the shell around the dorsoventral axis, first turning 90° in one direction then 90° in the opposite direction. The shell-twisting is sometimes accompanied by rapid pedal waves similar to those of a galloping gait. These movements are usually effective in breaking the starfish's hold.

The escape response is elicited only by and matched to particular predator species, and contact with other starfish species (or other stimuli) tends to cause these animals to clamp their shell down rather than raise it up (Bullock, 1953). The starfish species that elicits escape is intertidal and the response is only produced in intertidal animals; sublittoral species do not produce it.

Escape responses specific to predator starfish are also shown by the opisthobranch *Tritonia diomedia* (Figure 8-17). As we have seen the swimming movements of the nudibranch *Tritonia diomedia* consist of violent ventral and dorsal flexions. These maneuvers are accompanied by extension and flattening of the lateral edges of the mantle, the oral veil, and the foot so as to increase the surface area during powerful ventral flexions (Willows, 1971). This response is apparently a reaction to saponins (noxious chemicals) that are released by the tube feet of the starfish (Willows, 1973).

Aspects of the neural circuit of this escape-swimming in *Tritonia* have been analyzed (Figure 8-19). Escape-swimming consists of two components: (1) reflex withdrawal and (2) swimming, which consists of alternating ventral and dorsal flexions of the body, each produced by separate populations of motor neurons, the ventral and dorsal flexor neurons. The initial reflex withdrawal is produced by concomitant activity in both populations of motor neurons. This is followed by swimming movements produced by alternating bursts of spikes in the two populations. The program for swimming is centrally located (Dorsett, 1968, 1974; Willows, 1971).

Getting (1976, 1977) has found that the full escape response is triggered by a cluster of 60–80 neurons in the pleural ganglia sensitive to chemical and mechanical stimuli. These cells activate motor neurons in the pedal ganglia by means of three classes of interneurons located in the cerebral ganglia: reflex interneurons, a command interneuron for swimming, and swim interneurons. These interneurons switch the animal from reflex withdrawal to centrally programmed swimming (Figure 8-19).

Aplysia californica also shows an escape response upon contact with the starfish *Astrometis sertulifera* (Dieringer and Koester, unpubl. observ.). However, the stimulus appears to be mechanical rather than chemical, resulting from the pinching action of the grasping pedicellariae of the starfish. The response, a quick gallop, consists of rapid pedal waves accompanied by weak parapodial flapping (Figure 8-20). The sequence for this gallop appears similar to that of routine crawling, only it is faster (Hening *et al.*, 1976). As first shown for *Tritonia diomedia* by Willows (1971), the responses in *Aplysia* can also be elicited by salt (Wachtel and Impelman, 1973; Hening *et al.*, 1976).

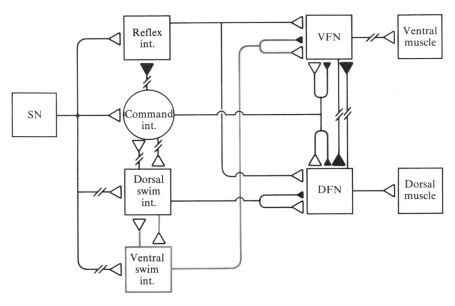

FIGURE 8-19.
Postulated neural circuit in *Tritonia diomedea* controlling escape-swimming. [After Willows, 1971; Getting, 1975, 1976, 1977.]

Sensory neurons activate the motor neurons by means of three classes of interneurons. The *reflex withdrawal interneurons* receive direct excitatory synaptic connections from the sensory neurons (SN) and make direct excitatory connections with both the dorsal- and ventral-flexion motor cells (DFN, VFN) These interneurons mediate the withdrawal phase. The second class of interneurons, the *command interneurons* (the symmetrical identified cells C2), are thought to provide the central command for swimming. They are also excited by the sensory neurons. They switch the behavior from withdrawal to swimming by inhibiting the reflex-withdrawal interneurons and concomitantly triggering the swimming pattern-generator. The C2 interneurons are dual-action cells that make many inhibitory and some excitatory connections with the contralateral dorsal- and ventral-flexion motor neurons, but they fire in phase only with the dorsal-flexion motor cells. The third group of interneurons, the *swim interneurons,* has two subgroups: the dorsal swim interneurons, which have been identified, and the ventral ones, which are inferred. They are thought to be excited by the mechanoreceptors and make indirect reciprocal excitatory connections with the C2 interneurons. These dorsal cells make excitatory and ferent dorsal-flexor neurons. The ventral cells are presumed to make similar connections to the ventral-flexor neurons. The dorsal swim interneurons burst in phase with the dorsal-flexor neurons and with C2. The positive feedback between C2 and the swim interneurons is thought to be important for bringing both groups of cells up to their maximum during each swim cycle. Gray lines indicate postulated cells and connections; broken lines indicate indirect connections. Squares indicate identified cell populations, circles identified single cells.

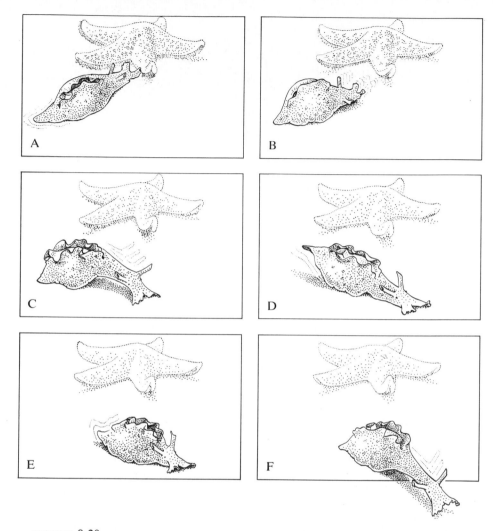

FIGURE 8-20.
Escape response of *Aplysia* to a starfish predator, *Astrometis sertulifera*. On contact with the starfish (**A**), the animal withdraws (**B**), turns away from the starfish (**C**), and escapes with rapid pedal waves (**D–F**). [From Kandel, 1976.]

Feeding

Feeding is one of the most complex and fascinating behavioral sequences in opisthobranchs and involves integrated reflex patterns as well as fixed-action patterns. The animal orients to the food by means of reflex patterns and consumes the food by means of fixed-action patterns.

Feeding has been studied in several species of *Aplysia* as well as in *Navanax, Pleurobranchaea, Helisoma,* and *Limax.* As a result we have a better understanding of the comparative aspects of feeding among gastropod molluscs than of any other class of behavior.

FEEDING BEHAVIOR IN APLYSIA

Aplysia is an herbivore and eats great quantities of seaweed (Eales, 1921; Carefoot, 1967a, c; 1970; Kupfermann and Carew, 1974). At one time it was thought that *Aplysia* changed its seaweed diet at various stages of development (Garstang, 1890; Eales, 1921). Thus, Eales thought that immature *A. punctata* ate the red alga *Delesseria* and mature animals ate green or brown alga, *Ulva, Laminaria,* or *Fucus.* This earlier idea was based on the assumption that sublittoral and littoral life represented different stages in the migratory life cycle of an individual and that the different seaweeds grew in different environments. But the findings that at least some animals do not migrate (Chapter 2) and that postmetamorphic animals can be reared in the laboratory on one species of seaweed (Chapter 7) indicate that *Aplysia* need not change its diet but can feed upon the same seaweed species throughout its life.

Littoral and sublittoral specimens of *A. punctata* eat different species of seaweed. Sublittoral populations feed predominantly on the red algae *Plocamium coccineum* and *Heterosiphonia plumosa* (Carefoot, 1967a). Littoral populations feed on the green alga *Enteromorpha.* However, both littoral and sublittoral *A. californica* tend to eat red seaweed, particularly *Laurencia* and *Plocamium* (Figure 8-21 and Table 8-1).

TABLE 8-1.
Seaweeds eaten at one location by Aplysia californica. *(From Kupfermann and Carew, 1974, based on studies at Scripps Oceanographic Institute, La Jolla, California.)*

	Genus	Sublittoral	Intertidal
Red seaweeds	*Laurencia**	Very frequent	Very frequent
	Gigartina	Frequent	—
	*Plocamium**	Infrequent	—
Brown seaweeds	*Macrocystis*	Infrequent	—
	Laminaria	Infrequent	—
	Eisenia	Infrequent	—
	Colpomenia	Infrequent	—
Green seaweeds	*Ulva*	Infrequent	—
	Codium	Infrequent	—

*At Santa Catalina Island, *Laurencia* is rare and *Plocamium* is consumed frequently (D. Willows and T. Audesirk, personal communication).

Red Algae

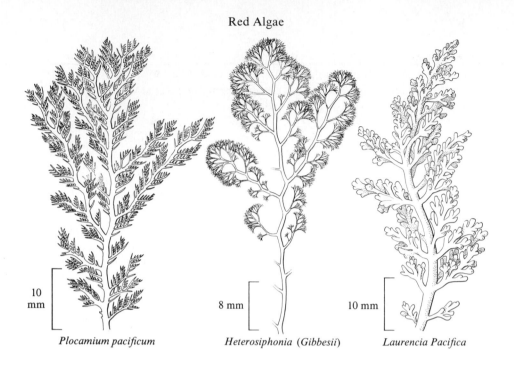

10
mm

8 mm

10 mm

Plocamium pacificum   *Heterosiphonia* (*Gibbesii*)   *Laurencia Pacifica*

Sea (eel) Grass   Green Algae

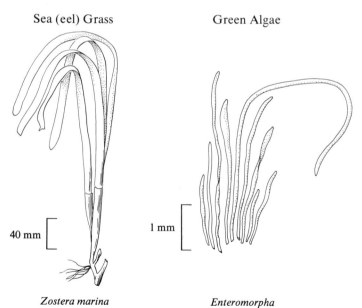

40 mm

1 mm

Zostera marina   *Enteromorpha*

FIGURE 8-21.
Some of the major classes of seaweed and sea grass on which *Aplysia* graze.
[After Dawson, 1956.]

There is general agreement with Eales's suggestion (1921) that color variations within a species are mainly determined by food pigments. Although *Aplysia* has few clearly identified predators, it would seem that one of its best defenses is adaptive coloration due to the color of the seaweed it grazes upon. Eales claims that the resemblance is sometimes so great that the animal cannot be distinguished from the seaweed unless the animal is moving.

Aplysia species have clear food preferences (Carefoot, 1967a). Animals will not eat some species of seaweed and may even prefer a seaweed to which they are normally not exposed over one upon which they normally feed. Thus, sublittoral *A. punctata* prefers *Enteromorpha,* which it rarely encounters, over *Plocamium,* its regular food. It grows almost as well on the rare seaweed as on its usual diet (Figure 8-22A).

The food preference of young animals can be behaviorally manipulated (Carefoot, 1967c). Sublittoral *A. punctata* normally prefer *Plocamium* to *Ulva,* but after 60 to 80 days of eating only *Ulva* in experimental conditions, they selected *Ulva* over *Plocamium* for several days. In addition, animals seem to become less selective in their diet with age (Kupfermann and Carew, 1974). These findings are based on a small number of animals and on experiments that did not use controlled behavioral techniques. It will be important to examine food preference in a more controlled way.

NUTRITION IN APLYSIA

Carefoot (1967a) studied the nutrition and absorption of food in *A. punctata* and found that growth occurred at a fairly high and uniform rate. For example, in 80 days a 2-g *A. punctata* increases its dry weight twentyfold, from 0.1 g to 2.0 g. The growth efficiency decreases as the animal gets larger; a 20-g damp dry *Aplysia* is only 30 percent as efficient as a 3-g animal in converting the absorbed digestion products of *Plocamium* into body weight (Figure 8-22B).

Carefoot (1970) has also studied two tropical species, *A. juliana,* an eater of the green algae *Ulva* and *Enteromorpha,* and *A. dactylomela,* an eater of the green alga *Cladophora* and the red alga *Centoroceras.* In the laboratory *A. juliana* grew most rapidly on *Ulva,* which is abundant and preferred in its own habitat. This alga also produced the highest growth efficiency. By contrast, *A. dactylomela* chose *Enteromorpha,* which is found only rarely in its own habitat. This confirmed Carefoot's earlier finding on *A. punctata* that regardless of their previous eating habits in nature, a number of *Aplysia* species prefer green algae to red in the laboratory, particularly the algae *Enteromorpha* and *Ulva* (Carefoot, 1967c). Even though *A. dactylomela* selected *Enteromorpha,*

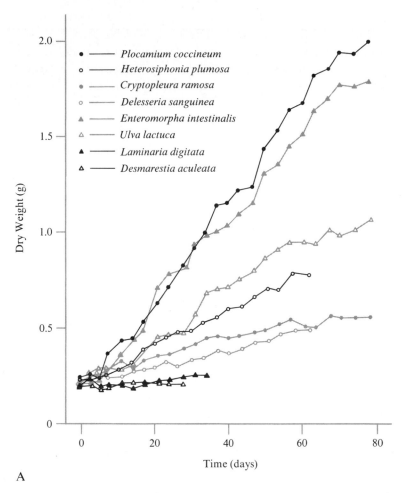

A

FIGURE 8-22.
Food preferences of *Aplysia*.

A. The growth of *Aplysia punctata* feeding on eight species of algae at 15°C. Each trace represents the mean of six individuals. The mean starting weight of each is 2 g (damp dry) or 0.244 g (equivalent dry weight at 110°C). [From Carefoot, 1967a.]

the highest growth efficiency was yielded by *Cladophora,* its natural food. Thus, independent of food preference in the laboratory, both *A. juliana* and *A. dacty-lomela* grow most efficiently on seaweeds they eat in nature. This suggested to Carefoot (1970) that in nature *Aplysia* eat the most nutritious food available to them.

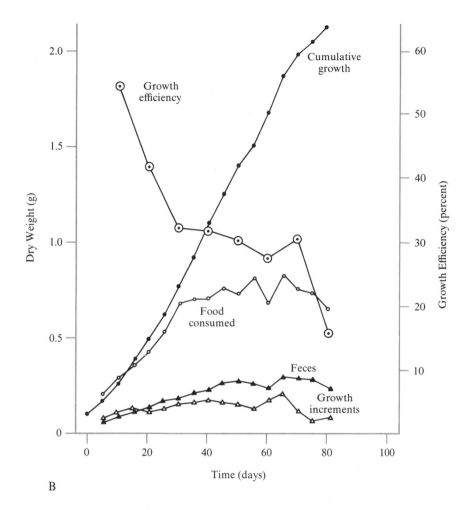

B. The growth, food consumption, fecal production, and growth efficiency of *Aplysia punctata* feeding on the red alga *Plocamium coccineum* at 15°C. Each trace represents the mean of nine animals. Growth efficiency is expressed as the percentage of dry-weight food absorbed that is converted into dry-weight body tissue, including spawn. [From Carefoot, 1967c.]

Carefoot's studies are a valuable bridge between the naturalistic and experimental studies of feeding in *Aplysia*. In addition, these studies are among the few comparative behavioral studies of *Aplysia* species. For example, he examined to what degree morphological adaptation may relate to food preference. He found that *A. juliana*, which eats *Ulva* naturally, will eat *Enteromorpha* in the laboratory, although it does not usually eat it in nature where it is also available. *A. juliana* belongs to the subgenus *Aplysia* and has a sucker on the posterior part of its foot with which it attaches to the substrate while eating. By means of the sucker *A. juliana* can attach itself to the broad thalli of *Ulva* and to the rocks on which this alga grows. But it cannot readily use its sucker to attach itself to the sand-covered limestone rock where *Enteromorpha* grows. In the laboratory, however, *A. juliana* can attach itself to the side of the tank, and readily feeds on *Enteromorpha* (Carefoot, 1970).

These studies illustrate why meaningful interpretations of behavioral findings require knowledge of the comparative structures of species of *Aplysia* as well as their natural histories and life cycles. In turn, comparative studies of behavior can help explain differences in the structure of *Aplysia* species. All species of *Aplysia* adhere to their substrates with the posterior part of the foot, but only in some species of the subgenus *Aplysia* does this lead to the development of a posterior sucker that permits animals to attach effectively with less exertion to relatively firm substances. Carefoot's studies point out the selective advantages of this evolutionary specialization. This characteristic allows *A. juliana* to graze more effectively on some seaweeds than on others.

Correlation between morphological adaptation and food preference has also been described among certain other opisthobranchs and particularly among nudibranchs (Thompson, 1964, 1976). For example, some nudibranchs that feed on a flattish vegetation tend to have a broad flattened oval foot, broad radula, and weak mandibles. Other nudibranchs that feed on erect plants are smaller and elongated; their foot is narrow and better adapted to this food. Voracious, cannibalistic nudibranchs are frequently brightly colored and do not seek concealment.

ORIENTING, FOOD-SEEKING, AND INGESTION IN APLYSIA

Feeding in *Aplysia* consists of a relatively variable appetitive food-seeking component—the orienting response—and a more stereotypic consummatory component—the biting response (Figure 8-23; for discussion see Kupfermann, 1974a).

Appetitive response.   Preston and Lee (1973) have examined food-seeking in *A. californica* and provide evidence for chemical detection at a distance and for chemotaxis (see also Jordan, 1917, and Audesirk, 1975). While searching for food the animal waves its head and anterior tentacles (oral veil), a behavior that facilitates sampling the composition of the chemical food gradient on each side of its head. When first placed in a small chamber in the absence of food,

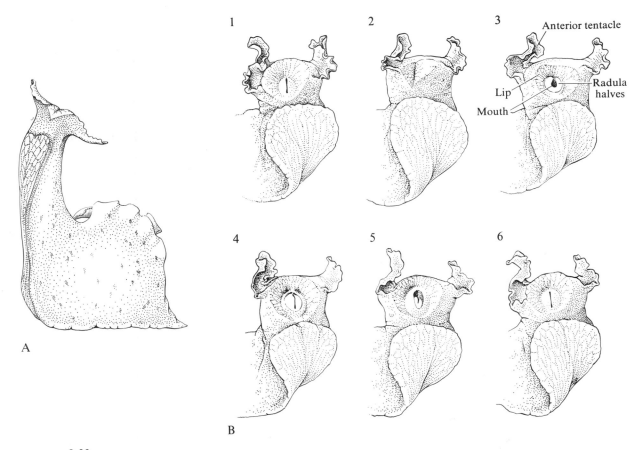

FIGURE 8-23.
Orienting and feeding responses of *Aplysia*. **A.** The orienting and searching (appetitive) component of feeding. **B.** The ingestive (consummatory) component. [From Kandel, 1976; based on Kupfermann, 1974a.]

an animal typically either crawls along the vertical walls, circling the perimeter of the tank, or rests. When food is floated on the surface of the water in the center of the chamber, the animal begins to wave its head and leaves the side of the tank and approaches the food by alternating forward locomotion and head waving. Having reached a position below the food, the animal waves its head for a variable length of time before contacting the food and eating it. Even with the food only one or two body lengths away, an animal typically does not go directly to the seaweed but approaches it slowly, with much trial and error. The time between food detection and initial contact is usually 30 minutes; about 40 percent of this time is spent head-waving, involving a mean of 3.5 separate stationary positions (head-waving loci) per hour (Figure 8-24).

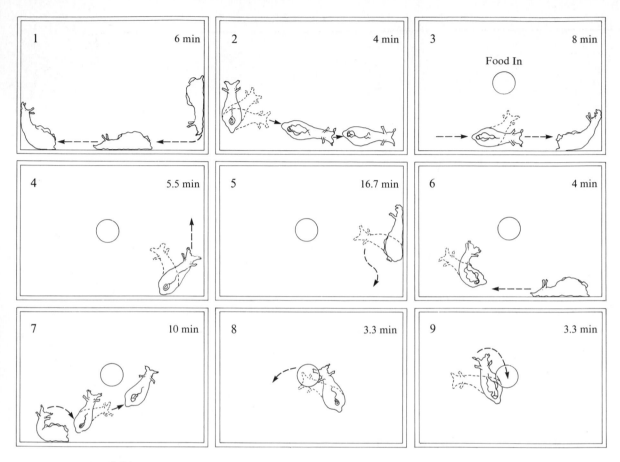

FIGURE 8-24.
Food detection by *Aplysia* at a distance. Response of an individual animal to seaweed placed on the surface of the seawater tank (view is from top of the tank). Prior to food presentation **(1, 2)** the animal locomotes around the periphery of the tank. When food is placed in the center of the tank on the water surface **(3)** the animal turns toward it and by means of head-waving and locomoting seeks out and finds the food **(4–9)**. The times given in the upper right corner of each frame indicate the duration of behavior in that frame. Dotted heads indicate sites of head-waving. The tank used for this timing was about five feet long. [After Preston and Lee, 1973.]

Further support for the hypothesis that *Aplysia* finds food at a distance using chemical gradient clues is the finding that an animal can select the correct arm of a T- or Y-maze that contains food (Figure 8-25; Preston and Lee, 1973; see also Jahan-Parwar, 1972a; Audesirk, 1975). Audesirk used this technique to examine the role of the anterior tentacles and the rhinophores in distance reception. She removed the distal portion of each organ surgically and ex-

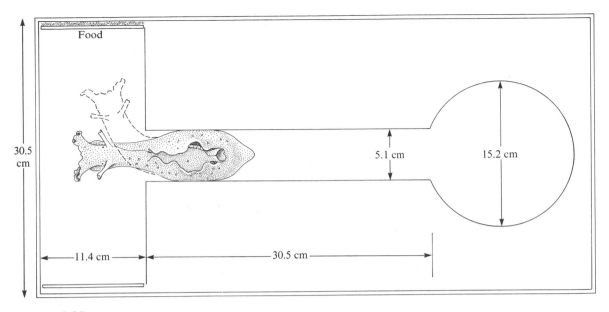

FIGURE 8-25.
T-maze used to determine chemoreception of food substances at a distance by *Aplysia*. The circular area at right is the starting area. The animal proceeds left to the choice point (just before the branching of the maze). The food may be in the right arm, as shown, or the left. [After Preston and Lee, 1973.]

amined the animals' ability to find food in a Y-maze. Removal of the rhinophores alone did not reduce performance, whereas removal of anterior tentacles alone led to an increase in the number of errors. By contrast, concomitant removal of rhinophores and tentacles led to a large decline in performance. Audesirk suggests that the rhinophores, which are carried high above the substratum, detect odors carried by currents and help recognize the presence of food upstream. But because they are close together they are perhaps not very accurate for localizing the precise direction of food. The more widely separated tentacles are also closer to the substratum, and therefore can locate food more easily. The tips of both the anterior tentacles and the rhinophores contain a darkly pigmented groove packed with ciliated sensory (presumably chemosensory) cells that send processes to the cluster of neurons lying underneath the epithelium (Audesirk, 1975; Emery, 1976).

The swimming species *A. brasiliana* has an additional orienting response: it bobs its head in search of food while swimming. As the neck bends, stretches, and lifts, the oral veil emerges temporarily at the water-air interface. This maneuver seems to be designed to increase the likelihood of encountering food at the surface, much as is head-waving of nonswimming animals (Aspey, Cobbs, and Blankenship, 1977).

Consummatory response.   The head movements of the orienting response are followed on contact with food by the biting response: a sequential protraction and retraction of the radula and buccal mass (Figure 8-23). The animal closes its lips to hold the food in place, opens its mouth, and protrudes its odontophore and the two radular halves. The odontophore and radular halves first diverge and then converge to grasp food. The odontophore then retracts and rotates backward to bring the food into the buccal cavity toward the anterior portion of the esophagus (Howells, 1942; Winkler, 1957; Frings and Frings, 1965; Lickey and Berry, 1966; Kupfermann and Pinsker, 1968; Lickey, 1968; Kupfermann, 1974a).

Aplysia punctata combines the gripping action of the radula halves with the biting action of its lips and mouth and eats a large string of seaweed, just as children eat spaghetti. With each backward movement of the radula, a 1–3 mm length of the seaweed may be drawn in. When approximately 1–2 cm of seaweed has been drawn into the buccal cavity, the lips tighten so that on the next backward movement of the radula the weed is torn. In this way relatively large strings of seaweed can be taken into the crop (Howells, 1942). Some *Aplysia* species may also grasp seaweed with the anterior end of the foot and manipulate the free end of the seaweed into the mouth.

The swallowing movements are also due to small odontophore protraction and retraction movements and do not depend upon tactile or chemical stimulation of the lips. Swallowing appears to be triggered by food applied to the inner surface of the mouth or buccal cavity. In addition to odontophore protraction (biting) and swallowing, the buccal muscles also produce a rejection response—reverse peristalsis—to inappropriate, inedible, or distasteful objects taken into the buccal cavity (Kupfermann, 1974a).

The head movements of the orienting response are mediated by the pedal ganglia; these movements are eliminated by cutting the cerebropedal connectives. By contrast, cutting the cerebrobuccal connectives largely eliminates the consummatory protraction and retraction of the odontophore. These movements are therefore primarily under the control of the cerebral and buccal ganglia (Kupfermann, 1974b).

Although the orienting response has not yet been analyzed, Kupfermann and his colleagues have analyzed aspects of the mechanisms of the consummatory response. The antagonist protraction–opening and retraction–closing movements of the radula are produced by cholinergic motor cells in the buccal ganglion that connect directly to the buccal muscle (Figure 8-26; Kupfermann

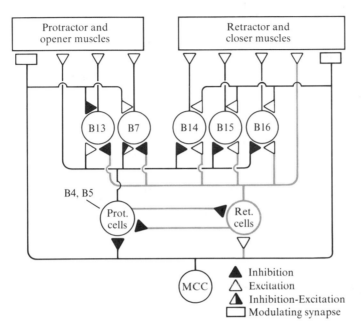

Inhibition
Excitation
Inhibition-Excitation
Modulating synapse

FIGURE 8-26.

Postulated neural circuit controlling feeding in *Aplysia*. The muscles that protract the radula and open its two halves and those that retract it and close the two halves are innervated by motor cells that are not interconnected. The motor cells are innervated by command cells. The command cells for protraction are two identified multiaction cells, B4 and B5. The retractor command cells are not yet identified, but are postulated to also be dual-action cells. The protractor and retractor command cells make mutually inhibitory connections. The dual-action protractor and retractor command cells are innervated by higher-order dual-action metacerebral cells. These excite the retractor command cells and motor cells and inhibit the protractor command cells and some of the protractor motor cells. The metacerebral cells also innervate the buccal musculature, causing a facilitation of motor neuron action by means of direct enhancement of contraction as well as by facilitating the synaptic actions of the motor neurons. The firing of the metacerebral cells is only weakly effective in initiating the feeding cycle; but the activity of these cells greatly enhances the contraction produced by the motor neurons. This enhancement is in part due to their central action. But it is also due, in larger part, to the action of these cells on the contractile process of the muscles. Unidentified cells and their connections are drawn in light lines. [Based on data in Kupfermann and Cohen, 1971; Weiss, Cohen, and Kupfermann, 1975, 1978.]

and Cohen, 1971; Cohen, Weiss, and Kupfermann, 1974; Cohen *et al.*, 1978).[5] The motor cells are driven by a set of protractor and retractor command elements. The program for the repetitive protraction–retraction sequence is thought to be centrally determined (Rose, 1972), but can be modified by sensory feedback from stretch receptors in the buccal mass (Kupfermann and Cohen, 1971; Laverack, 1970; Jahan-Parwar, 1976). The symmetrical metacerebral cells, a set of dual-action cells, serve as higher-order modulatory command cells that act centrally (on command cells and motor neurons) and peripherally (Figure 8-26). As will be discussed in Chapter 9, the metacerebral cells send axons to both retractor and protractor muscles and may act as general excitors for all phases of the consummatory component of feeding (Weiss and Kupfermann, 1978).

FEEDING BEHAVIOR OF OTHER OPISTHOBRANCHS

An informative contrast with the herbivorous *Aplysia* is the carnivorous cephalaspid *Navanax inermis*. Like *Aplysia californica*, *Navanax* lives off the California coast. A voracious eater, it feeds primarily on other opisthobranchs (often on *Hermissenda, Haminoea, Bulla,* and occasionally on small *Aplysia*) and sometimes fish, swallowing the prey whole (Paine, 1963b, 1965). *Navanax* also shows an appetitive response to chemical contact with food. When *Navanax* encounters the mucus trail of potential prey, it will follow the trail (Paine, 1963b; Murray, 1971). *Navanax* then stalks its prey, using two symmetrical head shield organs to contact the trail (Figure 8-27). If one organ loses contact with the trail the animal corrects its course until it is again on center. If the trail is lost, *Navanax* will orient by swinging its head back and forth, much like *Aplysia* searching for food.

Unlike *Aplysia*, *Navanax* lacks distance chemoreceptors (Paine, 1963a, b; Murray, 1971). It must contact a trail before it can follow it. The discovery of the trail is therefore accidental. During unoriented locomotion, *Navanax* keeps its head shield on the substratum in search of potential prey. Following the

---

[5]Cohen *et al.* (1978) have studied in detail the motor system controlling the accessory radular closer muscle, which brings the two radular halves together and seems to aid in grasping the seaweed. This muscle is innervated by three or four motor neurons, of which two, B15 and B16, are readily identified. Both of these cells are cholinergic and make direct connections to the muscle, but the cells differ in the size of their EPSP and rate of facilitation and posttettanic potential. Above a certain threshold the contraction of the muscle, which does not produce action potentials, is directly dependent upon the degree of depolarization produced by the summated EPSP. As a result, changes in the size of the summated EPSP translate directly into changes in the force of contraction.

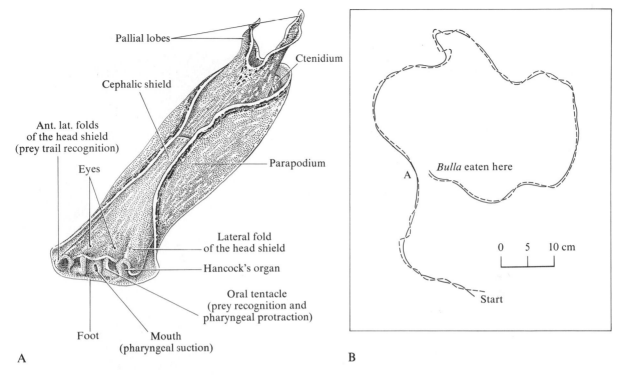

FIGURE 8-27.

Feeding in the carnivorous cephalaspid *Navanax inermis.*

A. Prominent external features of the animal. [From Murray, 1971.]

B. Stalking of prey by *Navanax.* The animal (dashed line) follows the mucous trail of *Bulla* (solid line). At point A, although within 2–3 cm of its intended prey, *Navanax* must continue to follow the cue of the trail. [From Paine, 1963b.]

trail of the prey involves specific chemical recognition; *Navanax* will not follow the trails of molluscs it does not eat, not even gastropods (such as prosobranchs). These trails are consistently ignored (Paine, 1963a, b). *Navanax* will, however, often track and consume members of its own species and may deplete an entire population of opisthobranchs in an area (Paine, 1963a).

Once *Navanax* has contacted its prey it suddenly protracts its pharynx, directs it at the prey, and then rapidly expands the pharynx, thereby sucking the prey into it (Figure 8-28). Because *Navanax* has no teeth, radula, or jaws, the prey is swallowed whole.

Murray (1971) found that *Navanax* rarely protracts its pharynx while following the trail of its prey. In fact, the pharangeal response does not depend on

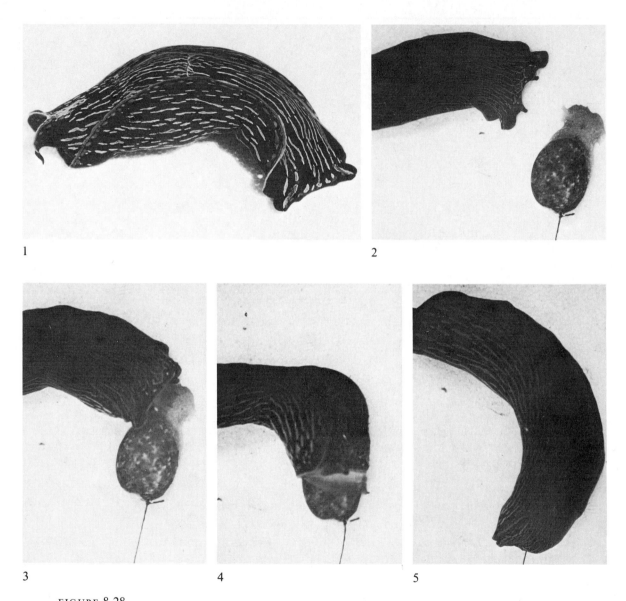

1

2

3

4

5

FIGURE 8-28.
Feeding behavior of *Navanax*. The first photograph shows the animal at rest. The succeeding photos show the animal attacking and swallowing its prey, *Bulla*. [Courtesy of Spira and Bennett.]

tracking; the head area need only contact the prey. Thus, the appetitive (tracking) response and the consummatory (pharyngeal) response are independent. Using tethered, suspended *Navanax,* Murray found that three independent receptor areas mediate the three components of the feeding behavior. Receptors for recognizing and following the trail are restricted to the anterolateral folds of the head shield. Receptors within these folds (the phalloform organs) signal whether or not the mucus of the prey is aligned with respect to the head. Receptors on the lips and on the pharynx recognize the prey and lead to protraction of the pharynx.

Large, presumably old individuals frequently fail to respond to a trail. Paine (1963a) suggests that these mature individuals may devote most of their efforts to reproduction. Following this idea, Murray (1971) found that some *Navanax* may follow the trail of another *Navanax* and copulate with it upon catching up. Thus, the ability to follow mucus secretions serves as a means of tracking prey for young animals, while at other times in the life cycle, or under other circumstances, it serves as a means of tracking a mate. Under these circumstances the mucus secretions serve as a pheromone for sexual activity. Tracking of members of the same species, in addition to propagating the species, may also limit it, by means of cannibalism (for the relationship between cannibalism and sexual activity see below).

The neural control of the consummatory response in *Navanax* has been studied by Bennett and his colleagues. There are three phases to the feeding cycle, each of which appears to be reflexively initiated. The first phase, protraction of the pharynx, occurs in response to chemical stimulation from food. The second phase, rapid expansion, is triggered by sensory feedback from protraction, produced by a group of mechanoreceptors that innervate the lip and the anterior third of the pharyngeal wall (Spray and Bennett, 1975a). It is mediated by the synchronous firing of several pharyngeal motor cells in the buccal ganglion that are electrically coupled to one another (Levitan *et al.,* 1970; Spira and Bennett, 1972). As the ingested prey is pushed into the esophagus, the stimulus of pharyngeal inflation produces in the motor cells inhibitory postsynaptic potentials that functionally uncouple the motor cells, allowing them to fire independently, by increasing their input conductance (Spira and Bennett, 1972). Uncoupling is thought to be produced by the activity of a second group of mechanoreceptor sensory neurons (located in the buccal ganglion) that respond to both tactile stimulation and stretching of the pharynx and make inhibitory connections on the motor neurons (Spray and Bennett, 1975b). Although independent activity in the motor cells following uncoupling has not yet been demonstrated, it could in principle produce the third phase of feeding: the wave of pharyngeal peristalsis that conveys the prey down to the midgut.

The feeding behavior of the notaspid opisthobranch *Pleurobranchaea californica* resembles that of both *Aplysia* and *Navanax*. Like *Aplysia*, *Pleurobranchaea* has distance chemoreceptors that are used for food-seeking (Lee, Robbins, and Palovcik, 1974). But, like *Navanax*, *Pleurobranchaea* is carnivorous and makes rapid ingestive movements. Upon contacting food odors, the animal rears up on its foot and extends its proboscis in searching movements that appear similar in function to head-waving in *Aplysia*. Davis and Mpitsos (1971) and Lee and his colleagues (1974) describe four phases in the feeding behavior of *Pleurobranchaea* (Figure 8-29).

1. Food detection. This behavior is elicited by chemical tactile stimulation with food of the oral veil, rostral foot, and tips of the rhinophores.

2. Proboscis extension. The highly mobile proboscis with a muscular buccal sphincter at its tip is extruded through the mouth for 1 cm or more. This extrusion can be rhythmic (in phase with biting movements), tonic (in orienting), or continuous (extending for many biting cycles).

3. Biting. The radula opens the teeth on the odontophore, grasps the food, and closes again.

4. Proboscis withdrawal.

Despite the basic difference in the diet of the two species, food detection in *Pleurobranchaea californica* is quite similar to that of *A. californica* (Preston and Lee, 1971, 1973; Lee *et al.*, 1974). The main difference is that *Pleurobranchaea* locomotes more continuously in search of food and makes fewer, less ranging head movements than does *Aplysia*. Without even crawling, *Aplysia* can use its flexible neck to sweep large areas while testing for chemical clues, whereas *Pleurobranchaea*, lacking a neck, needs to crawl to perform the same type of testing. In both animals the same parts of the body—the oral veil (anterior tentacles), anterior foot, and tips of the rhinophores—are most sensitive to chemical stimulation (Figure 8-30).

The neural control of feeding in *Pleurobranchaea* has been studied by Davis and his colleagues (Figure 8-31; Davis, Siegler, and Mpitsos, 1973; Gillette and Davis, 1977; Kovac and Davis, 1977). As in *Aplysia*, the alternating movements are due to alternate bursts of activity in antagonist motor neurons. The motor cells generating the motor program in *Pleurobranchaea* are located in several ganglia: the buccal, the cerebropleural, and perhaps others. Coordina-

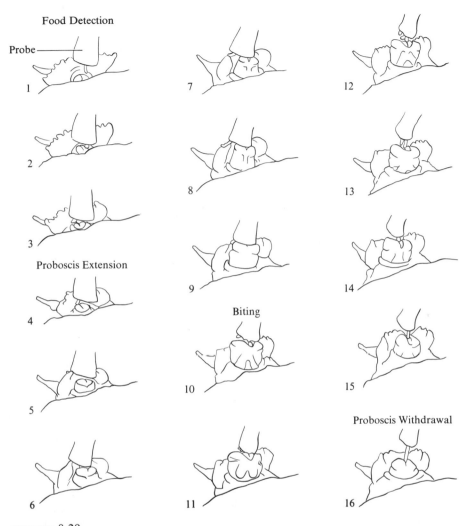

Food Detection

Probe

Proboscis Extension

Biting

Proboscis Withdrawal

FIGURE 8-29.
Feeding response of carnivorous opisthobranch *Pleurobranchaea*. The feeding behavior has four phases, as shown. [After Davis *et al.*, 1971.]

tion between ganglia is thought to be achieved by two types of interneurons (located in the buccal ganglion) that link the activity of cells in the several ganglia by conveying to them both an exact (efference) copy and a rough copy (a corollary discharge) of the efferent pattern of motor activity (Davis *et al.*, 1973; Kovac and Davis, 1977).

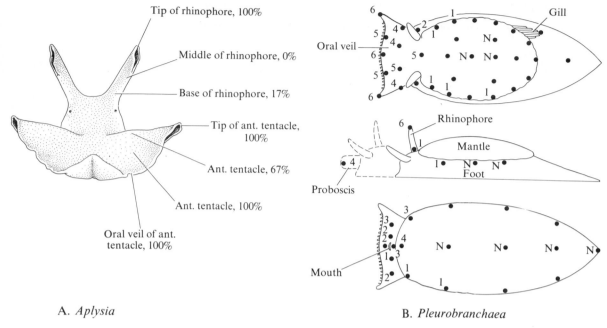

Tip of rhinophore, 100%

Middle of rhinophore, 0%

Base of rhinophore, 17%

Tip of ant. tentacle, 100%

Ant. tentacle, 67%

Ant. tentacle, 100%

Oral veil of ant. tentacle, 100%

Oral veil

Gill

N

Rhinophore

Mantle

Foot

Proboscis

Mouth

A. *Aplysia*

B. *Pleurobranchaea*

FIGURE 8-30.

Comparison of chemosensitive areas of *Aplysia* and *Pleurobranchaea*. **A.** View of *Aplysia* head indicating percentage of responses to food elicited in various regions. **B.** Dorsal, lateral, and ventral views of *Pleurobranchaea*. Dots indicate sites tested for chemosensitivity. Numbers indicate the number of animals showing feeding responses, with degrees of chemosensitivity ranging from six out of six responses (highest) to one in six (lowest); N means no response. [After Lee *et al.*, 1974.]

As with *Aplysia*, the metacerebral cells in *Pleurobranchaea* have a modulatory function in feeding. During feeding there is tonic burst activity in the metacerebral cells of *Pleurobranchaea*. Simulating this tonic firing with intracellular pulses in the metacerebral cells increases the frequency of the biting sequence. Another group of neurons, the paracerebral neurons, are also capable of initiating feeding (Gillette, Kovac, and Davis, 1978). As with the metacerebral cells, the paracerebral neurons are modulated cyclically with the feeding rhythm by repetitive sequences of alternating excitatory and inhibitory synaptic potentials. Some of this synaptic input is from command interneurons and motor cells. Thus the cyclical discharge of the metacerebral and paracerebral cells results from feedback to them from the interneurons mediating corollary discharge and from the motor cells. This mutual synaptic interaction between metacerebral and paracerebral cells, on the one hand,

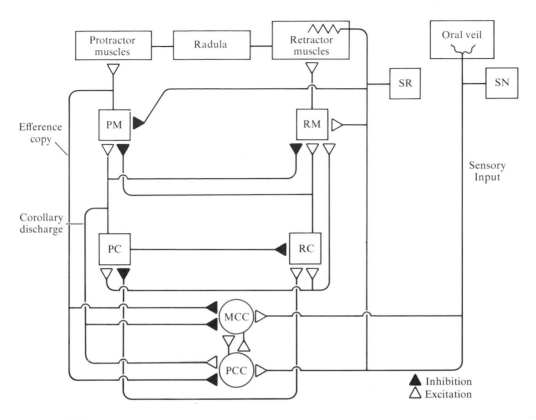

FIGURE 8-31.
Hypothetical circuit controlling feeding in *Pleurobranchaea*. The circuit illustrates the possible role of motor feedback in coordinating the activities of the buccal and cerebral ganglia. Mutually inhibitory dual-action command cells, PC and RC, are postulated to drive protraction and retraction, respectively, by means of their reciprocal innervation of the motor cells. Branches of the protractor motor cells (PM) are thought to produce an exact (efference) copy of the motor output from the buccal ganglia and convey it to the cerebral and pleural ganglia. Branches of the protractor command cells convey a rough copy (corollary discharge) of the motor output of the buccal ganglia and convey it to the meta- and paracerebral cells as well as to other cells of the cerebral and pleural ganglia (not shown). The interneurons conveying the efference copy are thought to fire in exact synchrony with the motor neurons, whereas the interneurons conveying the corollary discharge are thought to be active only during roughly the same phase of the feeding cycle as the motor neurons. Both types of interneurons send axons to buccal and cerebropleural ganglia by means of the cerebrobuccal connectives. These feedback copies of the motor output of the buccal ganglia serve to coordinate the activity of the cerebral and pleural ganglia with that of the buccal ganglia. Cutting the connectives dissociates the rhythms in the buccal and cerebropleural ganglia. The metacerebral cells and the paracerebral cells are higher-order modulatory command cells. Although the metacerebral cells can initiate a single coordinated protraction–retraction sequence in a quiescent preparation, their main function seems to be to increase the frequency of the protraction–retraction sequence by enhancing protraction at the expense of retraction. The metacerebral cells are in turn excited by chemical and mechanical sensory input from food applied to the oral veil and are inhibited by other neurons of the feeding circuit, including the cells that mediate corollary discharge. Boxes indicate populations of postulated cells. [After Davis *et al.,* 1973; Gillette and Davis, 1977.]

and the motor neurons for feeding, on the other, is thought to provide the sustained oscillations that maintain the cyclicity of feeding movements.

The feeding behavior of the closely related *Pleurobranchus membranaceus* has been studied by Thompson and Slinn (1959).[6]

FEEDING BEHAVIOR OF PULMONATES

Feeding has also been studied in the pulmonate *Helisoma trivolvis* by Kater and his colleagues.[7] Contact with food is followed by a fixed-action ingestive response similar to that of *Aplysia* and *Pleurobranchaea*. It consists of back-and-forth oscillation (retraction and protraction) of the buccal mass resulting in rasp-like movements of the odontophore and radula. Contraction of the protractor muscles of the buccal mass propels the radula forward. In *Helisoma* the radula then scrapes across the seaweed and shovels the food into the buccal cavity, or mouth, by contraction of the retractor muscles of the buccal mass (Figure 8-32).

The scraping action of the radula is produced by the protraction and retraction of the buccal mass. These movements result from the action of seven bilateral pairs of protractor muscles and four pairs of retractor muscles as well as by eight pairs of minor muscles. There are five pairs of bilateral and electrically coupled protractor motor neurons. These fire in phase with each other and 180° out of phase in relation to eight bilateral retractor neurons, which are also electrically interconnected and synchronously active. The antiphasing of the protractors and the retractors is timed by a population of electrically coupled dual-action command cells that excite the retractors and inhibit the protractors. Inhibition of the protractors is followed by rebound excitation, which drives the protractor muscles. The phase relationship is characteristic for each protractor motor neuron; some motor cells invariably rebound from inhibition before others, (Figure 8-33).

Command cells produce the central program for feeding (Kater, 1974; Kaneko *et al.*, 1978). Individual cells produce antagonistic out-of-phase movements by exciting the retractor motor cells and inhibiting the protractor motor cells. Bursts within this network of electrically coupled dual-action

---

[6]The pleurobranchs are almost unique in having a large acid-secreting gland attached to the anterior digestive tract to aid in digestion. This group also has the capability of releasing extraordinarily strong (pH 3.3) sulfuric acid from skin glands when distressed (Thompson and Slinn, 1959; Blankenship, personal communication).

[7]Kater and Fraser-Rowell, 1973; Kater, 1974; Kaneko, Kater, and Merickel, 1978; Merickel, Eyman, and Kater, 1977.

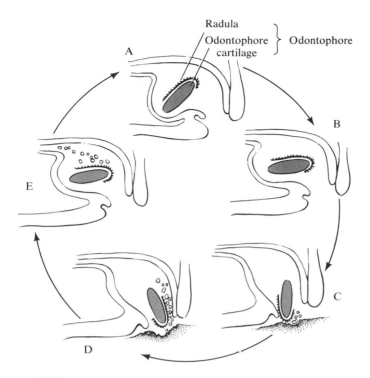

FIGURE 8-32.
The feeding cycle of *Helisoma trivolvis*. **A.** Rest position of the buccal mass, radula, and odontophore cartilage. **B.** Buccal mass in initial protraction. **C.** Buccal mass fully protracted; independent protraction of odontophore and radula. **D.** Slight retraction of the odontophore, as well as independent radular movement (these independent movements of mass and radula are not found in *Aplysia*). **E.** Completion of retraction of the buccal mass, radula, and odontophore. [From Kandel, 1976; based on Kater, 1974.]

command cells underlie the alternating cyclical movements for feeding. Because this population of cells has both command and pattern-generating functions, Kater (1974) has called it the *cyberchron network,* a term derived from the Greek words *Kybern* (to direct) and *chronos* (time). The feeding program generated by the cyberchron network is influenced by two known sources. One, it is modulated by sensory feedback from stretch-receptor neurons in the buccal mass—feedback that regulates the firing frequency and duration of the motor neuron burst (Kater and Fraser-Rowell, 1973). Two, it is also modulated by the metacerebral cells. These can initiate as well as modulate the biting sequence. They are thought to do so by acting on members of

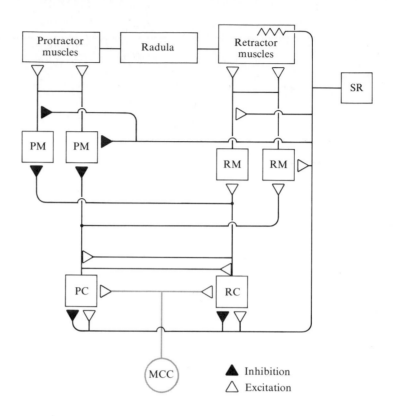

FIGURE 8-33.

Hypothetical circuit controlling feeding in *Helisoma* (PC = protractor command cells; RC = retractor command cells; SR = stretch-receptor cells, unidentified; PM, RM = protractor and retractor motor neurons). Boxes indicate populations of cells. Stretch receptors excite the retractor motor neurons and inhibit the protractors. They also excite and inhibit the command cells. Whether these opposite synaptic actions are mediated directly by individual cells is not clear (Kater, 1974). The stretch receptors appear to be activated by contraction of the retractor muscles, and thus send a positive feedback to the retractor motor neurons, causing them to fire until the buccal mass is fully retracted. The protractor motor neurons are excited by protractor command cells during the second phase of the cycle, and receive from the stretch receptors inhibitory feedback that persists until optimal retraction is assured, thereby relieving the stretch on the stretch receptors. As is the case for feeding in *Aplysia* and *Pleurobranchaea*, the central program does not reside in the intrinsic properties of individual cells. In *Helisoma* the program resides in the electrically coupled command cells (called the cyberchron network). Two command cells in the cyberchron population receive chemical as well as electrical connections from the other members of the network. The increase in conductance produced by chemical input to these two cells is presumed to uncouple (by shunting) the electrical connections between the constituent command cells of the population, thereby terminating the burst (Kaneko *et al.*, 1978). The electrically coupled cyberchron network is in turn thought to be activated by the metacerebral cell on each side that acts as a higher-order command cell capable of both initiating and modulating the feeding rhythm. [Based on data from Kater and Fraser-Rowell, 1973; Kater, 1974; Kaneko *et al.*, 1978); Granzow and Kater, 1977.]

the cyberchron network, but there are as yet no direct data on this point (Granzow and Kater, 1977).

The neural control of feeding behavior of *Helisoma* resembles that of *Aplysia* and of *Pleurobranchaea,* on the one hand, and that of *Navanax,* on the other. In *Helisoma,* as in *Aplysia* and *Pleurobranchaea,* dual-action command cells provide reciprocal motor controls and, as in *Aplysia,* the command cells are electrically coupled. In *Helisoma* this coupling seems to be essential for generating the feeding pattern. These cells thus resemble the ink motor cells and the bag cells in *Aplysia* (which mediate egg-laying, to be discussed below). As is the case in *Navanax,* uncoupling of the synchronous activity of electrically coupled cells by chemical synaptic transmission is thought to be important for terminating a phase of feeding in *Helisoma.*

A beginning has also been made in the analysis of feeding of a closely related pulmonate, the terrestrial slug *Limax maximus,* by Gelperin and his colleagues (Gelperin, 1975a; Prior and Gelperin, 1977; Gelperin *et al.,* 1978). Prior and Gelperin (1977) have found that the motor neurons to the salivary ducts are electrically coupled to the protractor motor neurons and are inhibited by the metacerebral cells, which are themselves inhibited during protraction and excited during retraction. As a result, the salivary gland motor cells are phase-locked to the motor sequence for feeding primarily during the protraction phase of the cycle.

As is evident in the analysis of feeding in the three opisthobranchs and two pulmonates considered above, there is no unique neuronal circuit for feeding among these two classes of gastropods. However, the individual neuronal circuits share a number of common features. Thus a surprisingly small number of organizational principles account for the biting responses of the several animals studied.

HIGHER-ORDER BEHAVIOR

Mating Patterns in *Aplysia*

Aplysia is hermaphroditic. Any individual can act as male or female for others. Animals cannot copulate with or fertilize themselves, although a pair of animals can inseminate each other simultaneously (J. Blankenship and T. Audesirk, personal communication). If more than two animals are available, a coupling chain is formed (Eales, 1921; and see Figure 4-15, p. 90 in Kandel, 1976).

Copulation in general and the coupling chain in particular are the highest forms of social behavior in *Aplysia.* When two individuals couple, one animal acts as a female and attaches itself to the substrate. The second animal acts as a male, crawling over the visceral hump of the first and positioning itself so that its penile aperture is level with the common genital aperture of the first animal. The male then grasps the mantle of the female with the anterior part of its foot while the posterior part embraces the tail of the female. Meanwhile, the female holds the body of the male by opening its parapodia. The male then inserts its penis into the vaginal segment of the common genital aperture. If three or more animals are involved, a third animal attaches itself in the same way above and behind the second. In this manner ten or more animals can copulate, with the last member of the chain serving only as a male. However, in some cases the last member will close the chain and be female to the first so that all animals behave bisexually simultaneously.

Coupling between two individuals and in chains lasts for hours and sometimes days, but the actual passage of sperm may take place within a few minutes (Eales, 1921). It is not unusual for animals that have served as females to extrude their eggs during prolonged (i.e., more than five hours) coupling or soon thereafter. But egg-laying need not be triggered by copulation and occurs spontaneously in individuals kept in isolation for up to 3–4 months. (MacGinitie, 1935). Typically, these eggs are not fertilized.

The neural and endocrine factors involved in the searching, foreplay, and coupling behavior of *Aplysia* have not been well studied. A beginning in this direction has been made by Newby (1972), who observed that the copulatory behavior of *A. dactylomela* has a circadian rhythm. Copulation occurs most frequently in the early morning and only rarely after 12:30 P.M. Under various laboratory conditions, e.g., lower light intensity, copulation was observed after noon but not very late in the day (see also Thompson and Bebbington, 1969, and Kupfermann and Carew, 1974). Newby found that when copulation occurred in pairs, the larger animal assumed the female role. Very small individuals (<40 g) were not observed to copulate, presumably because they were still sexually immature. Beeman's study of oogenesis and spermatogenesis in *Phyllaplysia taylori* indicates only very slight gametogenic activity in small animals (Beeman, 1970; Smith and Carefoot, 1967).

Mating Patterns in Other Opisthobranchs

In *Navanax,* another hermaphroditic opisthobranch, Paine (1965) found that sexual behavior and cannibalism are interrelated. In the mating ritual animals make initial contact by biting one another. In small animals the biting is particularly fierce. Biting evokes a typical response: the bitten partner withdraws its head and extends its parapodia, thereby increasing its diameter and reducing the likelihood of being swallowed by an overly eager mate. Since *Navanax* can only swallow its prey whole, this defensive maneuver is successful only if the bitten partner is heavier than 40 g. If the bitten animal is not swallowed, it is quickly released and assumes its normal shape. Then both animals circle each other, release a great deal of mucus, and begin to mate reciprocally.

The sexual behavior of *Pleurobranchaea californica* and *Pleurobranchus membranaceus* consists of four stereotypic phases: (1) mate detection; (2) approach; (3) body orientation (side by side but facing in opposite directions); and (4) reciprocal copulation (Davis and Mpitsos, 1971). During the last three phases the animal assumes a posture unique to courtship; the right edge of the mantle curls upward, exposing the gill and the sexual organs, which then partially or completely extend. This posture is presumably mediated by chemosensory cues (perhaps mucus) because it can be elicited in an isolated animal placed in seawater that previously contained mating animals.

In the nudibranch *Tritonia* copulation is also reciprocal, two animals meeting head-to-tail with their right flanks apposed (Thompson, 1961); each animal acts as both male and female for the other. Rarely, however, one animal acts only as the male while the other serves as female. Peaceful copulation often continues for hours without either individual appearing to move. *Tritonia* copulates frequently throughout the breeding season with or without egg-laying.

Elaborate courtship patterns have been described in many pulmonate molluscs. Courtship in the slug *Arion ater* begins with one animal pursuing the other, often eating its mucus trail. Once the pursuer has caught up with its partner, the pair circle each other, each producing masses of mucus. Finally, they evert their genital atria and start to copulate. In *Limax maximus* the courtship is truly dazzling. The pair start on a tree branch or on the top of a wall and follow each other in a tight circle for about an hour, caressing each other with their tentacles and secreting large amounts of mucus. The mucus forms a string, which may extend 45 cm, on which the animals lower themselves while they entwine their bodies. The penis sacs become everted, intertwine, and transfer of sperm occurs. In *Limax redii* the everted penile sacs are 70 mm long and courtship takes anywhere from 7 to 24 hours (Gerhardt, 1934; Chace, 1952; for review see Runham and Hunter, 1970).

SPAWN PRODUCTION AND EGG-LAYING IN APLYSIA

Aplysia is one of the most prolific egg producers of benthic (marine bottom) invertebrates. An individual *A. californica* weighing 2,600 g was recorded to have laid about 500 million eggs at 27 separate times in less than five months (MacGinitie, 1934; see Table 7-2, p. 222, for a comparison of the egg-laying capabilities of four *Aplysia* species).

Considering that only one larva from each egg mass needs to reach fertility and spawn to maintain the population in a steady state, there is a staggering loss of eggs and larvae during development. *Aplysia* eggs are small (50–100 μm) and contain a small amount of yolk. The small egg size and long larval period (about 45 days from fertilization) are two factors that are frequently associated with low survival rates among marine invertebrates (Thompson, 1958).

In Treaddur Bay, Anglesey (Wales), *A. punctata* breed from May, when the sea temperature is 9.5°C, to October, when it is 14°C. Smith and Carefoot (1967) found they could speed up gonadal maturation in animals taken from the sea in February by keeping them in 15°C seawater for periods of up to 28 days.

Carefoot (1967c) examined the spawn production of *Aplysia punctata* on different diets and found that in an 80-day period animals eating certain species of seaweed (*Plocamium* and *Enteromorpha*) produced seven times as many eggs as those eating others (*Crytopleura*). Spawn production represented 55 percent of the total growth increase for those animals feeding on *Enteromorpha*. The color of the spawn of *A. punctata,* like that of *A. californica* (Kriegstein, unpubl.), is in part determined by the algal diet, as is the color of the parent animal.

Egg-laying appears to be initiated by contractions of the muscles of the ovotestis that cause it to release mature oocytes (Coggeshall, 1971). This contraction is triggered by a polypeptide hormone with a molecular weight of about 6,000 daltons (Toevs and Brackenbury, 1969; Arch, 1972a, b; Loh, Sarne and Gainer, 1975). The hormone is secreted by two clusters of bag cells, a group of neuroendocrine cells located at the rostral margins of the abdominal ganglion (Frazier *et al.,* 1967; Kupfermann, 1967, 1970, 1972; Strumwasser, Jacklet, and Alvarez, 1969). In the mature animal each cluster contains about 400 cells. Stimulation of the connectives from the head ganglia causes the bag cells to fire in a long decelerating burst (that lasts up to 40 minutes). During the burst all the cells in a cluster fire in tight synchrony. Indirect evidence suggests that this is due to electrical coupling between the cells within each cluster. The number of stimuli necessary to initiate the burst is variable, but once triggered, an all-or-none pattern is released and the time course of the burst is essentially insensitive to further stimulation (Kupfermann and Kandel, 1970; Dudek and Blankenship, 1976). Because of the stereotypic burst pattern and the highly synchronous firing of the cells, it seems reasonable to assume that the bag-cell clusters release a relatively fixed amount of hormone each time they are activated (Kupfermann and Kandel, 1970).

The available evidence suggests that the motor program for the long-lasting repetitive discharge in the neurosecretory cells resides in the cells themselves. For example, the burst capability is unaffected when the bag cells are isolated from the rest of the abdominal ganglion. It is not yet known, however, whether the ability to generate a prolonged burst is due to reverberation of activity between the neuroendocrine cells (by means of their electrotonic connections) or, more likely, to endogenous properties of the cells' membranes that cause them to fire repetitively to a brief stimulus.

Because the natural stimulus that causes release of bag-cell hormone is not known, it has not yet been possible to describe the stimulus–response curve for egg-laying. Casual observations of *Aplysia* indicate that the egg masses laid by a given individual at a given point in its life cycle vary in size over only a rather restricted range. This suggests that egg-laying probably is an all-or-none fixed act that has a rather steep stimulus-response curve.

HORMONAL MODULATION OF BEHAVIOR

Since injections of extracts of the bag cells cause egg-laying within one hour (Kupfermann, 1967, 1970), it is possible to examine other behavioral responses that precede and accompany egg-laying (Strumwasser *et al.*, 1969; Arch and Smock, 1977). Thus, Arch and Smock (1977) found that 10 minutes after injection of the bag cell hormone the animal begins to pucker its mouth; at 15 minutes the genital groove begins to swell. Next, head waving begins. This behavior starts slowly but soon becomes the animal's dominant behavior, and increases even more during egg-laying. Just prior to egg-laying the animal tends to stop crawling. As the egg strand emerges (from the common genital groove near the base of the right tentacle) the animal uses the side-to-side head-waving movements to deposit the eggs on the substratum. Since the hormone has been isolated and partially purified, egg-laying provides an interesting model for a cellular analysis of hormonal regulation of behavior (see Mayeri and Simon, 1975).

Another instance of egg-laying that seems interesting to study at the cellular level is that of *Navanax*. *Navanax* deposits its egg mass in a highly regular geometric pattern, much as the web spun by a spider (Koester, personal communication).

RESPONSE HIERARCHIES AND BEHAVIORAL CHOICE

In nature animals are often exposed to a variety of stimuli, which may tend to elicit contradictory responses. Moreover, an animal may be exposed to the same set of stimuli while in different motivational states (see Chapter 9). Thus, not only do animals rarely emit contradictory patterns of responses, they also often do not emit the same pattern of responses in different internal states. Even when stimuli are not contradictory, the elicitation of some responses may be more essential for survival than others. Therefore there must be mechanisms for selecting among competing stimuli and assuring that only one of a family of possible responses is emitted. As Sherrington (1906) pointed out, hierarchy and choice are intrinsic to the integrative action of the nervous system. He suggested that when two or more reflexes utilize the same motor neurons, the "singleness of action" necessary for effective behavior is achieved because only one of a family of competing responses will prove dominant and be expressed (Tinbergen, 1951; Kovac and Davis, 1977).

The key to the integration of behavior in response to conflicting and varying demands is accomplished by coordinating the different behaviors emitted by an animal. This involves, among other things, reciprocal control over antagonist responses and coordination of priorities in a behavioral hierarchy so that only one of a family of competing responses is expressed at a time.

The effects of competing stimuli have been examined in *Navanax* and *Pleurobranchaea.* In *Navanax* Murray (1971) found that readiness for copulation does not preclude feeding; a copulating individual can have the best of both worlds and can follow and ingest food applied to its anterior lateral folds. But copulating individuals cannot be encouraged to track prey. Thus, some aspects of feeding behavior are not altered by sexual activity while others are inhibited.

As a result of matching a variety of (supramaximal) stimuli and responses in *Pleurobranchaea californica* Davis, Mpitsos, and Pinneo (1974a) found that various behavioral responses in this animal can be ranked in a hierarchy. Unfortunately, this type of study is difficult to carry out because comparisons of choice are predicated on the assumption that alternative stimuli are "equivalent" in some sense. In practice, however, it is hard to compare intensities for stimuli that differ in modality, e.g., mechanical versus chemical stimuli. Nevertheless, Davis and co-workers suggest that individual acts within the behavioral repertory of *P. californica* might be organized into a system of priorities in which some behavioral responses are dominant over others (Figure 8-34). For example, escape swimming is thought to have top priority; swimming suppresses egg-laying, which in turn suppresses feeding. Davis and his colleagues propose that inhibition of feeding by egg-laying may be designed to prevent an animal from eating its own eggs while they are still in close proximity. Feeding in turn is thought to have priority over certain other types of behavior. Typically, given a choice between feeding and righting, an animal will feed. Similarly, animals will feed rather than mate or withdraw defensively in response to weak tactile stimuli (Davis and Mpitsos, 1971). Since these stimuli normally elicit responses in the absence of food, it must be assumed that recognition

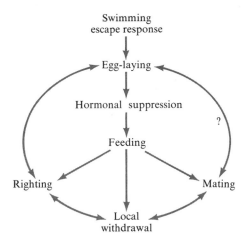

FIGURE 8-34.
Summary of the behavioral hierarchy of *Pleurobranchaea*. Unidirectional arrows from one behavioral act to another signify dominance of the former over the latter. Bidirectional arrows signify mutual compatibility between the behavioral acts, i.e., they can occur simultaneously. Escape behavior is dominant over all other behaviors. The question mark between egg-laying and mating means that no relationship has been demonstrated in *Pleurobranchaea*. However, in *Phyllaplysia*, a gastropod having a reproductive system similar to that of *Pleurobranchaea*, egg-laying and mating can occur simultaneously (Beeman, 1970). By analogy, Davis and Mpitsos suggest that egg-laying and mating may be mutually compatible in *Pleurobranchaea*. [After Davis *et al.*, 1974a, b.]

of food and initiation of feeding suppress the expression of these other behavioral responses.

Kovac and Davis (1977) have begun to analyze food-induced suppression of the withdrawal reflex to weak tactile stimuli. They have found that the two symmetrical corollary-discharge interneurons, which convey to the cerebral ganglia a rough copy of the efferent motor pattern of feeding (see above), also inhibit the central neurons controlling defensive withdrawal. Thus, the same set of cells that inform the central nervous system about the feeding program also assure that competing responses are suppressed.

SUMMARY AND PERSPECTIVE

Although opisthobranchs provide excellent opportunities for comparative studies of behavior, only a few of the studies so far carried out were specifically designed to be comparative. With the exception of some studies on several species of *Aplysia,* the opisthobranchs that have been examined are only distantly related. As a result, it is difficult to draw meaningful comparisons between the various groups. But a few suggestions about the relationship between structure and function have emerged. One is that the presence of a shell is not necessarily a powerful determinant of the behavior of molluscs. The behavioral capabilities of the shelled pulmonate snail *Helix* are not dramatically different from the shell-less pulmonate slug *Limax.* Similarly, diet does not seem to be as powerful an influence on behavior as one would have thought, although carnivorous and herbivorous feeding habits are clearly associated with distinctive behavioral types. Herbivores, like the anaspid *Aplysia,* tend to be bigger and slower moving than carnivores, like the notaspid *Pleurobranchaea.* Moreover, herbivores tend to spend a much greater part of their day eating than do carnivores. Nevertheless, despite a basic difference in diet and distribution of their activities during the course of a day, *Pleurobranchaea* and *Aplysia* share a number of similarities in feeding behavior.

An interesting and readily analyzable determinant of behavior is environment. For example, members of the same species of *Aplysia* living in intertidal zones are much less responsive to mechanical stimuli than are animals living in the sublittoral zone. Likewise, the natural alternation of light and darkness exerts a powerful control of behavior, acting through several photoreceptors and circadian clocks.

In order to move beyond these broad and vague generalities it will be necessary to carry out studies designed to examine specific questions in comparative behavior. For example, it will be worthwhile to examine inking and noninking species of *Aplysia,* burrowing and nonburrowing ones, swimmers and non-swimmers. This would allow us to determine what aspects of the environment encourage selection for a particular behavior and what differences in the basic neural machinery of the species allow for the generation of that behavior. Since a variety of behaviorally different species can coexist, these studies should not be difficult. In addition, *Aplysia* are hardy and can be shipped over great distances, thus making them easily accessible to researchers throughout the world. From these studies we might be able to reconstruct the behavioral patterns common to all the *Aplysia* species and thereby infer the possible behavioral repertory of the common ancestor for the genus. By extension, a comparison of *Aplysia* with the primitive anaspid *Akera* and with the even more primitive cephalaspid *Acteon,* thought to be a more direct descendant of the opisthobranch ancestor, might shed light on the origin of these opisthobranch orders.

Whereas the data necessary for a systematic comparative analysis of the behavior of opisthobranchs are still not at hand, the techniques for obtaining those data are now well developed. Major advances have been made in studying the behavior of opisthobranch and pulmonate molluscs, which previously had been relatively ignored. Considering the short time in which systematic behavioral research has been carried out on *Aplysia* and related opisthobranchs, a surprising amount has been learned about the five major types of naturally occurring behavioral responses that constitute the animal's total behavioral repertory. The five types are (1) defensive responses; (2) locomotor and escape behavior; (3) feeding; (4) autonomic regulation; and (5) sexual behavior and egg-laying. These behaviors range in complexity from elementary reflex acts (defensive withdrawal) to complex feeding responses and higher-order social behavior (mating). In each case a beginning has been made in analyzing the cellular architecture of the behavior. As a result, in some opisthobranchs and pulmonates one can now analyze a small but significant fraction of an animal's *total* behavior on the cellular level. This background knowledge makes it possible to explore the mechanisms of motivation, arousal, and behavioral choice, which we will consider in the next chapter. These mechanisms determine the coordination of the various component behaviors of an animal's total behavioral repertory and provide for an effective and integrated response to complex stimuli.

SELECTED READING

Bullock, T. H. 1953. Predator recognition and escape responses of some intertidal gastropods in the presence of starfish. *Behaviour*, 5:130–140. One of the few systematic comparative studies of species-specific behavior in gastropods.

Willows, A. O. D. 1973. Learning in molluscs. In *Invertebrate Learning*, Vol. II, edited by W. C. Corning, J. A. Dyal, and A. O. D. Willows. Reviews the behavioral capabilities of molluscs.

CHAPTER 9

Learning, Arousal, and Motivation in Opisthobranchs

As we have seen in the last chapter, in certain opisthobranchs and pulmonates one can now study aspects of an animal's total behavioral repertory on the cellular level. This makes it possible to move from a consideration of individual behavioral responses to an examination of higher order interrelationships, such as arousal, motivation, and learning, that pertain to the activity of the whole animal. For example, a given stimulus may evoke a family of responses at one time but not at another, suggesting that the behavioral thresholds of a variety of responses can change together. Changes in responsiveness that cannot be attributed to fatigue, injury, or maturation have been ascribed to two classes of so-called *intervening variables:* (1) *learning,* a change in behavior attributable to the previous history of exposure to the stimulus; and (2) *motivation* or *drive,* a change in behavior attributable to an alteration in internal homeostatic state of the animal (satiated versus hungry, hydrated versus thirsty).

Intervening variables cannot be measured directly. They are theoretical constructs, such as hunger, designed to explain the changes that occur in the brain and that alter the relationships between a given stimulus (food) and response (eating). To study learning and motivation more directly it will be necessary therefore to specify the loci and mechanisms of these postulated brain processes. This specification is best done at a cellular level. In the limit a cellular analysis may allow one to replace these hypothetical behavioral constructs with more precise neurobiological ones. A particularly useful beginning has come from the cellular analysis of *arousal,* a process thought to be a component of motivation as well as of certain types of learning.

This chapter reviews insights into learning, arousal, and motivation in opisthobranchs and pulmonates obtained from behavioral and cellular studies. Comparative studies of these processes, although only beginning, are likely to be important for two reasons. One, these studies can elucidate the relationships between various learning and motivational capabilities on the one hand and types of neuronal organization on the other. Two, comparative studies allow one to test the generality of the cellular mechanisms found in a particular group of animals.

LEARNING CAPABILITIES OF
OPISTHOBRANCHS AND PULMONATES

In its broadest meaning learning refers to any modification in the behavior of an animal that results from the history of its interaction with its environment and that is not due to maturation or injury or fatigue. Learning can be classified in several ways. For example, Thorpe (1963) divided learning into six categories: habituation, classical conditioning, instrumental conditioning, imprinting, latent learning, and insight learning. A simpler classification is to divide learning along two dimensions: complexity and time course. Learning varies in complexity according to the stimulus pattern necessary to initiate the behavioral modification. *Simple forms* of learning, such as habituation and sensitization, are nonassociative—they do not depend on a specific temporal pairing of stimuli or of stimulus and response. *Complex forms* of learning, such as classical and operant conditioning, are associative—they require a specific pairing of two stimuli or of a stimulus and a response. Learning also varies in the duration for which the learned change is retained. Thus learning is often arbitrarily divided into *short-term* learning, lasting minutes or hours, and *long-term* learning, lasting days or weeks and sometimes a lifetime.

There is general agreement that gastropods show simple, nonassociative types of learning of both short- and long-term duration. Recent studies suggest that gastropods, specifically opisthobranchs and pulmonates, also have associative capabilities, although they are apparently less well developed. In this chapter I will consider the behavioral evidence for nonassociative short- and long-term forms of learning in gastropods and then outline several insights into these processes that have been gained from cellular studies. In particular,

I will illustrate some conceptual relationships between sensitization (a form of prolonged nonassociative reflex modulation), arousal, and motivation. I will then consider associative learning.

NONASSOCIATIVE LEARNING

Of the several types of intervening variables that we will consider in this chapter, nonassociative learning has proved to be the most tractable for studies aimed at achieving cellular explanations. Since one form of nonassociative learning, sensitization, is related to arousal, insights from the cellular studies of this type of learning have also guided aspects of the research on motivation, a construct related to arousal.

Habituation and Sensitization of Elementary Behavior in *Aplysia*

Habituation, perhaps the simplest form of nonassociative learning, is a progressive decrease in the strength of a behavioral response to a relatively weak natural stimulus of constant intensity when that stimulus is repeatedly presented. Sensitization, the mirror-image process of habituation, is a slightly more complex form of learning; it is the enhancement of a reflex response to one stimulus as a result of the presentation of another stimulus, usually a strong or novel one. Sensitization thus resembles classical conditioning in that activity in one pathway facilitates reflex activity in another. Unlike classical conditioning, however, the reflex facilitation is not associative and does not require specific temporal pairing of the two stimuli.

Many defensive reflexes in molluscs show habituation and sensitization. These two forms of nonassociative learning have been extensively studied in the defensive withdrawal reflexes of the siphon and gill in *Aplysia* (Figures 9-1 and 9-2).[1] Two interesting features have been found to characterize these behavioral modifications. One is that the behavioral modifications can be both short and long term (Figures 9-1B and 9-2B). The second is that the time course is determined not by the number of stimuli but by the pattern of stimulation (Figure 9-3). For example, with an interstimulus interval of 30 sec a single training session of 10 stimuli produces short-term habituation that lasts at most several hours. Four repeated training sessions separated by as little as 1.5

---

[1]Local responses, such as the gill-pinnule or siphon responses in *Aplysia* and siphon withdrawal in the clam, also undergo habituation (Chapter 6).

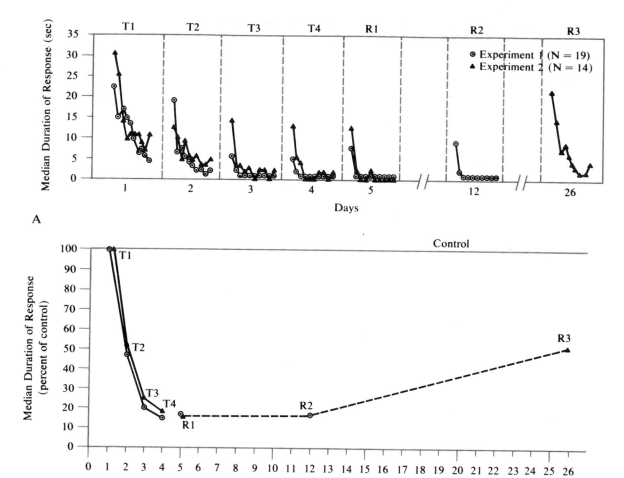

FIGURE 9-1.

Long-term habituation and sensitization of the siphon component of defensive reflex in the intact animal. [From Kandel, 1976; based on Carew *et al.*, 1972.]

A. Build-up of habituation during training for four days (T1 to T4) and retention after 24 hours (R1), one week (R2), and three weeks (R3). Data from two experiments are presented. In experiment 1 retention was tested at 24 hours and one week (data is pooled from two identical replications). In experiment 2 retention was tested at 24 hours and three weeks. Each data point is the median duration of siphon withdrawal for the population in a single trial (10 trials in each training session). A 30-sec interstimulus interval was used.

B. Time course of habituation, based on the data in part A. The score for each daily session is the median of the sum of 10 trials. Control duration (100 percent) is the response time during the first day of training. Experimental animals tested at 24 hours (R1), one week (R2), and three weeks (R3) showed significantly greater habituation than the controls. A Mann-Whitney U test was used for intergroup comparisons.

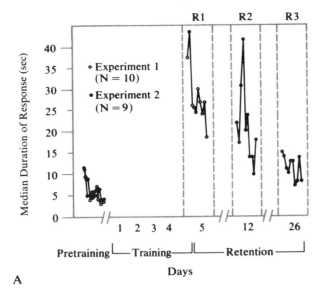

A

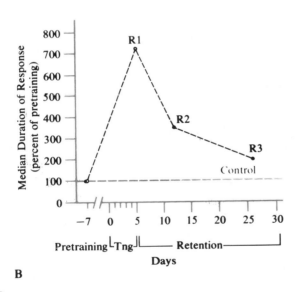

B

FIGURE 9-2.
Long-term sensitization of the siphon component of the defensive reflex in the intact animal. [From Kandel, 1976; based on Pinsker *et al.*, 1973.]

A. The median duration of siphon withdrawal is shown for each trial of a 10-trial block (minimum intertrial interval was 30 sec). The results of two independent experiments are shown. Experimental animals were given four electrical shocks per day for four days; the controls received no shocks. In experiment 1 retention of dishabituation was tested one day (R1) after the last shock. In experiment 2 retention was tested one week (R2) and three weeks (R3) after the last shock. Significant sensitization was evident in the one-day and one-week retention tests, while there was almost complete recovery in the three-week retention test.

B. Time course of sensitization, based on data from part A. Responses for each daily session have been summed and expressed as a single (median) score. Twenty-four hours (R1) and one week (R2) after the end of training the experimental animals showed significantly longer siphon withdrawal compared to their own pretraining score (expressed in the graph as 100 percent) and the score of a control group (not shown). (The Mann-Whitney U test was used for intergroup comparisons.) In the three-week retention test (R3) there was no longer a significant difference between the experimental group and control group (not shown) but the responses of the experimental animals were still significantly prolonged compared to their own pretraining levels (Wilcoxon matched pairs signed-ranks test; $p < 0.005$); the control animals were unchanged.

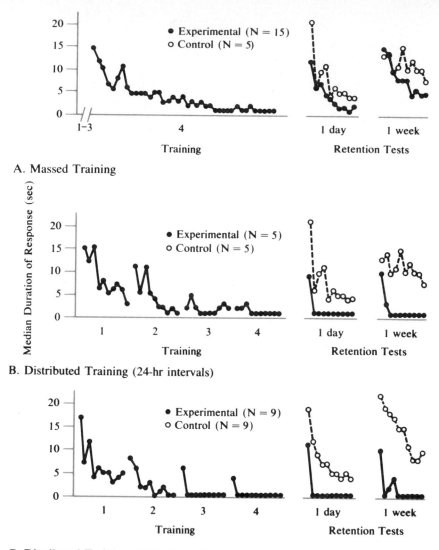

A. Massed Training

B. Distributed Training (24-hr intervals)

C. Distributed Training (1.5 hr intervals)

FIGURE 9-3.
Comparison of massed and distributed training of habituation of siphon withdrawal in *Aplysia*.

A. Massed training. Experimental animals received no stimulation for three days. On the fourth day 40 consecutive stimuli (trials) were applied to the siphon (intertrial interval = 30 sec). Control animals received no training. Retention was tested one day and one week after training, and was not significantly different from control. [From Kandel, 1976; based on Carew *et al.*, 1972.]

B. Distributed training (24 hours between training sessions). Experimental animals were given four sessions of training (10 trials per session); control animals received no training. Mann-Whitney U tests revealed that animals receiving spaced training exhibited significantly greater retention of habituation one day ($p < .01$) and one week ($p < .01$) after training than animals that received massed training. Studies illustrated in parts A and B were carried out concurrently; control animals were the same in both experiments. [From Kandel, 1976; based on Carew *et al.*, 1972.]

C. Distributed training (1.5 hours between training periods). Experimental animals were given training; controls received no training. Mann-Whitney U tests revealed that experimental animals exhibited significantly greater habituation than controls both one day ($p. < .001$) and one week ($p < .001$) after training. [From Kandel, 1976; based on Carew and Kandel, 1973.]

hours or more produce long-term habituation that lasts for up to three weeks (Figure 9-3C). If the four training sessions are presented consecutively, the training does not produce as pronounced long-term habituation (Figure 9-3A; Carew *et al.*, 1973). A similar sensitivity to a pattern of stimuli seems to apply to sensitization (Pinsker *et al.*, 1973).

Cellular Analysis of Short-Term Habituation and Sensitization

The mechanisms of short-term habituation and sensitization of the gill-withdrawal reflex have been specified on the subcellular level (for review see *Cellular Basis of Behavior*). With repeated sensory stimulation at rates that produce habituation in the intact animal (once every 10 sec to once every 3 min), the monosynaptic excitatory connections between the mechanoreceptors (from the siphon skin) and the motor neurons and interneurons undergo a depression due to a decrease in transmitter release by the sensory neurons. A sensitizing stimulus produces rapid facilitation of the same set of connections due to an increase in the amount of transmitter released per impulse (Figure 9-4; see Castellucci and Kandel, 1974, 1976).

The acquisition and retention of the synaptic depression and facilitation that accompany short-term habituation and sensitization are not altered when protein synthesis is inhibited by 95 percent (Schwartz *et al.*, 1971). This finding suggests that habituation and sensitization do not involve an alteration in macromolecular synthesis but might be due to an alteration in the levels of a small molecule—perhaps an intracellular second messenger like cyclic AMP. A second messenger could then act to alter either the probability of transmitter release or the distribution of the transmitter from one compartment in the presynaptic terminal to another.

To test this idea, Cedar, Kandel, and Schwartz (1972) and later Klein, Castellucci, and Kandel (unpubl.) examined changes in the level of cAMP in the abdominal ganglion. They stimulated the connectives that mediate the synaptic facilitation accompanying behavioral sensitization and found that strong stimulation produced a twofold increase in cAMP. The increase seemed to be synaptically mediated; it was blocked by high concentrations of Mg^{++}. This synaptically mediated increase in cAMP is simulated by serotonin, dopamine, and octopamine (Cedar and Schwartz, 1972; Levitan and Barondes, 1974). With a brief exposure to biogenic amines the increase in cAMP lasted about 40 min and therefore resembled in its time course the synaptic facilitation produced by a strong sensitizing stimulus (Figure 9-5).

338

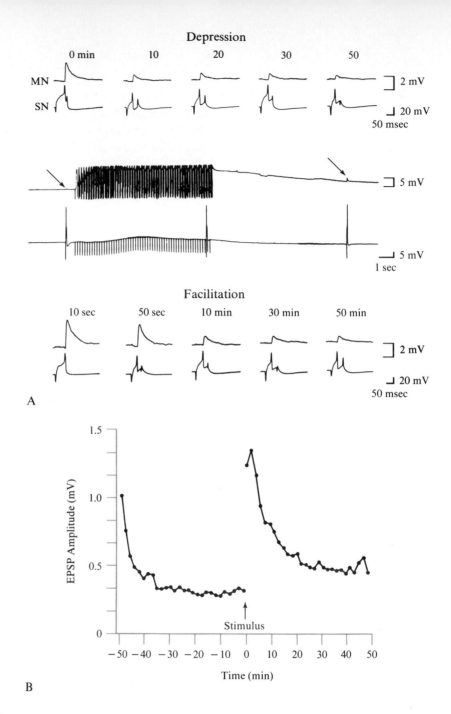

A

B

FIGURE 9-4.
Synaptic depression and facilitation accompanying short-term habituation and sensitization. [From Castellucci and Kandel, 1976. Copyright 1976 by the American Association for the Advancement of Science.]

A. Facilitation of a monosynaptic EPSP after a strong stimulus. Selected samples of EPSPs during synaptic depression produced by repeated stimulation every 10 sec. Arrows in the middle traces indicate the last EPSP before the facilitating stimulus to the left connective and the first EPSP after the stimulus (SN, sensory neuron; MN, motor neuron).

B. Time course of facilitation. Data were obtained from an experiment similar to the one illustrated in part A. Each point represents the average amplitude of 10 successive evoked EPSPs. Facilitating stimulus to the left connective was presented at time 0. The EPSPs are facilitated beyond the initial control amplitude.

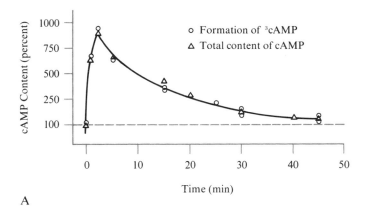

A

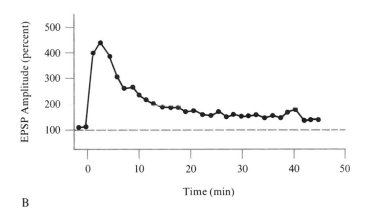

B

FIGURE 9-5.
Comparison of the time courses of the increase of cyclic AMP in the abdominal ganglion and presynaptic facilitation accompanying sensitization.

A. Increase of cAMP content. The abdominal ganglia from animals weighing an average of 130 g were incubated in 0.2 mM serotonin for 5 min. Cyclic AMP was determined at the times indicated on the graph. Each value is from two or more determinations that were normalized to the amount of cAMP in the tissue 5 min after onset of incubation with serotonin. [After Cedar and Schwartz, 1972.]

B. Presynaptic facilitation of a monosynaptic EPSP. A sensory neuron was stimulated once every 10 sec. Each point represents the average of 10 successive evoked EPSPs. Facilitation (time 0) was produced by stimulating the left connective for 10 sec (6 Hz). The EPSP amplitudes have been normalized in relation to the average of the last 20 EPSPs preceding the facilitation, which was taken as control (100 percent). [After Kandel *et al.,* 1976.]

These several findings suggested that sensitization involves one or more neurons that release a biogenic amine (serotonin, dopamine, or octopamine). These neurons might then act on the terminals of the sensory neurons to increase cAMP and thereby enhance transmitter release. To test the first of these two ideas, Brunelli and co-workers (1976) exposed the ganglion to serotonin, dopamine, and octopamine in turn, and found that only serotonin (in concentrations of 10^{-6} to 10^{-4} M) enhanced synaptic transmission between the sensory and motor neurons (Figure 9-6). Furthermore, the presynaptic facilitation produced by a sensitizing stimulus was blocked by a serotonin antagonist, cinanserin. Brunelli and co-workers next exposed the ganglion to the cAMP analog, dibutyrl cAMP, and found that it produced synaptic facilitation at the synapses made by the sensory neurons on the motor neurons but not at two other nonsensory synapses examined. Moreover, intracellular injection of cAMP into the sensory neurons also produced the facilitation (Figure 9-7). Finally, agents that increase the intracellular level of cAMP—either by direct stimulation of the enzyme adenyl cyclase, which synthesizes cAMP, or by inhibition of the degradative enzyme phosphodiesterase—lead to facilitation of the synapse between sensory and motor neurons (Klein, Castellucci, and Kandel, unpubl.). In turn, electrical stimulation of the pathway from the head that mediates sensitization, or incubation with serotonin, or intracellular injection of cAMP, all produce a prolonged increase in the Ca^{++} conductance normally activated by the action potential in the sensory neuron. The resultant increased Ca^{++} influx seems to account for the facilitated transmission (Klein and Kandel, 1978).

FIGURE 9-6.
Simulation of presynaptic facilitation by serotonin. [From Kandel *et al.*, 1976; Kandel, 1976.]

A. Effect of serotonin on monosynaptic EPSP. A sensory neutron (SN) was stimulated once every 10 sec and produced a monosynaptic EPSP in a gill or siphon motor neuron (MN). Between the 15th and 16th action potential there was a 2.5-min rest during which the ganglion was bathed with 10^{-4} M serotonin.

B. Summary graphs of experiments with serotonin. In each experiment the EPSP amplitudes were normalized to initial control EPSP. In the control group 15 stimuli produced synaptic depression; after a rest of 2.5 min (and a slight recovery of the EPSP) a second group of 15 stimuli was applied to the sensory neurons, producing further depression. In two experimental groups the 15 initial stimuli produced similar synaptic depression. There is complete overlap of the control and experimental curves during the initial EPSP depression. But after treatment with serotonin, both experimental curves are significantly higher than the control curve. To determine the statistical differences between the experimental and control curves, the amplitude of the responses to each of the 15 stimuli of a run were summed to obtain a single score, and a one-tailed Mann-Whitney U test was performed ($p < 0.016$, ●—●, and $p < 0.028$, ○—○).

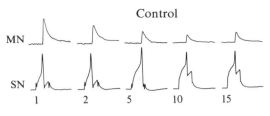

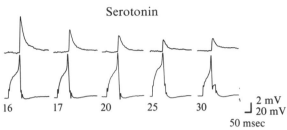

A

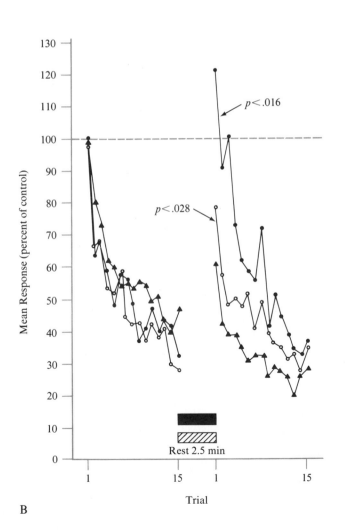

B

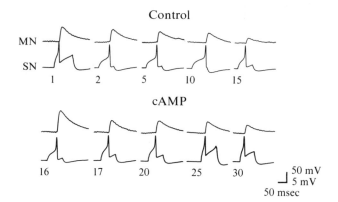

Control

MN

SN

1   2   5   10   15

cAMP

16   17   20   25   30

50 mV
5 mV
50 msec

A

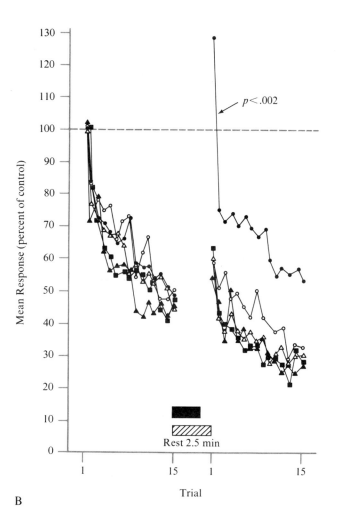

- cAMP, N = 8
- cGMP, N = 4
- 5'AMP, N = 3
- Hyperpolarizing pulses
  (K⁺ citrate), N = 6
- Rest, N = 5

$p < .002$

Mean Response (percent of control)

Rest 2.5 min

Trial

B

FIGURE 9-7.
Simulation of presynaptic facilitation by cyclic AMP. [From Brunelli *et al.*, 1976, copyright 1976 by the American Association for the Advancement of Science; and Kandel, 1976.]

A. Effect of cAMP on monosynaptic EPSP. **(1)** A sensory neuron was first stimulated once every 10 sec for 15 stimuli (1 to 15). During a 2.5-min rest cAMP was introduced by iontophoresis into the cell body of the sensory neuron. Hyperpolarizing current pulses (1 sec long) of 3×10^{-7} to 4.5×10^{-7} A were applied every 2 sec for 2 min from one barrel of a double electrode filled with 1.5 M cAMP; the second (recording) barrel was filled with 3 M potassium citrate. Thirty seconds after the end of the injection a second series of 15 stimuli was given (16 to 30). The amplitudes of the evoked EPSPs of the second series are higher than the amplitudes of comparable EPSPs evoked in control experiments.

B. Summary of cAMP effect and comparison with four control groups. The experimental group (●—●) is compared with a first control group (■—■) in which the rest (2.5 min) was simply followed by 15 additional stimuli without any current injection. In a second group (△—△) a hyperpolarizing current pulse comparable to that used to inject cAMP was injected into the sensory neuron but the pulses were applied through an electrode filled with 3 M potassium citrate. In a third group (○—○) cGMP (1.5 M) was injected intracellularly. In a fourth group (▲—▲) 5′-AMP was injected. The five curves generated during the initial EPSP depression produced by the first 15 stimuli overlap completely. But after injection of cAMP the experimental group was significantly higher than either the first control group ($p < .002$, one-tailed Mann-Whitney U test) or the three other control groups (for potassium citrate, $p < .001$; for cGMP, $p < .014$; for 5′-AMP, $p < .012$).

Based on these findings Kandel and co-workers (1976) and Klein and Kandel (1978) have proposed a molecular model for the mechanisms of sensitization and habituation (Figure 9-8). In this model, a certain number of voltage-sensitive Ca^{++} channels are capable of being activated at rest and can mediate normal transmitter release by permitting a certain inflow of Ca^{++} with each action potential. Stimulation of the sensitizing pathway leads to the release of serotonin, which activates a serotonin-sensitive adenylate cyclase in the membrane of the sensory neuron terminal. The resulting increase in cAMP leads, through a series of intermediate steps, perhaps involving phosphorylation of membrane proteins, to the activation of an increased number of voltage-sensitive Ca^{++} channels, allowing a greater influx of Ca^{++} with each action potential than during rest and consequently greater transmitter release. The increased Ca^{++} influx could be produced in several ways, including opening up new Ca^{++} channels, increasing the conductance of preexisting Ca^{++} channels, or reducing their activation threshold. By analogy, the repeated activation of the reflex pathway that accompanies habituation could lead to a progressive inactivation of Ca^{++} channels so that each action potential leads to progressively less transmitter release. In fact, preliminary experiments indicate that the repeated activation of the sensory neurons leads to a decrease in the Ca^{++} component of the spike (Klein and Kandel, unpubl.). (For a parallel analysis in another system in *Aplysia*, see Shimahara and Tauc, 1977.)

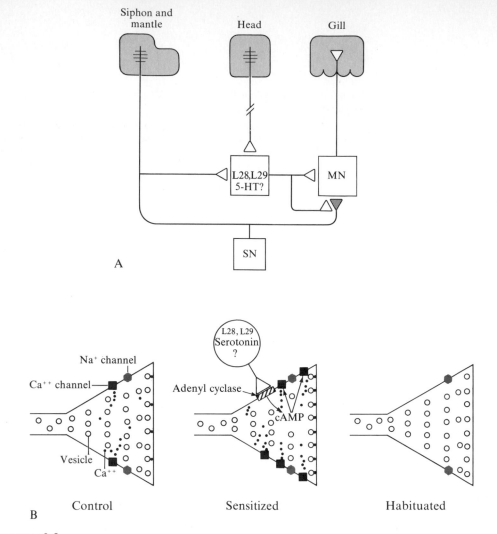

FIGURE 9-8.

Mechanisms for short-term habituation and sensitization of the gill-withdrawal reflex.

A. Postulated plastic changes in the circuit underlying habituation and dishabituation. For schematic purposes only the sensory neuron (SN) and motor neuron (L7) are illustrated. Similar processes are thought to be operative at all synaptic contacts made by the population of sensory neurons on the other motor neurons and interneurons. The terminal of a sensory neuron is the common locus of both the synaptic depression underlying habituation and the presynaptic facilitation underlying dishabituation. The pathway from the head mediating presynaptic facilitation acts on interneurons L28 and L29. These presumed serotonergic cells are postulated to synapse on the sensory neuron terminal.

B. Suggested molecular mechanisms for depression and facilitation based on Ca^{++} modulation. In the resting state the depolarization produced by each action potential opens up a certain number of Ca^{++} channels. Ca^{++} flows in and binds the vesicles to the membrane, allowing their release. Sensitization is postulated to be mediated by serotonergic cells. These act to increase cAMP, which is postulated to operate for prolonged periods to enhance transmitter release by increasing the Ca^{++} conductance. This could occur in one of three ways: **(1)** by opening up new Ca^{++} channels; **(2)** by increasing the conductance of preexisting Ca^{++} channels; or **(3)** by reducing the activation threshold of each Ca^{++} channel. It apparently does so by activating a voltage-dependent Ca^{++} channel that can be modulated by transmitter action. Habituation is postulated to be mediated by an inactivation of Ca^{++} channels.

To test this model further, Hawkins, Castellucci, and Kandel (1976) have found two cells, including one of the excitatory interneurons in the circuit for gill withdrawal (cells L28 and L29 in Figure 9-8), that form part of the population of neurons that mediate sensitization. These cells may be serotonergic since the facilitation produced by these cells is reversibly blocked by cinanserin. But more direct biochemical studies are required to establish this point.

Cellular Analysis of Long-Term Habituation

Studies of long-term habituation of the gill-withdrawal reflex by Carew and Kandel (1973) and by Castellucci and co-workers (1977 and unpubl. observ.) suggest that the long-term process may simply be a dramatic extension of the process underlying short-term habituation. During each of four ten-trial training sessions separated by 1.5 hours the complex EPSP produced in identified motor neuron L7 by electrical stimulation of the sensory input is progressively reduced and this reduction is retained for at least 24 hours (Carew and Kandel, 1973).

To determine whether these changes reflect a partial change in the effectiveness of all sensory connections or a more dramatic change in some connections, Castellucci, Carew, and Kandel (1978) examined the elementary synaptic potentials produced in motor cell L7 by individual sensory cells of animals one day and one week after long-term habituation training. They found that the training produced a profound functional disconnection (synapse inactivation) of most of the connections between sensory neurons and the motor cell. In control animals 90 percent of all sensory cells make strong excitatory connections to L7. These connections average 2 mV in amplitude (Figure 9-9). But one day and even one week after long-term habituation training only 30 percent of sensory cells show functional connections. Thus 60 percent of the previously existing connections have become inactivated by habituation training so that they cannot be resolved (less than 50 μV in amplitude). The remaining connections are only slightly reduced in effectiveness compared to control animals. Thus long-term habituation occurs at the same locus as does short-term habituation: the synapses made by the sensory neurons on the motor neurons. Moreover, long-term habituation seems to involve an electrophysiological mechanism somewhat similar to the one involved in the short-term process, namely a profound decrease in the functional effectiveness of the synapses. Whether inactivation of synapses following long-term habituation also represents an alteration in transmitter release, as it does in the short-term process, or a change in receptor sensitivity is not known.

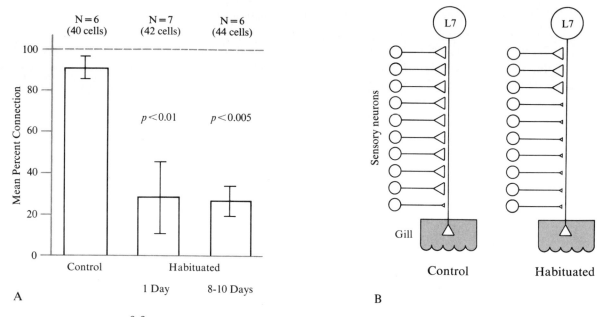

FIGURE 9-9.

Neuronal mechanisms underlying long-term habituation.

A. Incidence of monosynaptic connections between sensory neurons and motor cells in control and habituated animals. The habituated animals were tested one day and one week after long-term habituation training consisting of five sessions of 10 trials each. [Based on Castellucci, Carew, and Kandel, 1978.]

B. Functional disconnection model of long-term habituation. In control animals nine out of 10 sensory neurons connect to motor cell L7. Following long-term habituation only three out of 10 cells connect to L7; the synapses of six cells that were previously connected have been inactivated. The synapses were considered functionally disconnected if their synaptic action could not be distinguished from background noise. [Castellucci, Carew, and Kandel, unpubl.]

Effects of Different Environments on Habituation of Defensive Reflexes in *Aplysia*

Carew and Kupfermann (1974) found that long-term habituation may occur frequently in nature and may serve as an important adaptive mechanism for animals buffeted by the tides. Animals living in a calm sublittoral environment are initially very responsive to a novel tactile stimulus and show habituation that resembles that of normal, unperturbed control animals housed in the laboratory. By contrast, intertidal animals that are buffeted about by the ocean have a reduced responsiveness that resembles that of animals that have undergone long-term habituation (Figure 9-10). These changes in behavioral respon-

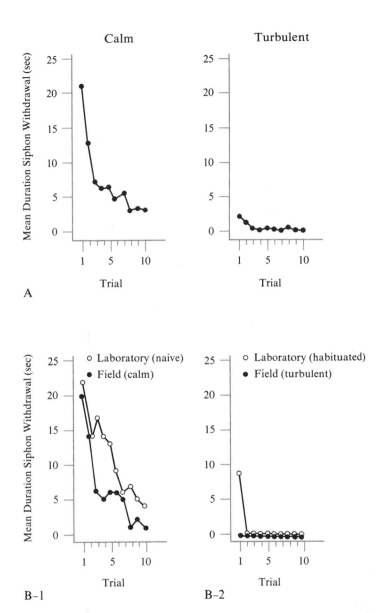

FIGURE 9-10.
Comparison of habituation of siphon withdrawal in animals living in different environments. **A.** Comparison of animals living in calm (N=11) and turbulent (N=10) environments. The reflexes of animals in the turbulent environment were significantly reduced compared to those in the calm environment ($p < .001$). [After Carew and Kupfermann, 1974.] **B. (1)** Comparison of habituation of naive animals from the laboratory with animals from a calm natural environment. **(2)** Comparison of habituation of siphon withdrawal in animals from the laboratory that had received four days of habituation training and of animals from turbulent environment. [After Carew, Pinsker, and Kandel, 1972; Carew and Kupfermann, 1974.]

siveness and in the kinetics of habituation are not simply related to exposure to air, which is a common occurrence in the life of intertidal animals. Animals that are constantly submerged in turbulent water show comparable reduced responsiveness and habituation kinetics.

Habituation of Complex Behavior in *Tritonia*

Habituation of a complex behavior, escape swimming, has been demonstrated in the opisthobranch *Tritonia diomedia* (Abraham and Willows, 1971; Willows, 1973). The behavior is elicited by contact with starfish (or salt crystals). When first presented, the stimulus elicits 3–10 swimming cycles that last 20–70 sec. But when the stimulus is repeatedly presented (at intervals of 10–20 min) the number of swimming cycles decreases to 50 percent of control. Forty-eight hours of rest leads to complete recovery. Briefly lifting the habituated animal out of the seawater produces sensitization. The interesting aspect of this modification is that whereas gill withdrawal is a graded reflex response and habituation leads to a graded decrease in responsiveness, *Tritonia* swimming is a complex fixed act in which the individual components are generated in an all-or-none fashion. Here habituation does not act on the amplitude of individual dorsal-ventral flexion swimming movements, but on the number of cycles in the swimming sequence. These are decreased in a quantized, all-or-none manner.

Sensitization of Complex Behavior in *Aplysia*

Sensitization, like habituation, is not limited to simple reflexes; it can also be induced in more complex fixed-action patterns. For example, exposure of *Aplysia* to food produces sensitization of the consummatory (biting) response: the latency for biting decreases and the number, speed, and amplitude of successive biting responses increases. In addition, there is a concomitant increase in heart rate (see p. 352).

The neural circuit controlling feeding as well as that controlling heart rate in *Aplysia* is now partially understood (Figure 9-11A). As noted earlier (Chapter 8), Weiss and Kupfermann (1976) have found that the symmetrical metacerebral cells of the cerebral ganglia have a particularly important role in feeding and seem to be critically involved in mediating sensitization of biting. The metacerebral cells are normally silent, but brief exposure of the animal

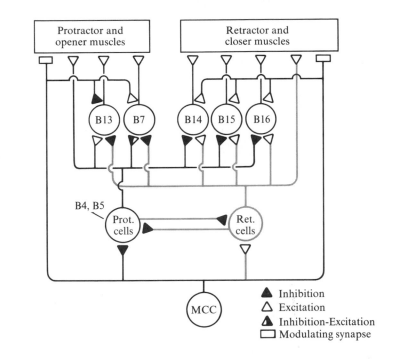

A

△ Inhibition
△ Excitation
△ Inhibition-Excitation
▢ Modulating synapse

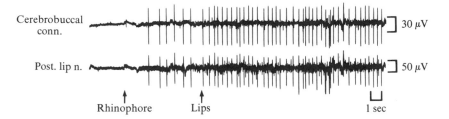

B

FIGURE 9-11.

The postulated neural circuit controlling feeding in *Aplysia*.

A. Postulated neural circuit for feeding. The muscles that protract the radula and open its two halves and those that retract it and close the two halves are innervated by motor cells that are not interconnected. The motor cells are innervated by command cells. The two command cells for protraction are identified multiaction cells (B4 and B5). The retractor command cells are not yet identified, but are postulated to be dual-action cells as well. The protractor and retractor command cells make mutually inhibitory connections. The command cells are innervated by one or more stretch receptors (SR) that excite the protractors and inhibit the retractors. The dual-action protractor and retractor command cells are innervated by higher-order dual-action metacerebral cells. These excite the retractor command cells and motor cells and inhibit the protractor command cells and some of the protractor motor cells. The metacerebral cells also innervate the buccal musculature, causing a facilitation of motor neuron action by means of direct enhancement of contraction as well as by acting on the connections between the motor neurons and the muscle. Unidentified cells and their connections are drawn as dashed lines.

B. Activation of the metacerebral cells by arousal stimuli. Recording from the axons of the metacerebral cells in the intact animal. The metacerebral cells show no spike activity before food presentation, but activity (large spikes) ensues when food is touched to the rhinophore (first arrow) and the lips of the animal (second arrow). [From Weiss, Cohen, and Kupfermann, 1978.]

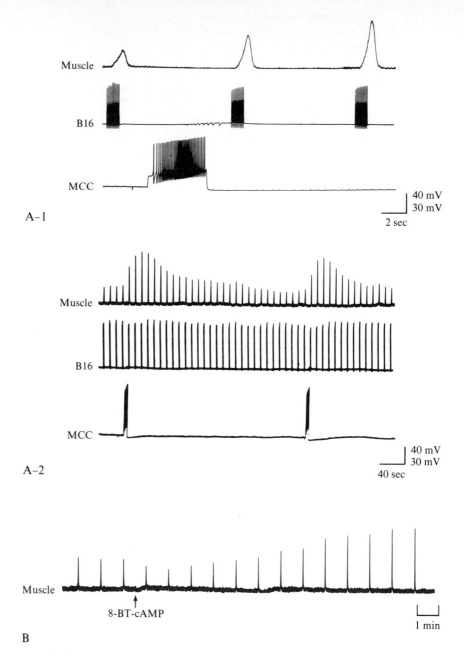

A-1

A-2

B

FIGURE 9-12.

Role of the metacerebral cells in the sensitization of biting in *Aplysia*.

A. Potentiation of buccal muscle contraction produced by an identified buccal motor neuron, B16. **(1)** Fast-sweep record showing muscle contractions produced by a fixed number and frequency of motor neuron spikes before and after stimulation of a metacerebral cell (MCC). **(2)** Slow-sweep record showing the time course of MCC potentiation. The MCC was stimulated twice to show the reproducibility of the effect. [From Weiss, Cohen, and Kupfermann, 1978.]

B. Potentiation of muscle contraction by an analog of cyclic AMP. Muscle contractions were produced by a constant burst of motor neuron spikes every 10 sec. At the arrow the preparation was perfused with 5×10^{-4} M 8-benzylthio-cAMP. [From Weiss *et al.*, 1976.]

to food activates the cells and causes them to fire (Figure 9-11B). The meta-cerebral cells in turn act on the muscle directly to increase its strength of contraction by enhancing excitation-contraction coupling (Figure 9-12A).

The metacerebral cells are serotonergic (Eisenstadt *et al.*, 1973; McCaman and Dewhurst, 1970), as may be the case for the cells producing sensitization of gill withdrawal. The metacerebral cells also mediate their sensitizing action on the buccal muscle by means of cAMP (Weiss *et al.*, 1976). Serotonin, applied to the intact muscle, or to cell-free homogenates, increases the synthesis of cAMP. Firing a single metacerebral cell can cause a marked increase in the synthesis of cAMP in the buccal muscle. In addition, analogs of cAMP simulate the enhanced contraction produced by the metacerebral cells (Figure 9-12B). A similar situation seems to apply to the sensitization of heart rate following the presentation of food. This increase in heart rate is largely mediated through the pericardial nerve, which carries the axon of LB_{HE}, the major heart excitor (Mayeri *et al.*, 1974). This cell is serotonergic (Liebeswar *et al.*, 1975) and pro-duces a prolonged speeding up of heart rate by means of an intracellular increase in cAMP (Koester, Weiss, and Mandelbaum, unpubl. observ.).

Thus, sensitization of gill withdrawal, biting, and heart rate in *Aplysia* appear to utilize a common molecular mechanism (Figure 9-13). It is tempting to speculate that in each case serotonin might act by means of cAMP to control the intracellular level of free Ca^{++}. In the case of gill withdrawal the control is exerted in the presynaptic terminals, where Ca^{++} could enhance transmitter release. In the case of heart rate the action is on the muscle directly where Ca^{++} could control the strength and frequency of contraction. In the case of biting the control is exerted at three points: (1) on the muscle; (2) on the motor neuron terminals; and (3) centrally, in an as yet unspecified manner, on the central program. Each of these actions again might be mediated by means of Ca^{++}.

AROUSAL AND SENSITIZATION

Groves and Thompson (1970) proposed the interesting idea that sensitization is a component of arousal. Since arousal is in turn a component of motivation, studies of invertebrate behavior provide an opportunity for attempting to translate into cellular terms several overlapping ideas.

Arousal is an intervening variable postulated to serve as an energizer that acts to alter at least three aspects of behavior: (1) the general level of activity—whether the animal is crawling or stationary, orienting for food or not, seeking

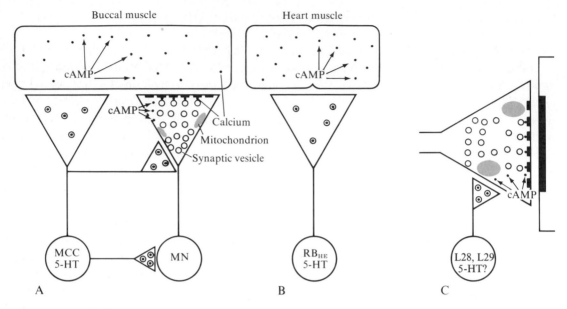

FIGURE 9-13.

Common features of the molecular mechanisms underlying the sensitization of biting, heart rate, and gill withdrawal. **A.** The metacerebral cells act on the terminals of the motor neurons and on the muscle to increase the level of cAMP. The cAMP level in turn defines the level of free Ca++. In the terminals the increase in free Ca++ enhances transmitter release; in the muscle it enhances excitation contraction coupling. **B.** The serotonergic heart excitor produces a prolonged increase in heart rate by increasing the level of cAMP, which in turn is proposed to act on the level of free Ca++ to produce a rate change. **C.** The presumed serotonergic interneurons produce presynaptic facilitation of the sensory neuron terminal to enhance gill withdrawal by increasing cAMP, which enhances Ca++ influx (see Figure 9-8).

a mating partner or fulfilled; (2) somatic reflex responsiveness—either enhancing or depressing several reflex response systems; and (3) autonomic responsiveness—heart rate, blood pressure, and respiration.

The concept of arousal was first developed to explain the well-known observation that most stimuli, particularly weak or familiar ones, have a restricted sphere of influence—they affect only a very limited range of behavioral response—whereas novel, noxious, or appetitive stimuli have more widespread influence and alter an animal's responsiveness to other stimuli. Moreover, the modulating effects can persist for minutes and even hours and share features of nonassociative learning.

The finding that some stimuli have a general modulatory influence on an organism's behavior has been interpreted by a number of theorists as indicating that these stimuli have an arousing or activating capability (Duffy, 1941, 1957;

Hebb, 1955; Malmo, 1959). For some of these theorists arousal represented a *single* behavioral system that varied along a single activity dimension ranging from sleep to wakefulness to states of high excitement or emotion.

The idea of a single arousal system, as originally proposed by Hebb and Malmo, seemed to be supported initially by the discovery of a single, diffusely acting neural system in the brain stem of mammals—the reticular formation—capable of producing arousal of the electroencephalogram, i.e., desynchronization of the electrical activity recorded from the surface of the skull or the cerebral cortex (Morruzzi and Magoun, 1949; French and Magoun, 1952; see Magoun, 1954). It seemed attractive, therefore, to assume that this neural system mediated behavioral arousal.

Although this formulation was first advanced more than 20 years ago, our understanding of the relation of the reticular formation to behavioral arousal in mammals has not advanced dramatically since then. In particular, there has been no clarification of the central question of whether arousal is due to a single neural system or to several independent systems. For example, recent anatomical and physiological studies indicate that different parts of the reticular system have distinct biochemical and functional properties (Jouvet, 1972; Ungerstedt, 1971). Each of these parts may have its own action on behavior. Whether one, several, or any of these regions is specifically related to behavioral arousal has not yet been determined. In addition, the behavioral concept of arousal is used in a variety of different ways (Andrew, 1974), which has made it difficult to compare studies and generalize from them.

However, the novel, noxious, or appetitive stimuli used to study arousal are similar to those commonly used to produce behavioral sensitization. Thus, the theoretical construct of arousal had much in common with the observed behavioral phenomenon of sensitization. Although sensitization typically refers to a nonassociative *enhancement* of a response, arousal stimuli may also produce nonassociative *depressions* of responses. For example, a noxious stimulus can cause a rat to freeze, thereby reducing locomotor behavior. It therefore seems useful to extend the term sensitization to include *all nonassociative modulation of the behavioral response to one stimulus by another, enhancement (positive sensitization) as well as depression (negative sensitization).* The depression and facilitation each need to be classified on the basis of operational criteria.

I would emphasize that although the consequences of sensitization are included in the construct "arousal," the two are not fully equivalent terms. Arousal is commonly defined as consisting of three components: general activity, somatic responsiveness, and autonomic activity. Sensitization often pertains only to the last two components. In addition, arousal can be produced by *internal* as well as external stimuli (see below, p. 365). Sensitization typically pertains only to changes produced by external stimuli, but this distinction does not seem very fundamental.

The advantage of reducing the study of arousal to that of reflex sensitization is that arousal is often used in a rather vague way so that it has different meanings in different contexts. By contrast, sensitization is used in a more restricted sense and is operationally well defined by Grether (1938). Sensitization is therefore reducible to clear, testable cellular hypotheses that can be effectively studied in invertebrates. By examining the effects of various novel, appetitive, and noxious stimuli on different response systems one can explore, on both a behavioral and a cellular level, whether arousal of different response systems depends on a common neural system or whether independent neural systems are involved.

The Multivariant Nature of Arousal in *Aplysia*

As is the case for vertebrates, a variety of appetitive or noxious stimuli—food, a potential sexual partner, or a predator—can arouse an otherwise quiescent opisthobranch. To determine whether different arousal stimuli have different consequences for different response systems in *Aplysia*, I will consider the consequences of two different types of arousal stimuli (appetitive and noxious) on three different response systems: (1) a variety of defensive responses (siphon withdrawal, inking, and escape); (2) an autonomic response (heart rate); and (3) an appetitive response (biting).

APPETITIVE STIMULI

As we have seen, food stimuli positively sensitize biting. Latency of the response decreases and the number of biting responses and the amplitude of each response increases with continued exposure to food (Figure 9-14A; Susswein, Kupfermann, and Weiss, 1976b). For example, if a hungry animal is touched on the lips with a small piece of food the median latency of the mouthing response to a second piece of food is much shorter (about 6 sec) than is the latency to the first piece (25 sec or longer) (Susswein *et al.*, 1978). This positive sensitization persists for half an hour, during which time the animal is said to be aroused. Food stimuli also increase heart rate (Figure 9-14A; Dieringer, Koester, and Weiss, 1978).

By contrast, defensive responses are depressed by appetitive stimuli such as food (Advokat, Carew, and Kandel, 1976). Animals tested immediately after a meal exhibit a significant depression (negative sensitization) of the graded

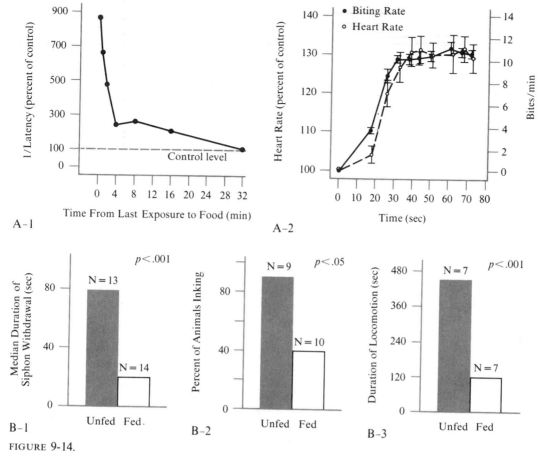

FIGURE 9-14.

Effects of appetitive stimuli on appetitive and defensive reflexes.

A. Sensitization of biting and heart rate. **(1)** Biting latency. Animals were presented with pieces of seaweed to the rhinophores until they oriented so as to make the lips visible. The lips were then stimulated until animals performed four biting responses. The time from the onset of stimulation of the rhinophore and lips until the first biting response was noted. Seven groups of animals were tested, with six animals in each group. Each was allowed a different period of nonstimulation (0.5, 1, 2, 4, 8, 16, and 32 min) and times needed for rhinophore and lip stimulation to elicit biting were again noted. Decline of sensitization is evident in the gradual increase in latency of biting over the time since the animals were last exposed to food. The control level (100 percent) is the mean latency of the initial response of all experimental animals together with a control group (dashed line) that was tested after a piece of seaweed was placed in the cage without touching it to the animal. [After Susswein, Kupfermann, and Weiss, 1976a and unpubl.] **(2)** Comparison of time courses for increase in heart rate and biting rate in response to appetitive stimuli. Starting at time zero a small piece of seaweed, held with a forceps, was kept in continuous contact with the lips. [After Dieringer, Koester and Weiss, 1978; Kupfermann, unpubl.]

B. Depression of defensive reflexes by appetitive stimuli. **(1)** Median duration of siphon withdrawal to 10 water-jet stimuli (interstimulus interval = 30 sec) in two groups of animals. One group (unfed) was examined immediately after eating, the other only 24 hours after its last meal. Animals that were tested without prior stimulation exhibited significantly greater reflex responsiveness than those tested after ingestion. **(2)** Inking to a noxious stimulus (shock) in two groups of animals. One group (unfed) received shock 24 hours after a daily meal; the other group (fed) received shock immediately following a meal. The unfed animals showed a greater tendency to ink than fed animals. These results demonstrate that prior feeding depresses responsiveness to noxious stimulation. **(3)** The effect of feeding on locomotion elicited by an aversive stimulus (salt crystals applied to the base of the siphon). The effect was determined by comparing the duration of locomotion (number of seconds elapsing until the animal remained stationary for a period of 60 sec) of animals who had just been fed and those who had not been fed for 24 hours. Fed animals crawled significantly less after the aversive stimulus. [After Advokat *et al.*, 1976 and unpubl.]

siphon-withdrawal reflex, the all-or-none inking reflex, and a complex fixed-action pattern, escape response, as compared to the animals tested 24 hours after feeding (Figure 9-14B). Eating is not critical for the reflex depression. Chemical stimulation of the lips with food produces similar effects (Advokat, Carew, and Kandel, 1976 and in preparation).

NOXIOUS STIMULI

As we have seen above, noxious stimuli can produce powerful enhancement of defensive siphon withdrawal (Figure 9-15A). By contrast, noxious stimuli depress the biting response. Kupfermann and Pinsker (1968) found that a strong electric shock to the head increases the latency for the biting reflex for at least 24 hours. The depression of the biting reflex is nonassociative; it is independent of whether the shock is paired with food or presented alone (Figure 9-15B).

Independent Positive Sensitizing System

The selective effects of appetitive and noxious stimuli indicate that arousal (defined as sensitization) is not a unitary process in *Aplysia*. Arousal produced by noxious or appetitive stimuli has opposite effects on different response systems (feeding and defensive withdrawal) and each of the arousal stimuli has opposite effects on the same system: noxious stimuli enhance siphon withdrawal and depress biting, whereas food stimuli depress siphon withdrawal, inking, and escape but enhance biting and heart rate. These results (Figure 9-16) are consistent with the physiological finding that separate nerve cells mediate positive sensitization of biting (Weiss *et al.,* 1978), gill withdrawal (Hawkins *et al.,* 1976), and heart rate (Mayeri *et al.,* 1974). In each case the neurons that mediate positive sensitization have turned out to be an intrinsic part of the neural circuit of the behavior. Cells such as L28 and L29 are excitatory premotor interneurons for gill and siphon withdrawal; the metacerebral cells are premotor interneurons in the feeding circuit; and the RB_{HE} cell is an excitor motor neuron for the heart. This anatomical arrangement suggests that *each family of behavioral responses may have its own arousal system.* Indeed, some families might conceivably have *two* independent systems, one for positive and the other for negative sensitization. Alternatively, negative sensitization might be mediated by inhibition of a tonically active positive system.

Although the neurons mediating positive sensitization of biting, heart rate, and siphon withdrawal are different and can be modulated independently,

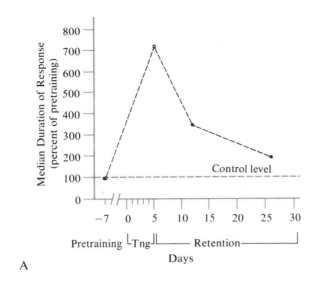

A

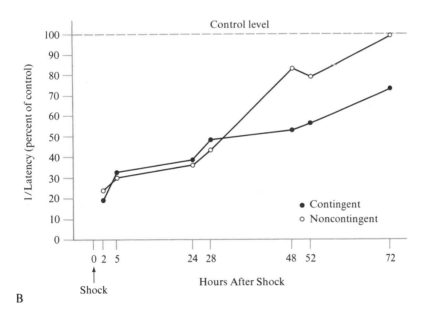

B

FIGURE 9-15.
Effects of noxious and appetitive stimuli on feeding. **A.** Positive sensitization of siphon withdrawal (see Figure 9-2B). **B.** Negative sensitization of biting. The data are mean latencies of the feeding response of shocked and unshocked (control) animals. The shock was either contingent or noncontingent upon a feeding response. The control level is the average latency for all unshocked trials in both experiments. [After Kupfermann and Pinsker, 1968.]

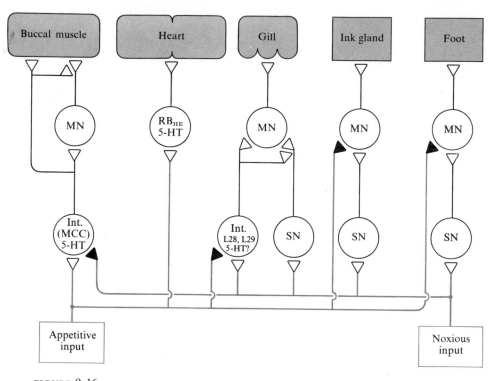

FIGURE 9-16.
A model of independent sensitizing mechanisms for gill withdrawal, heart rate, and biting (see text for details).

the molecular mechanism for sensitization may in each case be the same. In each case the positive sensitizing action seems to be produced by serotonergic neurons that act on intracellular Ca^{++} concentration by increasing cAMP in the target cells (see Figures 9-13 and 9-16; Weiss *et al.*, 1976; Brunelli *et al.*, 1976; Koester and Weiss, unpubl.). In view of the similarities of the mechanisms underlying positive sensitization, it becomes interesting to explore instances of negative sensitization and to determine how they are mediated.

MOTIVATIONAL STATE

The initial success in reducing aspects of the intervening variable arousal to sensitization, a variable that has been specified at the cellular level in invertebrates, has encouraged the attempt to apply a similar approach to motivation.

This task is more difficult, however, because motivation is not as well defined operationally. In the clearest cases motivation is used to describe homeostatic processes, e.g., temperature or food regulation, that are, as I will indicate below, perhaps best described in the language of regulatory biology before they can be analyzed in terms of cellular mechanisms.

Motivation refers to two interrelated aspects of behavior (Hebb, 1966). One, *arousal,* determines whether an inactive animal becomes active; the other, *cue* or *goal-directedness,* organizes the appropriate family of behavioral responses in relation to a stimulus. Even though motivation is only vaguely defined, psychologists have found it useful, since by using this concept it is often possible to relate several independent behavioral responses. For example, a hungry *Aplysia* crawls actively and shows orienting behavior (head-waving) in seeking out food, but this activity ceases once the animal is fed. The relationship between food deprivation, food seeking, and feeding can be explained by means of alterations in a single or related group of variables called *motivational* or *drive state.*

The concept of motivational state is most effectively applied to regulatory, goal-oriented behavior, such as feeding and drinking. These behaviors, like the even more basic regulatory processes of oxygen consumption and temperature regulation, are homeostatically governed. A set of regulatory mechanisms for each of these processes assures that the internal state is maintained within certain limits. Thus, if the body temperature increases beyond a certain *set point,* a family of processes (sweating, panting) designed to bring the body temperature back toward the set point is activated. As with these basic regulatory processes, the purpose of goal-oriented behavior is to regulate the internal environment so as to reduce deviations from the set point. Thus, in its simplest form motivation is the behavioral equivalent of the biological concept of homeostatic regulation.

Because alterations of motivational state are typically most dramatically seen in goal-directed behavior, some behaviorists define motivational state as goal-specific. They regard the associated behaviors as "motivated" and speak of a "hunger drive" or "thirst drive," as if the motivational state were unitary and restricted in its action to only that part of the animal's behavior directed toward the goal, e.g., feeding or drinking (for review see Brown, 1961; Hebb, 1966; and Bolles, 1967).

However, as we have seen, most stimuli (or deprivations) that produce alterations in motivational state act not only on a given goal-directed behavior; they also tend to produce nonspecific (arousing) effects on a variety of response systems not necessarily directly related to the goal-directed system being examined (see Brown, 1961). The nonspecific aspects of motivational state act as an intensity modulator for both the goal-oriented behavior (sometimes called "priming"—see Weiss *et al.,* 1976) and for other, often not directly related behaviors. Thus food arousal enhances biting (a related behavior) and depresses defensive reflexes (a not directly related behavior). As a result, the nonspecific arousal function of motivational state can energize an animal to act. In addition, this function can determine or initiate the choice of alternatives, so that appropriately synergistic behaviors (increase in heart rate) are expressed during feeding whereas inappropriate or antagonistic responses (defensive escape and withdrawal reflexes) are inhibited.

Goal-Specific Effects

Goal-oriented behavior typically involves three stages: (1) orientation for the goal (food, drink); (2) a consummatory response (feeding, drinking); and (3) a phase of quiescence following achievement of the goal (satiety).

Kupfermann and co-workers have examined motivational state in *Aplysia* by studying the control of feeding (for review see Kupfermann, 1974a; Kupfermann and Weiss, 1976).[2] Feeding is not simply related to the intensity of the eliciting stimuli, but is also related to the previous history of feeding (hunger versus satiation) and recent exposure to food. For example, food-deprived animals crawl extensively and show orienting and spontaneous biting responses (Figure 9-17A). By contrast, animals fed a large meal stop crawling and fail to bite in response to food—they are said to be satiated. Rather than approach food, they will avoid it (Figure 9-17B).[3]

The major signal for satiation in *Aplysia* is food bulk (Susswein and Kupfermann, 1975a). Animals that are first fed partial meals of nonnutritional bulk (paper) require less food to become satiated than do nonfed controls. The *total* bulk consumed by animals fed to satiation first with paper and then with seaweed does not differ from the amount of bulk consumed by animals fed to satiation on only seaweed (Figure 9-18). Satiation occurs when the anterior

---

[2]For parallel studies on motivation in the blowfly, see Dethier, 1964, 1966.

[3]Animals also have a circadian rhythm in their feeding behavior. The time an animal takes to get into a feeding position after exposure to food is shorter during the light period of a daily light-darkness cycle. Once a circadian rhythm has been established, this difference persists in experimental conditions of constant darkness, showing that the rhythm *per se* is responsible for the modulation of feeding behavior—light alone does not produce the effect. Animals show a shorter latency of response during the time of day when it would normally be light than they show in the periods when it would normally be dark (Kupfermann, 1974a).

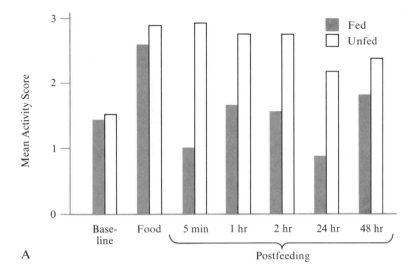

A

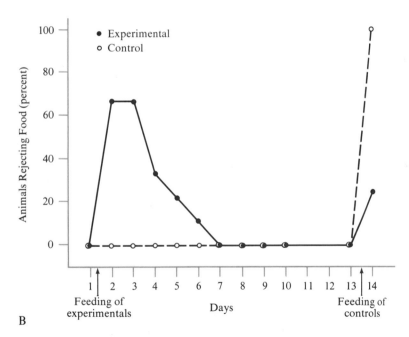

B

FIGURE 9-17.
Satiation in *Aplysia*. [After Kupfermann, 1974a.]

A. Activity before and after feeding. Each pair of bar graphs represents the median activity scores of the same group of animals before feeding (baseline) and at various intervals after the introduction of food into the water. Activity was rated from 0 to 3 based on a 30-sec period of observation. Zero equals no movement; 1 equals small head movements; 2 equals typical head-waving behavior; and 3 equals locomotion.

B. Number of animals failing to exhibit feeding responses in response to a 2-min exposure to seaweed. After testing on day 1 (zero rejection) experimental animals received a supramaximal meal of 4–8 g of seaweed; the control animals received their normal diet of 1 g. On day 2 all animals were tested using a blind procedure. On day 13 the experimental and control treatments were reversed.

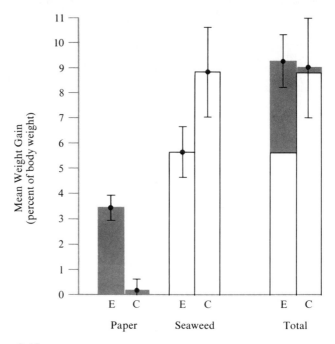

FIGURE 9-18.

Satiation in *Aplysia* is signaled by bulk. Mean weight gains (amount eaten) of experimental (E) and control (C) animals are shown. During the first part of a meal (shaded bars) experimental animals were fed paper and controls were induced to bite but were not fed. During the second part of the meal (clear bars) both groups were fed seaweed to satiation. The experimental animals ate less to reach satiation than did the controls, with the result that both the controls (which ate only seaweed) and the experimental animals (which first ate paper and then seaweed) gained the same amount of total weight in reaching satiation. The experiment was run in two sessions so that each of 10 animals would serve as both experimental and control. Vertical lines show standard error of the mean. [After Susswein and Kupfermann, 1975b.]

gut (esophagus, crop, anterior and posterior gizzard) is filled with bulk. Bulk injected in the body cavity of the animal does not lead to satiation. Presumably the action of bulk is mediated by stretch receptors in the wall of the viscera (Susswein and Kupfermann, 1975b).[4]

Susswein, Kupfermann and Weiss (1976a) have found that the amount of food ingested has a graded *inhibitory* effect upon biting, the consummatory

---

[4]A somewhat similar mechanism is operative in the blowfly, in which satiation is mediated by several types of stretch receptors (Gelperin, 1967; Dethier, 1976).

phase of feeding. As the animal is fed increasing amounts of food, the amplitude of biting decreases and the latency and the interval between repetitive responses increase progressively. Chemical stimuli from food, in turn, have a graded *excitatory* effect on all three aspects of biting (Figure 9-19). Thus, satiation in *Aplysia* seems to represent interaction between two competing

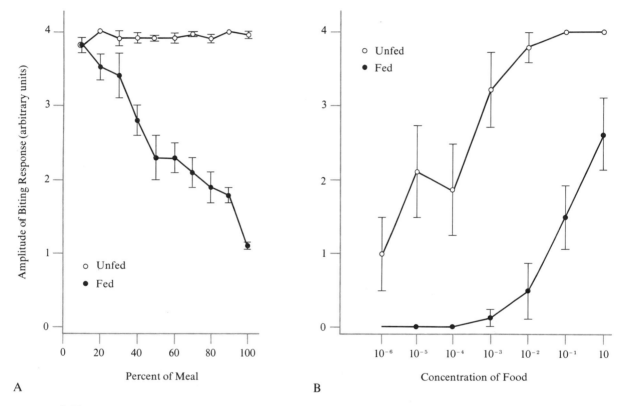

A   B

FIGURE 9-19.
Stimulus control of satiation. [After Susswein, Kupfermann, and Weiss, 1976b.]

A. Mean amplitude of the biting response rated on a scale of 1 to 4 in seven animals fed to satiation and seven yoked (unfed) control animals that were tested in a blind procedure along with the experimental animals. The total meal of each animal was divided into tenths and the mean amplitude of the biting response for each tenth was calculated. The mean amplitude and standard error for all seven animals are shown for each tenth of the meal. The amplitude can be seen to decrease as animals satiate; by contrast, no change in amplitude occurs in the unfed animals.

B. Amplitude of biting as a function of concentration of food before and after a meal. Both before and after a meal the response amplitude is a function of the nutrient concentration of the food. But after the meal the curve is shifted to the right so that higher concentrations are needed to elicit a response. Standard errors are shown.

stimuli: (1) the strength of the external stimulus (the concentration of nutrients in the seaweed extract), and (2) the strength of the internal stimulus (the consequences of the amount of food eaten on the gastro-intestinal tract). Satiation and hunger are therefore not discrete motivational states but extreme points along a motivational continuum (Susswein, Kupfermann, and Weiss, 1976b).

Motivational control of feeding has also been studied in *Pleurobranchaea* and *Navanax*, and interesting differences from *Aplysia* have been found. *Pleurobranchaea* and *Navanax* can ingest a large meal with a few swallows, so that they can be satiated within minutes, whereas *Aplysia* must eat for one hour or more before it is satiated (Davis and Mpitsos, 1971; Lee *et al.,* 1974; Kupfermann, 1974a).

Nonspecific Effects

As in vertebrates, motivational states in *Aplysia* have nonspecific as well as specific effects on the organism. These general (arousal) effects are reflected not only on goal-oriented behavior, such as feeding and sex, but also in simple reflexes. In both cases they can be studied by examining changes in the strength of a response to a constant stimulus.

Thus, as we have seen, hungry animals will show sensitization of biting produced by food stimuli. Animals satiated on one day prior to testing take up to 15 minutes before responding to food. By contrast, the biting latency of aroused animals is only six seconds. Although some aspects of enhanced biting are present even 40 minutes after exposure to food, the major components fall off quite rapidly. In fact, animals exposed to multiple successive food stimuli are rearoused every time, even if the interstimulus interval is only 10–15 seconds. Therefore, a partially satiated animal that for some reason stops eating may not start again because rearousal is more difficult to produce, even though it would have eaten more had it not stopped. A decrease in arousal may therefore contribute to termination of a meal (Susswein and Kupfermann, 1975a). Thus, the internal state—in this case inadequate bulk in the anterior gut—is not sufficient, by itself, to maintain the motivation for feeding; it is necessary to present the external stimulus, food.

As discussed above (p. 354), arousal produced by food depresses defensive siphon withdrawal, inking, and escape locomotion. Preston and Lee (1973) examined other aspects of arousal in *Aplysia* that had not been fed for 17 hours. Based upon the appearance of the animal, they defined six states of arousal: (1) *balled:* stationary animals having a round shape with contracted tentacles and neck; head buried in body mass; (2) *still:* stationary animals with neck and tentacles partially extended; (3) *alert:* animals with neck and tentacles fully extended and moving slightly; (4) *lip:* neck and tentacles fully extended, head-waving and spreading of lips; (5) *mouth:* as in the previous but with rhythmic movements of the mouth parts and occasional radula protrusion; (6) *moving:* locomotion. The first two states were regarded as nonaroused, while the other four were variations of arousal.

Preston and Lee presented a tactile stimulus to several body areas, of sufficient force to produce a slight withdrawal response of the stimulated part. They observed the responses, especially movements of the head and mouth, that followed the initial withdrawal. They found that movements of the head in response to touching the osphradium or side of the body were little affected by the degree of arousal, but the response to touching head structures was markedly affected. Whereas nonaroused animals (contracted or still states) tended to withdraw their head following the stimulus, animals in aroused states (alert, lip, mouth, and moving) tended to approach the stimulus (Figure 9-20). Dieringer *et al.* (1978) have found a gradient of heart rate values paralleling arousal. Quiescent animals had the lowest rate, active animals the highest.

RELATIONSHIPS BETWEEN SENSITIZATION, AROUSAL, AND MOTIVATION

As we have seen, sensitization can be considered a critical aspect of arousal. Since arousal is a component of motivational states it becomes essential to examine the relationship between sensitization and motivation. Clearly, some stimuli that alter motivational state resemble those used to produce sensitization (and arousal). However, whereas sensitization stimuli are typically exteroceptive, stimuli that alter arousal and motivation can be both exteroceptive and interoceptive. These conceptions of sensitization, arousal, and motivation can be brought into register by conceiving of the three processes as being mediated by a set of common neuronal systems that modulate behavioral responsiveness and are capable of being influenced by exteroceptive and interoceptive stimuli. Whereas the exteroceptive stimuli are neuronally mediated, the interoceptive stimuli are likely to be mediated by hormonal as well as neural means (Figure 9-21).

According to this scheme, motivational state includes both features of arousal: (1) the modulation of a variety of responses by interoceptive as well as

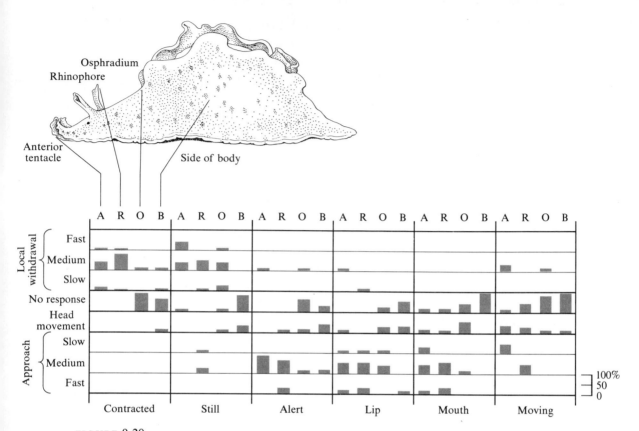

Effect of motivational state on the response to tactile stimuli. Each bar indicates the percentage of animals responding to a weak tactile stimulus applied to four different parts of the body (A = anterior tentacles; R = rhinophores; O = osphradium; and B = side of body) while the animal is in one of the six states indicated at bottom (contracted, or least aroused, to varying degrees of arousal: still, alert, lip, mouth, and moving). Two classes of response were noted: (1) local withdrawal response at the stimulated site and (2) body movement (no response, head movement, or approach). [After Preston and Lee, 1973.]

exteroceptive stimuli, i.e., sensitization, and (2) locomotion and other changes in activity level, such as orientation of the animal to the stimulus (the nonsensitizing component of arousal). Motivational state commonly also includes two additional functions. One is a goal or cue function, which is similar to the orientation aspect of arousal but is directed to the attainment of a specific physiological goal. This function leads to a temporal grouping of a set of behavioral responses into an appropriate sequence so that, for example, the appetitive response precedes consummation, which in turn precedes quiescence.

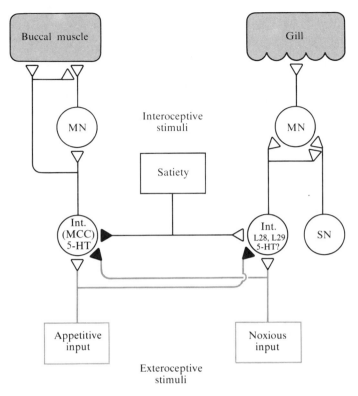

FIGURE 9-21.
A speculative neural model illustrating the various effects of sensitization, arousal, and motivational state on appetitive behavior (feeding) and defensive behavior (gill withdrawal). See text for details.

The other function controls the reinforcing value of stimuli and insures that a food stimulus will only serve as an effective reinforcer for a hungry animal. Thus, certain similarities between sensitization, arousal, and motivation may reflect the fact that arousal and motivation are in general somewhat similar constructs (motivation is essentially arousal *directed* toward a specific goal) and that in particular both constructs share a common process, sensitization.[5]

[5]An alternative view is to conceive of arousal as being primarily an energizing force without a directional component. Thus, several cups of coffee may wake one up from a drowsy state but this enhanced alertness is not goal-directed. According to this view, motivation provides the direction or specific drive to the energizing force. Thus hunger is specifically directed towards food, thirst towards drink. As this position nicely illustrates, the distinction between arousal and motivation is partly semantic, and confusion between the two terms exists primarily because of a failure to agree on the arbitrary definitions of each one.

This hypothesis is clearly speculative and requires better operational definition. Moreover, as I will argue below, until operational hypotheses are developed that can be tested on the cellular level, speculative discussions will not prove fruitful. Cellular studies of various opisthobranchs and pulmonates could however allow one to test this and other models that relate sensitization to arousal and motivation.

MOTIVATIONAL STATE AND THE HIERARCHY OF BEHAVIORS

Davis and his colleagues have attempted to rank a number of behavioral responses in *Pleurobranchaea* into a behavioral hierarchy in which feeding dominates over righting behavior and withdrawal responses of the head and oral veil (Chapter 8, p. 325). As we have seen above, motivational state (e.g., hunger or satiety) are postulated to account for alterations in the threshold of a behavior. This state can therefore also account for the hierarchical relationship between responses. Thus, feeding in *Aplysia* depresses and therefore dominates over various defensive responses—withdrawal, inking, and escape (Advokat *et al.*, 1976 and unpubl.).

Davis and co-workers (1977) first obtained these results in *Pleurobranchaea*. They satiated animals with raw squid and then compared the effects of food on righting and head withdrawal in hungry and sated animals. They found that food stimuli delay righting responses in sated as well as hungry animals, indicating that this component of the hierarchy is immune to satiation. By contrast, the dominance of feeding over head withdrawal is altered by satiation. Whereas food stimuli reduce head withdrawal in hungry animals, the dominance of feeding is abolished by satiation; food stimuli do not depress head withdrawal in sated animals.

The authors suggest that exposure to food suppresses righting—whether or not the animals eat—because the chemoreceptors of the oral veil act on the neurons controlling the righting behavior (Figure 9-22A). On the other hand, suppression of head withdrawal is contingent on feeding because the inhibitory actions of food on withdrawal neurons are indirect and may be mediated by the neurons of the feeding circuit (Figure 9-22B). When (in sated animals) the feeding neurons are not active, withdrawal is not suppressed. These several ideas have been incorporated by Davis and his colleagues into an overall neural scheme that suggests how feeding, righting, and withdrawal relate to each other (Figure 9-22C). As is evident from this diagram, behavioral choice and motivational states are alternate formulations of the same behavioral paradigm. In each case the animal's response to a given stimulus is determined

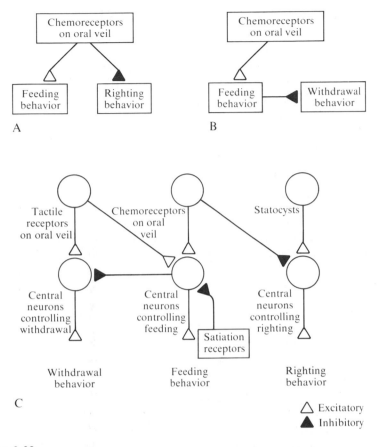

FIGURE 9-22.
Neurophysiological model of motivation and behavioral hierarchy in *Pleurobranchaea* according to Davis *et al.*, 1977.

A. Chemoreceptors that mediate food stimuli excite feeding and inhibit righting. This hypothesis is proposed because righting is delayed in sated animals by a food stimulus to the lips even though food does not usually cause feeding in sated animals.

B. Chemoreceptors excite feeding but do not directly affect withdrawal. This hypothesis is proposed because withdrawal is suppressed by application of food stimuli only if animals have fed.

C. Summary of these hypotheses in a simplified wiring diagram, with the addition of hypothetical inhibitory feedback from satiation receptors to feeding neurons to account for the elevation of feeding thresholds by satiation.

by an internal state that governs which of a family of responses will be expressed and which will be depressed. Moreover, the findings in *Pleurobranchaea* and in *Aplysia californica* are quite similar. In each case appetitive stimuli depress a variety of defensive and escape responses.

MOTIVATION, HOMEOSTASIS, AND NEUROBIOLOGICAL EXPLANATIONS

In certain cases the behavioral construct motivation may be reduced to a set of homeostatic mechanisms that can be examined on the cellular level. The need for this reduction comes from two sources. One is the existence of various alternative formulations for motivation, such as behavioral choice. The other is the fact that satiety—a key factor in the motivational state related to food can be accounted for by means of bulk in the gut and the fact that arousal can in part be accounted for by sensitization. As Dethier (1964, 1966), one of the first to extend the concept of motivational state to invertebrates, has recently (1976) argued:

> Whereas psychologists might speak of feeding behavior as characterized by drive (motivational state) leading to goal-directed behavior—which results in satiation (drive reduction) when the goal is achieved—physiologists would describe the same behavior as characterized by an internal state (hormonal, neural, metabolic, etc.) that initiates locomotor activity . . . , lowers central thresholds to specific stimuli that were until then ineffective, and releases these effects when the initiating internal states are altered.

The physiological formulation is not more elegant than the behavioral, but the former does focus attention more directly on regulatory processes that are likely to be causal in the behavioral sequence. Formulating the factors that control homeostatic behaviors, such as feeding, drinking, and temperature regulation, in descriptive biological terms facilitates the search for their neural mechanisms. With the discovery of the loci and the cellular mechanisms of the homeostatic processes for which the concept of motivational state was formulated, the concept may be reformulated so as to take on more precise meaning.

Constructs such as motivation were immensely useful to psychologists in the past because our understanding of the cellular basis of behavior was extraordinarily limited as little as 10 years ago. Thus, in 1966, Hebb, one of the most progressive and optimistic theorists, described the relationship between brain function and behavior in the following way.

> Until neurological theory is much more adequate, the psychologist has to take it with a grain of salt. But we must go further. It seems that some aspects of behavior can never be dealt with in neurological terms alone. We turn next to the necessary limitations of neurologizing, and the use of *psychological constructs:* conceptions for dealing with behavior that do not derive from anatomy or physiology.
>
> The essential point is that the simplest behavior of the whole animal involves a fantastic number of firings in individual neurons and muscle cells, as the ani-

mal moves, for example, out of the starting box in a maze, or as the student reads a line of this text. There is no possible way of keeping track of more than a few of these cells, and little prospect that it will become possible to do so in the future. To describe mental activity in such terms would be like describing a storm by listing every raindrop and every tiny movement of air.

We must have units on a larger scale for the description. To deal with the storm, the meteorologist speaks of showers or inches of rainfall (instead of counting raindrops), a moving weather system (extending over hundreds of miles) and so forth. For our problem, we can use neurological constructs such as a volley of impulses, the level of firing in the arousal system, or the occurrence of widespread summation in the cortex. But the intricacies of brain function are such that this still does not take us far enough, and we reach a point at which the use of psychological conceptions, on a still larger scale of complexity, becomes inevitable.

To discuss what goes on inside a rat's head as he runs the maze, for example, we use such terms as "hunger," "expectancy of food," "stimulus trace" and "the stimuli of the choice point." Such constructs have little direct reference to neural function. They were invented and subsequently refined in the context of studying behavior, and their use does not depend on first knowing how the brain functions. Instead, we can learn about how the brain functions *from the behavior,* beginning with these psychological constructs. . . .

When eventually we learn in detail what the neural processes are, we may still find that they are as complex and variable as the raindrops in the meteorologist's weather system, and that just as the meteorologist needs his large-scale construct of a weather system for convenience in thought as well as in communication, so we as psychologists will continue to need such [large-scale] constructs. . . .

Although the gap between moderately complex psychological processes, such as arousal and memory, and cellular explanations is still enormous in the study of vertebrates, it has been significantly reduced in the study of invertebrates. As a result, the advantages that these intervening variables once offered in interpreting behavioral data and for designing new experiments have also been reduced. The language that seems most useful now for interpreting data and designing experiments is that which most effectively focuses on the regulatory processes involved. From that point of view the descriptive language of regulatory biology ("set point," "feedback," etc.) may prove most useful.

ASSOCIATIVE LEARNING

The last intervening variable I want to consider is associative learning. This form of learning has always had a special interest for students of behavior because it includes a variety of complex forms commonly related to higher mental functions, such as insight learning and verbal learning. Clearly these more advanced forms of associative learning cannot be studied in opistho-branchs. But there has been encouraging progress in developing preparations in which lower-order types of associative learning can be investigated.

A variety of recent studies have indicated the existence of important biologi-cal constraints on learning (see Seligman and Hager, 1972). What associations an animal can learn to make is in good part determined by what is ecologically and adaptively important for it to learn. Thus comparative studies of associa-tive learning in closely related species could provide important insights into the cellular determinants of these biological constraints.

Early Attempts to Produce Classical Conditioning in Gastropods

In 1917 Thompson described her attempts to produce classical conditioning in the freshwater snail *Physa*. She paired a neutral tactile stimulus as a condi-tioned stimulus (CS) with food (which produces feeding movements) as an unconditioned stimulus (US). After repeated pairing the CS elicited feeding movements that persisted for 48 hours after pairing. As was the case for many early experiments on classical conditioning, Thompson did not run random or backward pairing controls that allow one to distinguish between positive sensi-tization (which had not yet been discovered) and associative (pairing-specific) learning.

Attempts to Produce Operant Conditioning in *Aplysia*

Lee (1969, 1970) has applied operant conditioning techniques to locomotor activity in *Aplysia*. *Aplysia* are often submerged, and Lee found that a low water level was aversive; animals preferred a high water level. He thus rein-forced animals for taking certain positions in the tank and found that animals tended to move more frequently to reinforced positions (where the water level was raised) than to nonreinforced ones (where the water level was lowered). To test for specificity, Lee removed experimental and control animals twice daily from their respective chambers for 10 seconds and found that after 24 hours of training, each of the experimental animals took less time to approach the photocell (which measured the reinforced behavior) than did the controls (Figure 9-23A).

Lee (1970) also used a chamber with photocells at different heights to record head-raising responses of the animal. Using high-water levels as a positive reinforcement, Lee trained an animal to assume, with increasing frequency,

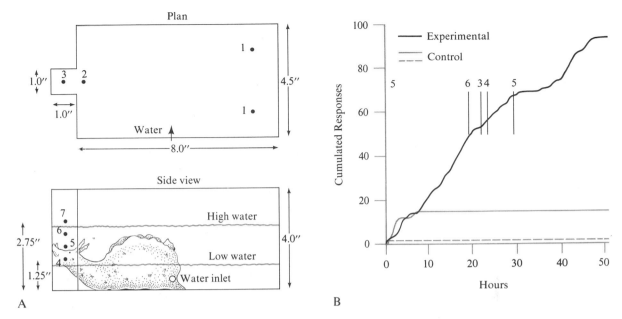

FIGURE 9-23.
Reinforcement of head level with water level in *Aplysia*. [After Lee, 1969, 1970.]

A. Plan and side view of two identical chambers used for experimental animals and yoked controls. The two chambers have common water inflow and outflow and each chamber has a vertically running tube at one end. The numbered dots indicate the locations of photocells. In the side view an outline of an *Aplysia* is shown with its head raised to the level of photocell 4 in the narrow chamber extension at left. The water level was raised by pumping water into the chamber and the water leaked out of a grid of small holes along the bottom of the side wall near photocell 1.

B. Cumulative performance of an experimental animal reinforced for assuming the head position at photocell 5 (see the side view in part A) and two control animals who were studied with the water continuously low. The vertical lines indicate short periods of reinforcement for head levels other than those at photocell 5.

certain head-holding positions only rarely shown by controls (Figure 9-22B). Lee interprets these two experiments as indicating that water level reinforcement will produce operant conditioning of locomotor activity and head position in *Aplysia*.

The use of water level as reinforcement of locomotor activity (Lee, 1969) has been questioned by Downey and Jahan-Parwar (1972) who found that the reinforcing effect of seawater levels was not due to submergence of the animal but to a decrease in temperature produced by the increase in water level. Downey and Jahan-Parwar suggest that the higher rate of response may be attributable to the (unconditioned) changes in the activity level of the animal as a result of changes in temperature produced by changes in water level.

The experiments by Lee and by Downey and Jahan-Parwar illustrate some of the difficulties of conditioning experiments, in particular the ability of certain reinforcing stimuli, such as those used by Lee, to produce unconditioned responses (body position). This makes it difficult to establish that learning occurred during training. To show that learning has occurred, it helps to demonstrate that training of the response takes less time during *retraining*, since most forms of learning show some retention. Neither Lee (1970) nor Downey and Jahan-Parwar (1972) have examined retention for this learning task. Moreover, to distinguish associative from nonassociative forms of learning, one needs to demonstrate temporal specificity between the response and the onset of reinforcement, and the evidence for this in the above experiments is also weak.

Perhaps head raising is not a particularly good operant response in *Aplysia*. Octopods—intelligent animals even by vertebrate standards—can be taught a variety of tactile and visual discriminations (see below), but they do not seem to learn well tasks dependent upon postural cues, i.e., those that require them to take their own body position into consideration (Wells, 1964, 1968). Similarly, octopods have difficulty learning mazes and can master only simple T-shaped ones. The inability to learn tasks based upon postural cues has been thought to be characteristic of most molluscs (Wells, 1964, 1968).

Sensory Interactions in *Hermissenda*

Alkon (1973, 1975a, b) has examined sensory-sensory interactions in the nudibranch *Hermissenda* by pairing light and rotation. Both light and rotation elicit behavioral responses. In response to repeated rotation *Hermissenda* rights itself and clings to the surface so as not to be shaken free. Light elicits two responses. On a daily schedule of 6.5 hours light and 17.5 hours darkness animals show a circadian locomotor rhythm; they move more in the light than in the dark. In addition, on this schedule *Hermissenda* shows a positive phototactic response to a light spot; after 10 minutes in darkness, animals exposed to a light spot will move toward the spot. However, the animals' response to light spots can be inhibited by repeated association of rotation with light (the light is turned on 45 seconds before the rotation and is maintained on throughout it). When light and rotation are paired for three hours animals are slower to move toward the light than animals receiving rotation alone, light alone, or darkness alone. The effect is not specific to a precise temporal pairing of light and rotation; it therefore resembles sensitization rather than classical conditioning. The effect only requires that the two stimuli be presented; animals exposed to unpaired light and rotation are as much inhibited as are animals receiving paired light and rotation.

The effectiveness of the pairing procedure appears to be affected by the light-dark schedule upon which animals are maintained. Whereas pairing of light and rotation inhibits positive phototaxis when animals are maintained on a schedule of 6.5 hours light and 17.5 hours darkness, this inhibition does not occur when the light-dark cycle is reversed and the animal is placed on a schedule of 18 hours light and 6 hours darkness.

To study the cellular mechanisms of this paradigm Alkon (1973, 1976) has undertaken an extensive study of the eye and statocyst and their interaction. Each of the two eyes of *Hermissenda* has only five photoreceptors: two type A cells and three type B cells. The A cells generate large spikes (45 mV), the B cells small spikes (15 mV). The type A and B photoreceptors make reciprocal inhibitory connections with one another. Each eye projects to an optic ganglion on the same side that contains 14 secondary neurons. Type B photoreceptors inhibit the second-order (optic ganglion) cells, type A do not (Alkon, 1973). Hair cells of the bilateral statocycsts respond to gravitational (mechanical) stimulation with excitation followed by inhibition. The hair cells in both statocysts are interconnected by chemical inhibitory and/or electrical synapses. The hair cells also respond to light, some cells giving "on" responses, others giving "off" responses. The excitation by light of the hair cells is thought to be caused by type A photoreceptors. In turn, in response to light the contralateral hair cells interrupt the firing of the ipsilateral type B photoreceptors. The type B photoreceptors are usually silent in the dark so that inhibition by hair cells would have little consequence for the activity in the optic ganglion in the dark. However, hair cell action on the photoreceptors (and consequently on the optic ganglion) might be important in the presence of light. Moreover, the combination of the light stimulus and the stimulus produced by change of orientation of the body could cause some hair cells to discharge that would not do so to one stimulus alone.

To test this idea Alkon (1976) developed a neural analog of the behavioral interaction in the isolated nervous system in which he paired visual and statocyst input (Alkon, 1975c, 1976). He found that this pairing produced synaptic inhibition, which reduced impulse activity in the type A photoreceptor response to light. After three pairings fewer impulses were produced to light alone. The type B photoreceptor was unaffected. This reduction in impulse frequency does not occur with repeated presentations of light or rotation alone (Figure 9-24). Moreover, unlike the behavior, it does not occur when the period of light

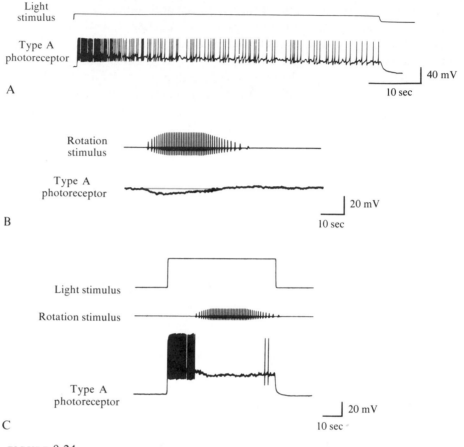

Light
stimulus

Type A
photoreceptor

A

40 mV

10 sec

Rotation
stimulus

Type A
photoreceptor

B

20 mV

10 sec

Light stimulus

Rotation stimulus

Type A
photoreceptor

C

20 mV

10 sec

FIGURE 9-24.
Responses of type A photoreceptors in the eye of *Hermissenda* to light stimulus alone **(A)**, rotation stimulus alone **(B)**, and paired light and rotation stimuli **(C)**. [From Alkon, 1976.]

is followed by a period of rotation. In addition, there was a change in the character of the impulse activity in the type A photoreceptors; the afterpotential decreased with repeated light steps and the duration of the action potential was increased following pairing of light and rotation. Both of these findings are thought to be due to increased inhibition (as a result of stimulus pairing) of type A photoreceptors by the type B photoreceptors. The time course of these changes is one to two hours and parallels roughly the behavioral changes that result from stimulus pairing. Thus the type of changes observed in the type A photoreceptors in this neuronal analog are of the appropriate sign and duration necessary to produce the behavioral change. Whether or not, or to what degree, they are causally related is however not clear.

This paradigm can also be used to study long-term memory. Thus, Crow and Alkon (1977) have found that repeated pairing of the light and rotational stimuli over several days leads to retention of the behavioral modification for at least four days.

Modification of Feeding in *Aplysia*

Both the appetitive and consummatory components of feeding are subject to training. Lickey and Berry (1966; and see Lickey, 1968) have studied the consummatory response. Somewhat different experiments on this behavior by Kupfermann and Pinsker (1968) were discussed above. Lickey (1968) presented *Aplysia vaccaria* with one of three stimuli: (1) *Food:* the lips were touched with the seaweed *Ulva,* held by blunt forceps; this produced radular grasping movements, after which the experimenter released the seaweed. (2) *Test.* the lips were touched with closed forceps that contained no food. (3) *Ambiguous stimulus:* the lips were touched simultaneously with forceps and seaweed. The seaweed was held deep in the forceps and not released, although the animal could tear the seaweed free.

A response was scored as *ingestion* if the animal grasped the stimulus with its radula; it was scored as a *rejection* if it withdrew all mouth parts from contact with the stimulus without first grasping it with the radula. Training was preceded by five food trials. The animals were divided between two training programs: (1) three ambiguous sessions each consisting of 10 *ambiguous* stimuli, i.e., forceps and food; or (2) three test sessions each consisting of 10 *test* stimuli (with the forceps only), alternating with 30 of presentation of food stimuli.

All the animals ingested the seaweed during the preliminary five food trials. Of the animals in the ambiguous training program, 41 percent rejected the ambiguous (forceps-and-food) stimuli whereas 68 percent of the animals in the test program rejected the test stimuli of forceps only. Repetition had no effect on responses to ambiguous trials; the probability of a rejection response remained the same. There was, however, a significant increase (to 95 percent) in the number of rejection responses to test trials during the three testing sessions on the first of two test days. This increase in rejection of the forceps persisted until the second test session four days later (Figure 9-25). The increase in the likelihood of forceps rejection was specific to the test procedure in which food and forceps were given in alternative trials. Lickey (1968) interprets these data as indicating the acquisition of a passive avoidance of food ingestion. However, as Lickey has pointed out, the results are equally consistent with habituation of the response to the forceps.

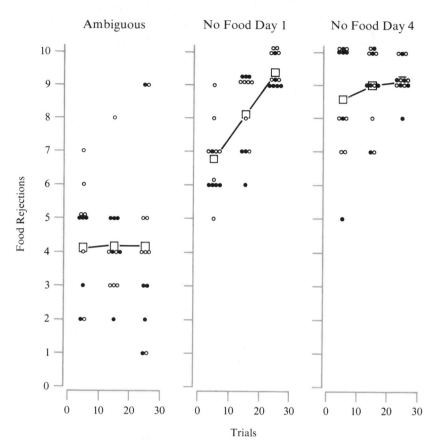

FIGURE 9-25.

Modification of feeding in *Aplysia vaccaria*. During training animals were presented with one of two types of training programs: ambiguous sessions and test sessions. In the ambiguous training program animals received three sessions, each containing 10 ambiguous stimuli (forceps plus food). In the test program each animal received three sessions of 10 test stimuli (forceps only) alternating with 30 food stimuli. The graphs plot the number of rejection responses during each block of 10 trials. (Open circles are individual scores from a group that received two test sessions followed by an ambiguous session; filled circles are individual scores from a group that received an ambiguous session followed by two test sessions; squares are the means of all subjects.) Whereas the animals receiving ambiguous stimuli did not alter their rejection rate with repeated stimulation, animals receiving test training showed an increase in their rejection rate that built up across the three sessions on the first day of training from 65 percent to 95 percent and persisted into the second day of testing. [From Lickey, 1968. Copyright 1968 by the American Psychological Association. Reprinted with permission.]

Modification of Feeding in Other Gastropods

AVOIDANCE TRAINING AND HABITUATION OF FEEDING IN NAVANAX

Feeding in *Navanax* can be modified in a variety of nonassociative ways. First, the pharyngeal retraction of *Navanax* habituates following about 15 repeated presentations of prey to the tentacles (Murray, 1971). In addition, the animal can be sensitized for many minutes following a single presentation of prey. During sensitization the animal shows an aroused or excited state in which it turns vigorously toward any stimulus applied to the head. Finally, a noxious stimulus that causes the phalliform organs of the anterolateral folds to be withdrawn inhibits the animal's search for prey (Murray, 1971).

CLASSICAL CONDITIONING OF PLEUROBRANCHAEA

Mpitsos and Davis (1973) have examined associative learning in the opistho-branch *Pleurobranchaea californica.* Naive animals withdrew from tactile stimulation of the oral veil and showed a feeding response to stimulation of the oral veil with food. Experimental animals received 20 training trials daily for seven days during which the oral veil was stroked with a glass probe (CS) coated with homogenized squid (US). With this procedure experimental animals showed a greater increase in the frequency of feeding response during training days (20 trials, rod coated with squid extract) and during extinction days (20 trials, rod alone) than did controls. The controls received 20 trials per day of the tactile stimulus (rod) alone or 20 trials of the tactile stimulus followed 3–4 hours later by food. This experimental protocol did not, however, dissociate specificity to pairing (the characteristic feature of classical conditioning) from sensitization, a nonassociative increase in the experimental groups. In particular, the control procedure of presenting the UCS alone is not comparable in intensity to that of paired CS and US. This control is essential since the controls do sensitize and the difference between sensitized and conditioned animals during extinction was only 30 percent (Mpitsos and Davis, 1973).

A more compelling example of conditioning in *Pleurobranchaea* has been provided by Mpitsos and Collins (1975). Prior to training *Pleurobranchaea* normally feeds when presented with food (squid juice) and normally withdraws from electric shock. Combining punishment training and avoidance conditioning, Mpitsos and Collins presented experimental animals with 10 conditioning trials of food paired with electric shock (intertrial interval one hour). Control animals were given alternating (unpaired) food and shock every half hour

1

4

2

5

3

6

A

FIGURE 9-26.

Avoidance learning in *Pleurobranchaea*. [From Mpitsos and Collins, 1975.]

A. Selected examples of a *Pleurobranchaea* showing avoidance learning. **(1)** A specimen of *Pleurobranchaea*, about 8 cm in length, crawling from left to right in the photograph. Structures pointing upward on either side of the head are the rhinophores; the oral veil, resembling a cowcatcher, is below and in front of the rhinophores; the mouth area lies below the oral veil. **(2)** Bite-strike response of naive animal to food: the proboscis (which extends beneath the outstretched rhinophores and oral veil) and the lifted foot are aimed toward the food. **(3)** Unconditioned withdrawal response to shocks applied by means of electrodes. **(4)** Shock-elicited withdrawal response during conditioning when food was present. **(5)** Approach-avoidance response. The animal's posture is between that for withdrawal and that for feeding: the rhinophores and oral veil are withdrawn, the foot has begun to swing away, and the proboscis is only slightly extended (the lips of the mouth eventually moved in a tiny bite). **(6)** A fully conditioned withdrawal response elicited by food (compare with 2).

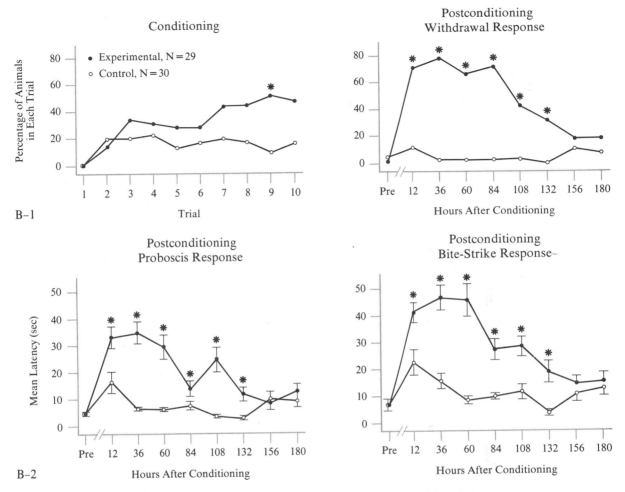

B. Pooled data from five replicate experiments. **(1)** Effects of conditioning on the withdrawal response. The asterisks indicate statistical differences between experimental and control groups significant at $p < .01$; chi-square test differences at other points are not significant ($p > .05$). **(2)** Effects of conditioning on feeding behavior. Latency of the proboscis and bite-strike responses was first measured at 12 hours after conditioning; preconditioning control values are also shown (at *Pre* on the abscissa). As in part 1, the asterisks indicate statistical differences between experimentals and controls significant at $p < .01$; differences at other points are not significant ($p > .05$, Mann-Whitney U test). Vertical bars indicate standard errors.

during the same 10-hour session. Experimental animals showed a statistically significant increase over control animals in the number of withdrawal responses to food and there was an increase in threshold and latency of feeding. The effect persisted for about six days (Figure 9-26). The greatest change occurred during extinction 12 hours after training. Since animals surgically operated on so as to expose their central ganglia still showed feeding responses, it may be possible to study associative conditioning on the cellular level in this animal.

FOOD-AVOIDANCE LEARNING IN LIMAX

Conventional classical and operant conditioning techniques may not be optimal for demonstrating associative learning in simple invertebrates. The recent behavioral literature provides many examples of biological constraints on learning. Animals of a species capable of learning specific tasks fail to learn other tasks that can be quite readily learned by another species because the stimuli involved in the task may not be biologically meaningful to the animal (see Seligman and Hager, 1972; Hinde and Stevenson-Hinde, 1973). Conversely, appropriate test situations may reveal mechanisms whereby animals of a species that may at first appear to have restricted learning capabilities (according to conventional classical and instrumental conditioning) will learn rapidly and retain what is learned for long periods. One powerful type of learning is exemplified by food-avoidance learning (bait-shyness), whereby an animal learns to associate food-related sensory cues with the negative internal consequences of ingestion (nausea, vomiting) (Garcia et al., 1972, 1974; Seligman and Hager, 1972).

This approach has been applied by Gelperin (1975) to the pulmonate slug *Limax maximus. Limax* is a polyphagous herbivore whose laboratory diet includes mushrooms, potato tubers, carrot root, and cucumber seed pods. Animals were fed potatoes once a day for seven days. On the eighth day animals were again given potatoes but followed four hours later by mushrooms. Half the animals were given toxic doses of CO_2 for five minutes as soon as they finished eating the mushrooms. To control for nonspecific effects of the CO_2, the other animals were exposed to CO_2 a full three hours after eating the mushrooms. For the next 10 days animals continued to receive potatoes followed by mushrooms, and CO_2 was given to the two groups of animals (either immediately following or three hours after eating the mushrooms). The combined feedings and CO_2 exposure depressed food intake in both groups, but animals that had been given CO_2 immediately after eating were much more depressed than animals given CO_2 three hours after eating (Figure 9-27). Food-aversion learning had a rapid onset. After one or two training trials, 40 percent of the animals avoided mushrooms for six days.

The neural circuit controlling feeding in *Limax* and another pulmonate, *Helisoma*, is being analyzed on the cellular level (see Gelperin et al., 1978; Kater and Fraser-Rowell, 1973; and see Chapter 8). This opens up the possibility for studying a potentially powerful type of associative learning.

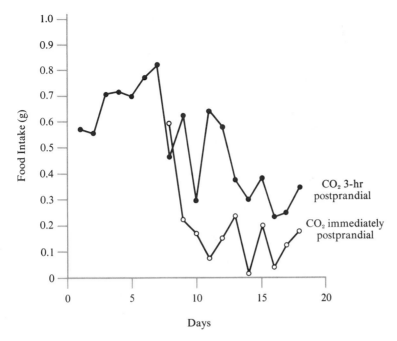

FIGURE 9-27.
Food avoidance learning in *Limax*. The graph measures the average daily intake of potato and mushroom. From the eighth day onward all animals were given a second meal of mushroom. Half the animals were then exposed to CO_2 immediately after the mushroom meal; the other half were given CO_2 three hours later. [After Gelperin, 1975. Copyright 1975 by the American Association for the Advancement of Science.]

SUMMARY AND PERSPECTIVE

Although the attempt to reduce intervening variables, such as arousal, motivation, and learning, in opisthobranchs to a cellular level of explanation has only recently begun, significant progress has already been achieved. A variety of behavioral responses have been found to undergo both short- and long-term nonassociative and associative learning. The cellular mechanisms of several types of nonassociative learning have been analyzed. Of particular interest is the converging evidence for a possible common molecular mechanism for positive sensitization based on studies in three different behavioral systems in *Aplysia:* gill withdrawal, heart rate, and biting. In all three cases sensitization seems to be mediated by serotonergic neurons that act on their target cells to enhance the response by increasing the level of cyclic AMP and perhaps thereby raising the concentration of free Ca^{++}.

A variety of arguments suggest that sensitization accounts for a significant component of arousal. Moreover, studies in *Aplysia* indicate that the sensitization component of arousal is not unitary. Each behavioral system seems to have its own positive sensitizing component and different arousing stimuli can selectively activate the various arousal systems. Thus, appetitive stimuli enhance biting and heart rate and depress defensive responses, whereas noxious stimuli depress feeding and enhance defensive responses. Since arousal is in turn a component of motivation, it now becomes possible to explain even aspects of this latter intervening variable in terms of cellular mechanisms.

In addition to the evidence for nonassociative learning in opisthobranchs, there is now encouraging and important evidence for associative learning in the opisthobranch *Pleurobranchaea* and in the gastropod *Limax*. A critical next step will be to develop behavioral systems in which both nonassociative and associative modifications occur. In such a system it will be possible to see whether classical conditioning is an associative extension of sensitization.

Even at this early juncture in the comparative study of learning in molluscs it is clear that the learning capability increases with the complexity of the nervous system and reaches its peak in the cephalopods. Cephalopods have long-term associative learning capabilities comparable to those of vertebrates (Wells, 1964).

There are clear similarities in these learning capabilities of molluscs and those of vertebrates. This similarity emphasizes an inherent conservatism in evolutionary processes that results in independent groups of organisms arriving at similar solutions to common problems of adaptation (Simpson, 1949). Thus, habituation, associative learning, and short- and long-term memory have evolved in molluscs and are well established in cephalopods. There is little question that the behavioral modifications of which octopods are capable are similar in behavioral complexity to certain forms of learning in vertebrates. The finding of remarkable learning capabilities in some classes of molluscs is reassuring for cellular biologists studying the behavior of molluscs.

Although this finding illustrates that the evolution of learning capabilities is as conservative as the evolution of other physiological functions, it is nevertheless important to be cautious about evolutionary conservatism. That phyla as different as molluscs and vertebrates have independently evolved complex brains does not mean that all molluscs have learning capabilities comparable to vertebrates or that similar mechanisms are present that produce the same result. The brains of *Aplysia* and other opisthobranch molluscs are less complex than those of the cephalopod molluscs. The learning capabilities of opisthobranchs may therefore be more restricted than those of vertebrates. However, an examination of the molluscan phylum illustrates clearly that *the relationship of brain complexity to learning is not a discontinuous process.* Learning did not evolve as an all-or-none event when a certain critical degree of neuronal complexity was reached. Varying degrees of neuronal complexity and behavioral modifiability are found in all molluscs.

SELECTED READING

Dethier, V. G. 1966. Insects and the concept of motivation. In *Nebraska Symposium on Motivation,* pp. 105–136. An early attempt to examine motivation in invertebrates.

Dethier, V. G. 1976. *The Hungry Fly.* Cambridge, Mass.: Harvard University Press. A beautifully written summary of feeding in the blowfly. Advances arguments against the use of intervening variables, such as motivation.

Kupfermann, I. 1974. Feeding behavior in *Aplysia:* A simple system for the study of motivation. *Behavioral Biology, 10:*1–26. Reviews the rationale for using *Aplysia* to study motivation.

Willows, A. O. D. 1973. Learning in gastropod molluscs. In *Invertebrate Learning*, Vol. II, edited by W. C. Corning, J. A. Dyal, and A. O. D. Willows. New York: Plenum. A good review of learning in gastropods.

CHAPTER 10

Epilog:
Cellular and Behavioral Homologies, Divergence and Speciation

Until recently the opisthobranch molluscs, in particular the genus *Aplysia*, had not been extensively used in comparative studies of behavior. However, because of the diversity among the various species of the genus (swimmers and nonswimmers, inkers and noninkers, burrowers and nonburrowers), *Aplysia* and the other opisthobranchs are likely to be very useful in the future for examining a number of problems, including the adaptive value of specific behaviors, the selective forces causing modifications of a specific behavior, the evolutionary trends among closely related groups, and the generality of neural mechanisms.

Comparative studies of behavior require the establishment of behavioral homologies between taxa and these in turn must be based on objective criteria. As I indicated in the Introduction, homology is basically a morphological concept. The more effectively one can describe behavior in morphological terms the better one can define homologous behavior. The most compelling evidence for homologous behavior would be the demonstration that responses are generated by homologous structures (Hodos, 1974), and ideally this would be done by delineating neural circuits made up of specific homologous cells. The identification of invariant nerve cells with specific behavioral functions in species affords a new and objective basis for identifying homologies between behavioral responses of closely related species.

Comparative studies of behavior, based on identified cells, can also analyze the functional significance of specific evolutionary changes. Such analyses can differentiate between adaptive and nonadaptive characteristics and are therefore more likely to evaluate correctly the selection pressures for evolutionary change than studies of structural changes alone. In the broadest sense, cellular studies of the evolution of behavior can indicate which properties of a neural network are most likely to change when behavior is modified by selection pressures. Are these changes different from those involved in ontogenetic behavioral modifications such as learning?

In this chapter I will consider, in a somewhat speculative way, how identified cells could be used to compare the functional neuroanatomy underlying the behavior of closely related species. I will first describe the use of identified cells in examining neuroanatomical homologies between the central ganglia of different opisthobranch classes, particularly the ganglia of the visceral complex. Next I will discuss the use of identified cells to establish behavioral homologies between *Aplysia* species and between *Aplysia* and closely related genera. Finally, I will review the role of behavior in speciation and suggest how this role might be examined on the cellular level. Since this last problem has not yet been studied in molluscs, I will illustrate my argument with examples taken from other invertebrates.

HOMOLOGIES BETWEEN THE
CENTRAL GANGLIA OF OPISTHOBRANCHS

Opisthobranchs are thought to have evolved from the prosobranchs (Fretter and Graham, 1962; Ghiselin, 1965). But malacologists disagree as to whether the various opisthobranch orders derive from a common mesogastropod prosobranch ancestor (an archetypal opisthobranch) or whether different orders derive from different prosobranch groups. The torsion and detorsion that characterize the phylogenetic history of prosobranch and opisthobranch molluscs is best reflected in the mantle and viscera and in the neuronal groupings that innervate them (Chapter 4). Comparative studies of the neural control of mantle organs in closely related species of mesogastropod prosobranchs and opisthobranchs could lead to further insights into the phylogeny of prosobranchs and opisthobranchs and the interrelationships (until now poorly understood) between the various opisthobranch orders.

The nervous system of primitive opisthobranchs, such as the cephalaspid *Acteon* or the primitive anaspid *Akera* (a close relative of *Aplysia*), resembles that of prosobranchs in being torted. These nervous systems of primitive opisthobranchs consist of four paired head (circumesophageal) ganglia that form a ring around the esophagus: the cerebrals, buccals, pleurals, and pedals (Figure 10-1). The cerebropleural ganglia give rise to the visceral (abdominal)

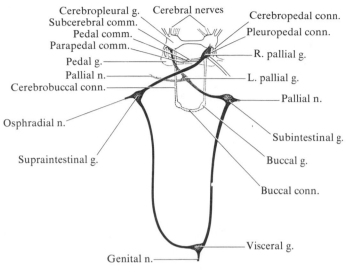

Primitive Cephalaspid
(*Acteon*)

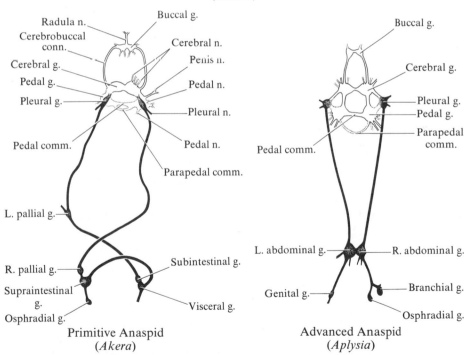

Primitive Anaspid
(*Akera*)

Advanced Anaspid
(*Aplysia*)

FIGURE 10-1.
Comparison of the nervous systems of a primitive cepahalaspid and examples of primitive and advanced anaspids. [*Acteon* after Bouvier, 1891; *Akera* after Hoffmann, 1939; *Aplysia* from Kandel, 1976.]

nerve cords (or connectives) that unite posteriorly. Along the torted course of the visceral nerve cords lie five ganglia: the two pallial ganglia, the left and right intestinal ganglia (supra- and subintestinal, respectively) and the single visceral ganglion.

Detorsion does not involve the head ganglia (see Chapters 1 and 4). In the various *Aplysia* species the four paired head ganglia are essentially in the same position as in *Akera* (and *Acteon*), and each ganglion innervates roughly the same region in the two groups. The cerebrals innervate the tentacles, mouth, eyes, rhinophores, and penis. The buccals innervate the buccal musculature, pharynx, salivary gland, esophagus, esophageal crop, and gizzard. The pedals innervate the head, foot, and parapodia. The pleurals give rise to the visceral nerve cords, called the pleuroabdominal connectives in *Aplysia*.

The major difference between the more primitive and torted anaspid *Akera* and the more advanced and detorted *Aplysia* is in the ganglia of the abdominal nerve cords (Figure 10-1). As in *Akera,* this complex in *Aplysia* supplies the mantle and its glands, the gill, osphradium, genital organs, kidney, and parts of the digestive apparatus. However, in *Akera* there are five ganglia (two pallials, supra- and subintestinals, and visceral) lying along the torted visceral (abdominal) cords, whereas in *Aplysia* there is only one large, asymmetrical, fused abdominal ganglion. This ganglion is considerably larger than the visceral ganglion of *Akera* and is thought to represent a fusion of several of the ganglia that lie along the visceral cords of more primitive opisthobranchs. According to Guiart (1901), the right hemiganglion is a fusion of the supraintestinal and the right pallial ganglia, while the left hemiganglion is a fusion of the subintestinal, the left pallial, and the visceral ganglia. Guiart supported his hypothesis with the finding that the left hemiganglion in *A. punctata* has two and occasionally three distinct divisions. He thought that the right hemiganglion had no divisions because the right pallial and supraintestinal ganglia were tightly fused.

As we have seen, in Chapter 4, an alternative view of the evolution of the abdominal ganglion has been presented by Hughes and Tauc (1963), who suggested that the left pallial ganglion (which contains LP1) fused with the left pleural, and the right pallial ganglion (which contains R2) fused with the supraintestinal ganglion. But these postulated homologies are based entirely on gross anatomical comparisons. Detailed embryological and comparative studies are needed in order to provide more convincing evidence for homologies.

In Chapters 4 and 7 I considered the embryological studies of Kriegstein (1977a) that partially support the Hughes and Tauc hypothesis. However, Kriegstein found that the abdominal ganglion is an embryological fusion of three separate ganglia (the subintestinal, visceral, and supraintestinal) and that the left pleural ganglion develops as a single ganglion. This would suggest that here ontogeny does not simply recapitulate phylogeny, but abbreviates and condenses steps of phylogenetic development. The left pleural and the supraintestinal ganglia seem to develop with the pallial homolog already fused to them. Comparative embryological studies could strengthen this hypothesis by showing that the abdominal ganglion is homologous to only four of the five abdominal ganglia of primitive opisthobranchs. This hypothesis would predict that in some groups there will be clear evidence that the fifth ganglion, the left pallial, originates independently and later fuses with the left pleural.

Comparative studies require markers that are characteristic for a given ganglion (or, in the case of ganglionic complexes, for different regions) and that can serve to distinguish homologous ganglia (or regions) in closely related species. Identified cells can serve as such markers. Thus, all the cells that can be identified in the abdominal ganglion of *A. californica* can be found in every member of the species. Most of these cells are also found in different species of the subgenus and in at least three different subgenera. For example, the abdominal ganglia of *A. (Varria) dactylomela* (Castellucci and Kandel, unpubl.), *A. (Varria) brasiliana* (Blankenship and Coggeshall, 1976), and *A. (Aplysia) depilans* (Kandel, unpubl.) contain most of the identified cells characteristic of *A. (Neoaplysia) californica* (Figure 10-2). It appears likely that many common identified cells will be found in every member of the genus *Aplysia*. It is important that the abdominal ganglia of the other Aplysiinae genus, *Siphonata*, and other Aplysiidae subfamilies (Dolabellinae, Dolabriferinae, and Notarchinae) also be mapped in order to see which of the identified cells are maintained throughout Aplysiidae, and which of these cells are even more widely distributed and occur in Akeridae, the other family of the order Anaspidea.

The family Akeridae forms an important transition group between the orders Anaspidea and Cephalaspidea. There are in fact a number of similarities between the torted nervous system of Akeridae, the most primitive anaspid family, and that of Acteonidae, the most primitive cephalaspid family. The acteonids are thought to be the most primitive of all living opisthobranchs and show the greatest evidence of a prosobranch (specifically, mesogastropod) ancestry (Fretter and Graham, 1962; Brace, 1977b). Although present-day acteonids are a specialized branch and clearly not ancestral to other groups of opisthobranchs, it is attractive to think that the ancestors of the acteonids are the common ancestors of the cephalaspids and anaspids and perhaps of all opisthobranch orders. A precise (cellular) determination of homologies between the ganglia of the nervous system of Akeridae and Acteonidae could be useful in exploring the phylogenetic relationships and origins of cephalaspid and anaspid opisthobranchs. By constructing a phylogenetic tree based on identified

392

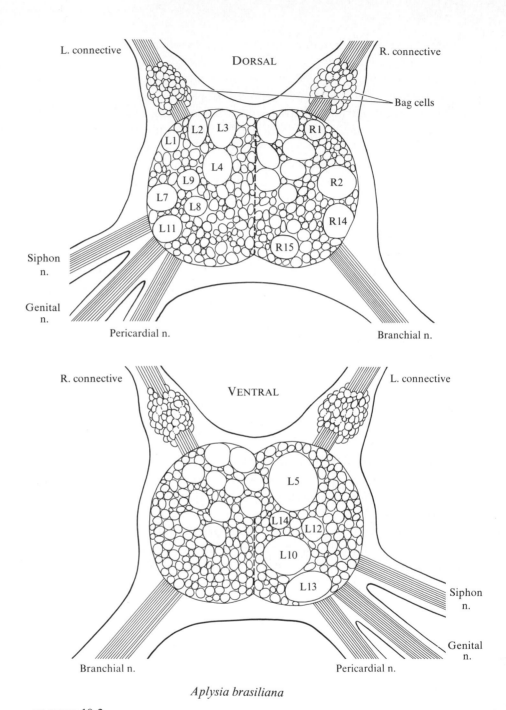

Aplysia brasiliana

FIGURE 10-2.
Cell maps of the abdominal ganglia of *A. brasiliana* [after Blankenship and Coggeshall, 1976] and *A. californica* [from Kandel, 1976] illustrate the extent of cellular homology in two closely related species.

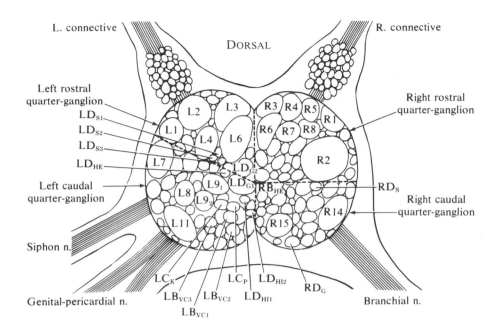

L. connective
DORSAL
R. connective

Left rostral
quarter-ganglion
LD_{S1}
LD_{S2}
LD_{S3}
LD_{HE}

Right rostral
quarter-ganglion

L2 L3 R3 R4 R5 R1
L1 L4 L6 R6 R7 R8
L7 R2
A LD_{G2}
Left caudal
quarter-ganglion
LD_{G1} RB_{HE} RD_S
L8 L9₁ L9₂
L11 R15 R14
Right caudal
quarter-ganglion

Siphon n.

LC_K LC_P LD_{HI2}
Genital-pericardial n. LB_{VC3} LB_{VC2} LD_{HI1} RD_G Branchial n.
LB_{VC1}

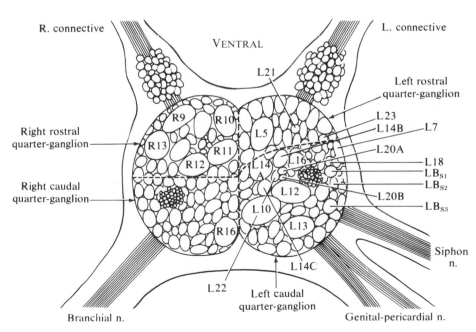

R. connective
VENTRAL
L. connective

L21
Left rostral
quarter-ganglion

L23
L14B L7
Right rostral
quarter-ganglion
R9 R10
R13 L5
R11
L20A
R12 L14 L16 L18
A LB_{S1}
LB_{S2}
L12 L20B
Right caudal
quarter-ganglion
L10 LB_{S3}
R16 L13 Siphon
n.

L14C
Branchial n. L22 Left caudal
quarter-ganglion Genital-pericardial n.

Aplysia californica

cells, one would find which identified cells (and neural circuits) are archetypal, i.e., those that belong to all (or at least most) families of anaspids and cephalaspids. This approach would provide a new perspective on molluscan phylogeny. Moreover, phylogenetic studies based on cellular homologies would allow one to relate evolutionary changes in morphology to adaptive changes in behavioral capabilities.

In principle at least, the insights that studies of identified cells could yield are not limited to one phylum. In certain advantageous comparisons identified cells might be used to explore selected problems in invertebrate phylogeny. Perhaps they could even be used to examine the relationship between the annelid worms and molluscs to their presumed common flatworm ancestors (Chapter 1).

BEHAVIORAL HOMOLOGIES BETWEEN CLOSELY RELATED OPISTHOBRANCHS

Following the work of the early ethologists Whitmann (1899) and Heinroth (1911), there have been efforts to trace homologous behavior back to a common ancestor, much as comparative anatomists had done for homologous structures. In some closely related species the morphological resemblance is so great that the species could not be distinguished until differences in their behavior were discovered. For example, Mayr (1942) found that one could distinguish more effectively between morphologically similar species of bank swallows (*Riparia riparia*) and barn swallows (*Hirundo rustica*) on the basis of their nest-building behavior than on the basis of morphological differences. Andriaanse (1947) used nest-building differences to discover a new species of digger wasp.

The usefulness of behavior as a taxonomic tool is not surprising. Behavior is not merely an additional taxonomic criterion; it is a qualitatively different one. Whereas morphology is static, behavior is kinetic and functional—behavioral analyses are based on examining an effector system as it operates adaptively in space and time. A functional and kinetic behavioral analysis can reveal features about a system that cannot be inferred from a static morphological analysis.

Despite the potential power of a behavioral analysis, the comparison *between* behavioral responses often is hampered by interpretive difficulties. With some exceptions (Figure 10-5), behavior is usually more difficult to treat quantitatively than is structure because it is transient and variable (Atz, 1970). These difficulties might now be partially overcome, and the utility of behavioral analysis expanded, by using the neural structure underlying and causing behavior as an additional taxonomic characteristic (see also Hodos and Campbell, 1969).

It should soon be possible to examine the complete neural circuitry underlying homologous behaviors in closely related species. In *Aplysia brasiliana,* for example, Blankenship and Coggeshall (1976) have compared the connections made by cell L10, a dual-action command cell in the heart circuit, to those made by this cell in *A. californica* (Koester *et al.,* 1974). Similarly, Carew (unpubl. observ.) has shown that a homologous cell (L14A) contributes importantly to inking in *A. brasiliana* and in *A. californica* and that the cells in the two species have similar properties.

The phylogenetic range for which identified cell markers can be used is surprisingly broad, extending in some cases to an entire subclass. For example, the dorsal metacerebral cells of the cerebral ganglia can be found in at least two different orders of opisthobranchs, Anaspidea and Notaspidea, and in two different gastropod subclasses, Opisthobranchia and Pulmonata. The metacerebral cells of the opisthobranch *Aplysia californica* are quite similar to those of its remote pulmonate relatives *Helix aspersa* and *Helix pomatia.* The similarity is found in various features, including axonal pathway (Figure 10-3), resting potential of the cell, biophysical properties (anomalous rectification), synaptic output and input, the pattern of firing of interneurons that converge upon them, and transmitter biochemistry (Kunze, 1918; Weiss and Kupfermann, 1976; Eisenstadt *et al.,* 1973; Weinreich *et al.,* 1973; Kandel and Tauc, 1965a; Cottrell and Osborn, 1970; Sakharov and Zs.-Nagy, 1968; Senseman and Gelperin, 1974; Gillette and Davis, 1977). In a few of the opisthobranch and pulmonate species examined, the metacerebral cells serve as modulatory cells controlling feeding behavior (Weiss and Kupfermann, 1976; Kater, unpubl. observ.; Davis, Siegler, and Mpitsos, 1973; Kovac and Davis, 1977). A detailed reconstruction of the neural circuit controlling feeding in primitive opisthobranch and pulmonate species might clarify the common origin of the untorted gastropods.[1]

Dorsett (1974) has studied neuronal homologies in the control of branchial tuft retractor motor cells in two species of *Tritonia: T. hombergi* and *T. dio-*

---

[1]There are however some interesting differences between the properties and connections of the metacerebral cells even among opisthobranchs. For example, in the opisthobranch *Phestilla* (*Acoela, Aeolidiiae*) the metacerebral cells are electrically interconnected and also innervate and move the lips (Willows, personal communication).

396

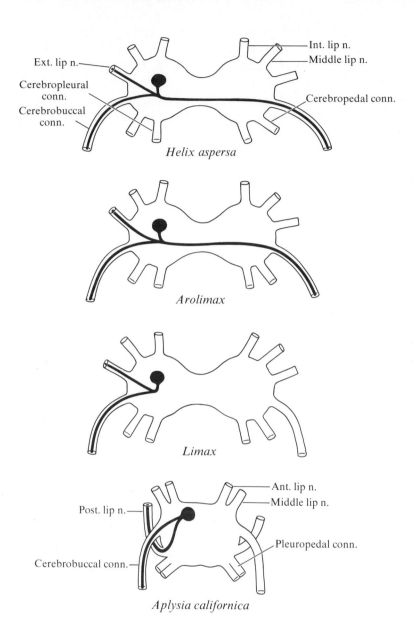

Ext. lip n.

Cerebropleural
conn.
Cerebrobuccal
conn.

Int. lip n.
Middle lip n.

Cerebropedal conn.

Helix aspersa

Arolimax

Limax

Post. lip n.

Cerebrobuccal conn.

Ant. lip n.
Middle lip n.

Pleuropedal conn.

Aplysia californica

FIGURE 10-3.
Comparison of the axonal pathways of the metacerebral cells in species of pulmonates (*Helix, Arolimax,* and *Limax*) and opisthobranchs (*Aplysia*). [After Kandel and Tauc, 1965a; Senseman and Gelperin, 1974; Weiss and Kupfermann, 1974.]

media. In both species two homologous motor cells in each pedal ganglion produce comparable ipsilateral contraction of the branchial tufts. By contrast there are differences in the cellular control of the bilaterally controlled movements in the two species. In *T. diomedia* only one pair of symmetrical cells (the giant pleural cell PL-1) is known to cause reliable bilateral contraction. By contrast, in *T. hombergi* PL-1 causes retraction of the rhinophore and contraction of the longitudinal muscle that causes the body to twist to the ipsilateral side. It would clearly be of interest to extend this analysis to a variety of *Tritonia* species to see where this divergence in cellular function occurs and what its behavioral consequences are.

Identified cells can also be used to study ganglionic homologies in other invertebrate phyla. In a systematic study of homologies Lent (1973 and in press) has examined the Retzius cells (symmetrical identified serotonergic cells that control mucus secretion) in four widely different species of leeches (*Haemoptis marmorata, Hirudo medicinalis, Macrobdella decora, Placobdella parasitica*) and found that in all four species the cells have very similar properties. They are always electrically coupled to each other and have similar branching patterns (Figure 10-4). Similarly, seven pairs of identified mechanoreceptor cells responding to touch, pressure, or noxious stimuli, two types of motor neurons involved in separate behaviors, and one interneuron had similar properties in the species examined. All told, Lent examined 30 homologous cells and found no differences in any cellular property: morphology, axonal branching pattern, membrane potential, connections, and behavioral function. As Lent points out, this high degree of evolutionary conservatism on the cellular level raises an interesting question that parallels the one raised by the studies on *Tritonia.* Different species of leeches have different types of behavior, different prey, different sign stimuli for mating, etc. At what level of neuronal organization do these differences arise? Lent suggests that perhaps the identified cells studied so far represent those that code for aspects of behavior common to all leeches. Species differences may arise because of the differences between other neurons that are less obvious and therefore not yet identified (as seems to be the case in the example of *Tritonia* considered above). On the other hand, species differences might arise by subtle alterations in the patterns of connectivity between identified cells.

The ability to relate behavior to a more fundamental, functional anatomical level (neural circuitry) is crudely analogous to relating phenotypic expression to its genotype. The analysis of the relationships of phenotype to genotype was greatly advanced when it became apparent that the amino acid sequences of proteins represented the primary "print" or direct translation of the correspond-

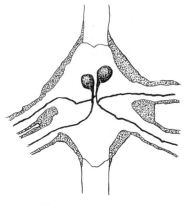

Haemopis marmorata

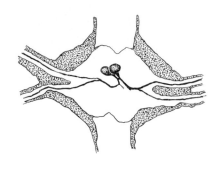

Hirudo medicinalis

250 μm

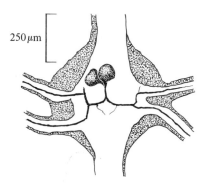

Macrobdella decora

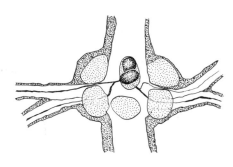

Placobdella parasitica

FIGURE 10-4.
Ventral views of segmental ganglia of the leeches *Haemopis, Hirudo, Macrobdella,* and *Placobdella* showing the orientation and geometry of their Retzius cells. The heavily stippled regions represent the volume between the neural sheath and neuropil, which is usually filled by packet cells and neuronal somata. The Retzius cell diameter of *Haemopis* is 65 μm, that of *Hirudo* 53 μm, *Macrobdella* 61 μm, and *Placobdella* 59 μm. The light stipple represents the glial packet cells, which are more spherical and compact than in other species. In four preparations examined no fine posterior branch was seen. [After Lent, 1973.]

ing stretches of genetic information in the DNA. Thus, comparable gene loci in different species could be examined by comparing the amino acid composition of a given protein in the different species (Anfinsen, 1973; Margoliash, 1972). As Francis Crick (1958) first pointed out:

> Biologists should realize that before long we shall have a subject which might be called "protein taxonomy," the study of the amino acid sequences of proteins of an organism and the comparison of them between species. It can be argued that these sequences are the most delicate expression possible of the phenotype of an organism and vast amounts of evolutionary information may be hidden away within them.

The relationship of behavior to its neural circuitry is by no means as powerful, or as direct, as that between an amino acid sequence and its corresponding DNA sequence. Nonetheless, by analyzing the neural circuits of homologous behaviors in different species it should be possible to determine whether the behaviors are mediated by a common basic circuit with secondary modifications, much as dispensable and indispensable amino acid sequences exist in homologous proteins. For example, some positions in the amino acid sequence of cythochrome c are invariant, other positions accept a few alternatives and remain otherwise unvaried over extensive taxonomic groupings of species, and yet others are commonly changed (Margoliash, 1972). The invariant component of the neural circuit can provide clues to the minimal structure necessary for the satisfactory execution of a particular action. The variations might give clues to successful variants that have evolved over time.

Comparing the products of single genes has provided important information about the evolution of particular organisms by allowing the construction of a phylogenetic tree for the gene product and therefore for the gene. By applying this strategy to neural circuits of homologous behaviors it might be possible to document the types and sequence of mutational events in a given neural circuit and thereby explain variations in homologous neural circuits in related organisms.

THE ADAPTIVE FUNCTION OF BEHAVIOR

In addition to clarifying the evolution of behavior, comparative studies of close relatives can clarify the adaptive function of a behavior, particularly when that function is not obvious. For example, *Aplysia californica* inks when seriously disturbed; by contrast, *Aplysia vaccaria,* which coexists with *A. cal-*

ifornica, has a reduced ink gland and secretes little ink. The adaptive function of inking is not known; presumably it does not simply serve to protect the animal from predators by camouflage because the ink does not form a very satisfactory screen. A comparative study of the two species could explain the function of inking in *A. californica.* A comparative analysis of the neural circuit controlling inking could also delineate differences in the cellular organization of the nervous system of two closely related animals. Inking in *A. californica* is controlled by three electrotonically coupled motor cells (Carew and Kandel, 1977a). Are these present in *A. vaccaria*? If so, are they coupled to one another?

THE ROLE OF BEHAVIOR IN SPECIATION

As is implied by their use as taxonomic characters, fixed-action patterns often differ among close relatives. Thus closely related species of birds emit slightly different mating songs and different species of crickets have different calling songs (Figure 10-5).

As Dobzhansky (1941) and Mayr (1942) first emphasized, subtle differences in behavior may play a role in the origin of a new species (speciation). Hybridization is disadvantageous for survival, and even closely related species usually do not interbreed in nature. Hybrids are often sterile or lack the full fertility of parent species. Even if they are fertile, hybrids rarely adapt as successfully to their environment as do the parent species. Each of the parents inherits a set of genes selected over many generations for survival in a particular ecological niche. The hybrid thus inherits a "compromise" of two sets of genes that usually does not equip it to survive as well in either ecological niche. As a result, there is a powerful selective advantage for mating with a member of the same species and for not mating with members of different species. Separation of two species, particularly those that coexist in the same terrain, requires that they be reproductively isolated. Yet close relatives tend to have very similar (often identical) sexual organs and mating methods. In fact, many species that readily hybridize in the laboratory will not mate in nature, despite the opportunity to do so. What mechanisms serve to reproductively isolate coexisting close relatives so as to assure the continued segregation of the species? The most important mechanisms in restricting random mating of close relatives are ethological, and usually take one of two forms: (1) exclusive food and habitat selection (potential mates do not meet) and (2) exclusive mate selection (potential mates meet but do not mate) (for discussion see Mayr, 1963).

Mate selection involves distinguishing between features during one or several steps of courtship behavior (calling, display, etc.). For example, two closely

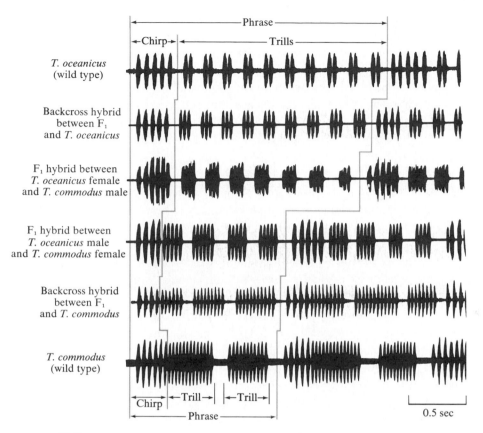

FIGURE 10-5.

Quantitative comparison of the song patterns of hybrids of the cricket *Teleogryllus*. The pattern shifts systematically in proportion to the ratios of the different wild-type genes inherited by the individual. The records of the hybrids are shown between those of two species (top and bottom). The records are aligned so that a complete phrase consists of a chirp followed by two or more trills. The phrase of *T. oceanicus* (top) is not only much longer than the phrase of *T. commodus* (bottom) but also distinctively different. For example, in each of the last three patterns, including that of *T. commodus,* the first trill is fused to the chirp. The patterns of the F_1 (first generation) hybrids of the two species are shown in the two middle traces. Again, they are quite different depending on which species served as the male parent and which as the female. The second and fifth traces were produced by backcrosses between the two different F_1 hybrids and members of the parent species. [From Bentley and Hoy, 1974.]

related and similar appearing species of grasshoppers (*Chorthippus bigluttus* and *C. brunneus*) coexist within the same territory, but the males of each species emit different calling and courtship songs. The female is attracted only to the call of the male of her species so that hybridization does not occur in nature. Hybrids can be established experimentally by deceiving the female of one species and attracting her to an alien male by means of tape recorded species-specific calling and courtship songs (Perdeck, 1958).

The importance of behavioral differences in maintaining sexual isolation in courtship behavior is supported by three lines of evidence (Mayr, 1963; Manning, 1965). First, the divergence in characteristics between the species is often most marked in calling songs and courtship behavior (or in pair-formation in birds). Second, behavioral divergence tends to be enhanced if the two species coexist and tends to diminish and disappear in species that become geographically isolated from one another. Third, the divergence is more marked in males than in females, indicating a selection for female preference. This is because the selective advantage of choosing the proper mate is greater for the female than for the male. Males of most species mate several times whereas females of some species may mate only once in a lifetime.

Food and habitat selection also lead to reproductive isolation, as well as reduce the competition for common food. Thus, whereas close relatives tend to occupy different ecological niches when they coexist, one of the species will occupy both types of habitats in areas where the relative is absent. When close relatives do occupy the same habitat they differ in their feeding behavior. Thus, behavior not only evolves from but plays a determining role in evolutionary trends.

It is commonly thought that one set of mechanisms for achieving speciation begins with physical isolation of one strain from the parent species by a geographical barrier (Dobzhansky, 1941; Mayr, 1942, 1963). Isolation reduces gene flow between the two groups and allows them to differentiate from each other. Once the two groups are isolated, both the selection and the genetic drift for each group will be different; this divergence in turn often leads to behavioral isolating mechanisms. Precursor behavioral isolating mechanisms are commonly found in isolated races of the same species (Mayr, 1963). When the two populations again coexist, the selection for behavioral mechanisms will be further increased and hybrid mating will become disadvantageous; sexual isolating mechanisms then pose a behavioral barrier that can substitute for the earlier geographic barrier. According to this view, behavioral isolating mechanisms arise as a secondary (often accidental) by-product of differences in selection pressures due to physical isolation. However, there is now compelling evidence in grasshoppers that speciation in the absence of physical isolation can come about through behavioral isolation resulting from changes in mating-song patterns (Perdeck, 1958). These findings raise the interesting possibility that in certain cases behavioral divergence, which usually is a later step in speciation, may itself initiate speciation (for discussion see Hinde, 1970).

The powerful role of behavior in determining reproductive isolation is illustrated in the mating song of male crickets. Crickets are classified into 3,000 species, in part on the basis of widely divergent song patterns (Alexander, 1962a, b). Despite this great divergence the transmitter and receiver of a song are matched so that the receiver, the female, responds only to the call of the male of her species. Findings by Alexander (1962a) and Bentley and Hoy (1972) suggest that song *reception* is as much genetically determined as is song emission. Indeed, the brain of the female cricket has at least part of the neural circuitry for sound production found in the male (Huber, 1962). The work of Alexander and of Hoy suggests that the same genes may be responsible for both song production by the male and song reception by the female of a species. This mechanism would insure assembly of appropriate genes on adjacent segments of homologous chromosomes for a synchronous evolution of both the transmitter and the receiver. If this is the case, a mutation in this gene assembly could readily initiate the behavioral step necessary for speciation.

THE CELLULAR BASIS OF SPECIATION

Since behavior plays a role in speciation and can be examined in terms of its cellular architecture, it should be possible, in properly selected cases, to study the cellular mechanisms of speciation. One is closest to doing so in insects such as the fruit fly *Drosophila* (see Manning, 1965) or certain species of crickets (Bentley and Hoy, 1974). Thus, the mating song of the male cricket of the species *T. oceanus* has two pulses per trill and that of *T. commodus* 14 pulses per trill (Figure 10-6). Hybrids of the two strains that have been backcrossed with a parental strain have between three and nine pulses per trill. The number of trills shifts systematically in proportion to the ratio of different type genes inherited by the male.

The trills are produced by opener and closer motor neurons that move the wings. The detailed neural circuit for singing has not been worked out, but preliminary work suggests that at least some of the synergistic cells within a group make excitatory connections with one another. In addition, the wing-openers make inhibitory connections to the wing-closer motor neurons. The rebound excitation following inhibition by the wing-closers apparently accounts for the precise spacing of impulse in the two groups of motor cells. The central program for singing is thought to be mediated by command elements. Stimulation of a single cell can release the entire song pattern (Bentley, 1973; Bentley and Hoy, 1974).

Bentley has examined an identifiable homologous motor cell in the two wild types, the F_1 cross, and the F_2 backcross. Although this cell represents only one member of the two populations of motor cells driving the wing, the firing pattern of this motor cell reflects the calling song. Typically, the motor cell fires once for each trill. Thus, the motor cell of the wild type *T. oceanus* fires twice corresponding to two trills, whereas the motor cells of the backcross between the F_1 and *T. oceanicus* fire three times corresponding to three trills. Thus, differences in the output activity of motor cells—in this case the difference between two and three impulses in each motor cell—contribute to distinctions in the calling song that can lead to reproductive isolation and speciation. Bentley has suggested that the difference in firing pattern of the motor cells could be due to slight differences in the two central programs caused by differences in the rate of firing of the command cell or by an increased effectiveness of the synapses made by the command cell in the motor neurons. Alternatively, the critical change could reside in the threshold of the motor cell.

Related species often have different behavioral thresholds for the production of homologous responses to the same stimulus (Manning, 1965). In extreme cases certain behavior patterns of the parent are normally not apparent in the hybrid but become manifest only under stressful or otherwise abnormal circumstances (Manning, 1965). Presumably, in the hybrid the neural mechanism for the behavior is intact, but the threshold for triggering it is dramatically elevated. Inking in *Aplysia* again provides an interesting opportunity for examining this question. Why is this behavior present and well developed in some species and poorly developed or absent in others? Is its absence related to a progressive increase in threshold? Does its disappearance have a role in establishing isolation between two coexisting species, such as *A. californica* and *A. vaccaria?*

In the near future it should be possible to move beyond inking and comparatively simple behavior to analyze in behavioral and cellular terms the mating signals between conspecifics. What are the behavioral signs (or pheromones) used by one animal to release mating behavior in another? Given that each *Aplysia* is hermaphroditic, what determines which animal will be male and which female? How does the neural circuitry for the signals of one species differ from that of another coexisting species?

As Mayr (1974) has pointed out, shifts in locomotor and feeding patterns are important for a variety of macroevolutionary changes, such as adaptive radiation and development of evolutionary novelties. Structures that evolve under new selection pressures in turn allow new behavioral responses to develop. Many new orders owe their origin to invasion of a new ecological niche. Behaviors that are useful for exploiting natural resources, such as feeding and locomotion, tend to be flexible so as to adjust to changes in environment, particularly expansion of the ecological niche or the opening of a new niche. As we have seen (Chapter 8) feeding and locomotion are now being extensively explored in various opisthobranchs. They are not, however, as yet being examined in species specifically selected for their evolutionary relationships.

It would be interesting to determine how the changes in behavior that lead to speciation are initiated by the nervous system. This problem will be most readily solved in genetically tractable animals such as *Drosophila,* but one might be able to gain some insights into some aspects of the problem in opisthobranch molluscs. Identified cells might now be used to examine homologies between the nervous systems of closely related species, homologous behaviors of closely related species, and the cellular mechanisms underlying ethological isolating mechanisms of species. The speciation question—how one species divides into two—one of the central and most intriguing problems of evolutionary biology, might now be examined with a new approach.

SELECTED READING

Campbell, C. B. G., and W. Hodos, 1970. The concept of homology and the evolution of the nervous system. *Brain, Behavior and Evolution,* 3:353–367. Discusses the several usages of the brain homology.

Manning, A. 1965. *Drosophila* and the evolution of behavior. *Viewpoints in Biology,* 4:125–169. Considers the role of behavioral threshold in divergence.

Mayr, E. 1963. *Animal Species and Evolution.* Harvard University Press. An authoritative treatment of the species concept and of the role of behavior in sexual isolation.

Mayr, E. 1974. Behavior programs and evolutionary strategies. *American Scientist,* 62:650–659. A simple introduction to evolutionary biological aspects of behavior.

References

Abbott, R. T.
   1958 The marine mollusks of Grand Cayman Island, British West Indies. *Monogr. Acad. Nat. Sci. Phila.*, No. 11.

Abraham, F. D., and A. O. D. Willows.
   1971 Plasticity of a fixed action pattern in the sea slug *Tritonia diomedia. Commun. Behav. Biol. Part A,***6:**271–280.

Advokat, C., T. J. Carew, and E. R. Kandel.
   1976 Modulation of a simple reflex in *Aplysia californica* by arousal with food stimuli. *Neuroscience Abstracts,* **2**:445.

Alexander, R. D.
   1962a Evolutionary change in cricket acoustical communication. *Evolution,* **16**:443–467.
   1962b The role of behavioral study in cricket classifications. *Syst. Zool.,* **11**:53–72.

Alkon, D. L.
   1973 Neural organization of a molluscan visual system. *J. Gen. Physiol.,* **61**:444–461.
   1974 Associative training of *Hermissenda. J. Gen. Physiol.,* **64**:70–84.
   1975a Neural correlates of associative training in *Hermissenda. J. Gen. Physiol.,* **65**:46–56.
   1975b A dual synaptic effect on hair cells in *Hermissenda. J. Gen. Physiol.,* **65**:385–397.
   1975c Responses of hair cells to statocyst rotation. *J. Gen. Physiol.,* **66**:507–530.
   1976 Signal transformation with pairing of sensory stimuli. *J. Gen. Physiol.,* **67**:197–211.

Alkon, D. L., and A. Bak
   1973 Hair cell generator potentials. *J. Gen. Physiol.,* **61**:619–637.

Anderson, J. A.
   1967 Patterns of response of neurons in the cerebral ganglion of *Aplysia californica. Exp. Neurol.,* **19**:65–77.

Andrew, R. J.
   1974 Arousal and the causation of behaviour. *Behaviour,* **51**:135–165.

Andrews, E., and C. Little.
   1971 Ultrafiltration in the gastropod heart. *Nature (London),* **234**:411–412.

Andriaanse, A.
   1947 *Ammophila campestris* Latr. und *Ammophila adriaansei* Wilcke. *Behaviour,* **1**:1–35.

Anfinsen, C. B.

1973 Principles that govern the folding of protein chains. *Science (Wash., D.C.)*, **181**:223–230 (Nobel Prize Lecture).

Antonini, E., and M. Brunori.

1971 "Hemoglobin and Myoglobin in Their Reactions with Ligands." *North-Holland Research Monographs: Frontiers of Biology*, Vol. 21. Amsterdam: North-Holland.

Arch, S.

1972a Polypeptide secretion from the isolated parietovisceral ganglion of *Aplysia californica. J. Gen. Physiol.*, **59**:47–59.

1972b Biosynthesis of the egg-laying hormone (ELH) in the bag cell neurons of *Aplysia californica. J. Gen. Physiol.*, **60**:102–119.

Arch, S., and T. Smock.

1977 Egg-laying behavior in *Aplysia californica. Behav. Biol.*, **19**:45–54.

Arey, L. B., and W. J. Crozier.

1919 The sensory responses of *Chiton. J. Exp. Zool.*, **29**:157–260.

Arvanitaki, A., and H. Cardot.

1941 Contribution à la morphologie du système nerveux des Gastéropodes. Isolement, à l'état vivant, de corps neuroniques. *C. R. Séances Soc. Biol. Fil.*, **135**:965–968.

Arvanitaki, A., and N. Chalazonitis.

1961 Excitatory and inhibitory processes initiated by light and infra-red radiations in single identifiable nerve cells (giant ganglion cells of *Aplysia*). In *Nervous Inhibition: International Symposium on Nervous Inhibition*, edited by E. Florey. New York: Pergamon, pp. 194–231.

Ascher, P.

1972 Inhibitory and excitatory effects of dopamine on *Aplysia* neurones. *J. Physiol. (London)*, **225**:173–209.

Aspey, W. P., and J. E. Blankenship.

1976a *Aplysia* behavioral biology: I. A multivariate analysis of burrowing in *A. brasiliana. Behav. Biol.*, **17**:279–299.

1976b *Aplysia* behavioral biology: II. Induced burrowing in swimming *A. brasiliana* by burrowed conspecifics. *Behav. Biol.*, **17**:301–312.

Aspey, W. P., J. S. Cobbs, and J. E. Blankenship.

1977 *Aplysia* behavioral biology. III: Head-bobbing in relation to food deprivation in *A. brasiliana. Behav. Biol.*, **19**:300–308.

Atz, J. W.

1970 The application of the idea of homology to behavior. In *Development and Evolution of Behavior: Essays in Memory of T. C. Schneirla*, edited by L. R. Aronson, E. Tobach, D. S. Lehrman, and J. S. Rosenblatt. San Francisco: W. H. Freeman, pp. 53–74.

Audesirk, T. E.

1975 Chemoreception in *Aplysia californica*, I: Behavioral localization of distance chemoreceptors used in food-finding. *Behav. Biol.*, **15**:45–55.

Bailey, C. H., V. F. Castellucci, J. D. Koester, and E. R. Kandel.

1975 Central mechanoreceptor neurons in *Aplysia* connect to peripheral siphon motor neurons: A simple system for the morphological study of the synaptic mechanism underlying habituation. *Neuroscience Abstracts*, **1**:588.

Bailey, C. H., E. B. Thompson, V. F. Castellucci, and E. R. Kandel.

1976 Fine structure of synapses of identified sensory cells which mediate the gill-withdrawal reflex in *Aplysia californica*. Society for Neuroscience, Sixth Annual Meeting, **2**:446.

Bailey, D. F., and M. S. Laverack.

1966 Aspects of the neurophysiology of *Buccinum undatum* L. (Gastropoda), I: Central responses to stimulation of the osphradium. *J. Exp. Biol.*, **44**:131–148.

Banks, F. W.
 1975 Inhibitory transmission at a molluscan neuromuscular junction. *J. Neurobiol.*, **6**:429–433.
Barnes, R. D.
 1968 *Invertebrate Zoology*, 2nd ed. Philadelphia: Saunders.
Bauer, V.
 1929 Über das Tierleben auf den Seegraswiesen des Mittlemeeres. *Zool. Jahrb.*, **56**:1–42.
Baur, P. S., Jr., A. M. Brown, T. D. Rogers, and M. E. Brower.
 1977 Lipochondria and the light response of *Aplysia* giant neurons. *J. Neurobiol.*, **8**:19–42.
Beach, F. A.
 1950 The snark was a boojum. *Am. Psychol.*, **5**:115–124.
 1960 Experimental investigations of species-specific behavior. *Am. Psychol.*, **15**:1–18.
Beeman, R. D.
 1963 Notes on the California species of *Aplysia* (Gastropoda: Opisthobranchia). *Veliger,* **5**:145–147.
 1968 The order Anaspidea. *Veliger,* 3(Suppl.):87–102.
 1970 An autoradiographic study of sperm exchange and storage in a sea hare, *Phyllaplysia taylori,* a hermaphroditic gastropid (*Opisthobranchia: Anaspidea*). *J. Exp. Zool.,* **175**:125–132.
Belon, P.
 1553 *De Aquatilibus,* Book II, Chapter 12. Paris.
Bennett, M. V. L.
 1973 Function of electrotonic junctions in embryonic and adult tissues. *Fed. Proc.,* **32**:65–75.
Bentley, D. R.
 1973 Postembryonic development of insect motor systems in *Developmental Neurobiology of Arthropods.* (D. Young, Ed.), pp. 147–177. Cambridge University Press, London.
Bentley, D. R., and R. R. Hoy.
 1972 Genetic control of the neuronal network generating cricket (*Teleogryllus gryllus*) song patterns. *Animal Behavior:* **20**(3):478–492.
 1974 The neurobiology of cricket song. *Sci. Am.,* **231**(2): 34–44.
Berg, C. J., Jr.
 1976 Ontogeny of predatory behavior in marine snails (Prosobranchia: Naticidae). *Nautilus,* **90**:1–4.
Bergh, R.
 1902 Malacologische Untersuchungen, Band 7, Vierte Abteilung, Vierter Abschnitt: Ascoglossa, Aplysiidae. In *Archipel der Philippinen,* edited by C. Semper, Wiesbaden: C. W. Kreidel.
Berrill, N. J.
 1971 *Developmental Biology.* New York: McGraw-Hill.
Bethe, A.
 1903 *Allgemeine Anatomie und Physiologie des Nervensystems.* Leipzig: G. Thieme.
 1926 Vergleichende Physiologie der Blutbewegung. In *Handbuch der Normalen und Pathologischen Physiologie,* edited by A. Bethe, B. V. Bergmann, G. Embden, and A. Ellinger. **7**:1, Blutzirkulation. Berlin: Springer.
 1930 The permeability of the surface of marine animals. *J. Gen. Physiol.,* **13**:437–444.
 1934 Die Salz- und Wasser-Permeabilität der Körperoberflächen verschiedener Seetiere in ihrem gegenseitigen Verhältnis. *Pflügers Archiv gesamte Physiol. Menschen Tiere,* **234**: 629–644.
Bevelaqua, F. A., K. S. Kim, M. H. Kumarasiri, and J. H. Schwartz.
 1975 Isolation and characterization of acetylcholinesterase and other particulate proteins in the hemolymph of *Aplysia californica. J. Biol. Chem.,* **250**:731–738.

Bidder, A. N.
  1966 Feeding and digestion in cephalopods. In *Physiology of Mollusca*, vol. 2, edited by K. M. Wilbur and C. M. Yonge. New York: Academic, pp. 97–124.
Blankenship, J. E., and R. E. Coggeshall.
  1976 The abdominal ganglion of *Aplysia brasiliana:* A comparative morphological and electrophysiological study, with notes on *A. dactylomela. J. Neurobiol.,* 7(5):383–405.
Blochmann, F.
  1883 Beiträge zur Kenntnis der Entwicklung der Gastropoden. *Z. Wiss. Zool.,* 38:392–410.
  1884 Die im Golfe von Neapel vorkommenden Aplysien. *Mitt. Zool. Stn. Neapel,* 5:28–49.
Block, G. D., D. J. Hudson, and M. E. Lickey.
  1974 Extraocular photoreceptors can entrain the circadian oscillator in the eye of *Aplysia. J. Comp. Physiol.,* 89:237–249.
Block, G. D., and M. E. Lickey.
  1973 Extraocular photoreceptors and oscillators can control the circadian rhythm of behavioral activity in *Aplysia. J. Comp. Physiol.,* 84(4):367–374.
Boettger, C. R.
  1955 Die Systematik der euthyneuren Schnecken. *Verh. Dtsch. Zool. Ges.,* 1954:253–280.
Bohadsch, J. B.
  1761 *De Quibusdam Animalibus Marinis.* Dresden.
Bolles, R. C.
  1967 *Theory of Motivation.* New York: Harper and Row.
Bonar, D. B., and M. G. Hadfield.
  1974 Metamorphosis of the marine gastropod *Phestilla sibogae* Bergh (Nudibranchia: Acolidacea). I: Light and electron microscopic analysis of larval and metamorphic stages. *J. Exp. Mar. Biol. Ecol.,* 16:227–255.
Borradaile, L. A., and F. A. Potts
  1963 *The Invertebrata: A Manual for the Use of Students,* 4th ed. rev. by G. A. Kerkut. University of Cambridge Press.
Bottazzi, F.
  1899 Ricerche fisiologiche sul Sistema nervoso viscerale delle *Aplisie* e di alcuni Cefalopodi. *Riv. Sci. Biol.,* 1:837–924.
Bottazzi, F., and P. Enriques.
  1900 Recherches physiologiques sur le système nerveux viscéral des Aplysies et de quelques Céphalopodes. *Arch. Ital. Biol.,* 34:111–143.
Boutan, L.
  1899 La cause principale de l'asymétrie des Mollusques gastéropodes. *Arch. Zool. Exp. Gén.,* 3rd Série, 7:203–342.
  1902 La détorsion chez les Gastéropodes. *Arch. Zool. Exp. Gén.,* 3rd Série, 10:241–268.
Bouvier, E. L.
  1891 Recherches anatomiques sur les Gastéropodes provenant des campagnes du yacht "L'Hirondelle." *Bull. Soc. Zool. de France,* XVI:53–56.
Boyden, A.
  1943 Homology and analogy: A century after the definitions of "homologue" and "analogue" of Richard Aven. *Q. Rev. Biol.,* 18:228–241.
Brace, R. C.
  1977a The functional anatomy of the mantle complex and columellar muscle of tectibranch molluscs (Gastropoda: Opisthobranchia), and its bearing on the evolution of opisthobranch organization. *Philos. Trans. R. Soc. London (B),* 277:1–56.
  1977b Anatomical changes in nervous and vascular systems during the transition from prosobranch to opisthobranch organization. *Trans. Zool. Soc. London,* 34:1–25.
  1977c Shell attachment and associated musculature in the notaspidea and anaspidea (Gastropoda: Opisthobranchia). *Trans. Zool. Soc. London,* 34:27–43.

Bridges, C. B.
1975 Larval development of *Phyllaplysia taylori dall*, with a discussion of development in the anaspidea (Opisthobranchiata: anaspidea). *Ophelia*, **14**:161–184.

Brown, A. C., and R. G. Noble.
1960 Function of the osphradium in *Bullia* (Gastropoda). *Nature (London)*, **188**:1045.

Brown, A. M., P. S. Baur, Jr., and F. H. Tuley, Jr.
1975 Phototransduction in *Aplysia* neurons: Calcium release from pigmented granules is essential. *Science (Wash., D.C.)*, **188**:157–160.

Brown, H. M., and A. M. Brown.
1972 Ionic basis of the photoresponse of *Aplysia* giant neuron: K^+ permeability increase. *Science (Wash., D.C.)*, **178**:755–756.

Brown, J. S.
1961 *The Motivation of Behavior*. New York: McGraw Hill.

Brunelli, M., V. Castellucci, and E. R. Kandel.
1976 Synaptic facilitation and behavioral sensitization in *Aplysia*: Possible role of serotonin and cyclic AMP. *Science* (Wash., D.C.), **194**:1178–1181.

Buchsbaum, R. M.
1965 *Animals without Backbones: An Introduction to the Invertebrates*, 2nd ed. University of Chicago Press.

Bullock, T. H.
1953 Predator recognition and escape responses of some intertidal gastropods in presence of starfish. *Behaviour*, **5**:130–140.

Bullock, T. H., and G. A. Horridge.
1965 *Structure and Function in the Nervous Systems of Invertebrates*, vol. 2. San Francisco: W. H. Freeman.

Byrne, J. H., V. F. Castellucci, T. J. Carew, and E. R. Kandel.
1978 Stimulus response relations and stability of mechanoreceptors and motor neurons mediating defense gill-withdrawal reflex in *Aplysia*. *J. Neurophysiol.*, **41**:402–417.

Byrne, J., V. Castellucci, and E. R. Kandel.
1974 Receptive fields and response properties of mechanoreceptor neurons innervating siphon skin and mantle shelf in *Aplysia*. *J. Neurophysiol.*, **37**:1041–1064.

Byrne, J., V. Castellucci, and E. R. Kandel.
1978 Contribution of individual mechano-receptor sensory neurons to defensive gill-withdrawal reflex in *Aplysia*. *J. Neurophysiol.*, **48**:418–431.

Byrne, J., and J. Koester.
1978 Respiratory pumping: Neuronal control of a centrally commanded behavior in *Aplysia*. *Brain Research*, **143**:87–105.

Cajal, S. R.
1911 *Histologie du Système Nerveux de L'homme et des Vertébrés*, vol. 2. Paris: Maloine. (Republished 1955 as *Histologie du Système Nerveux*, translated by. L. Azoulay. Madrid: Instituto Ramón y Cajal.)

Campbell, C. B. G., and W. Hodos.
1970 The concept of homology and the evolution of the nervous system. *Brain, Behavior and Evolution*, **3**:353–367.

Carazzi, D.
1900 L'embriologia dell'*Aplysia limacina* L. *Anat. Anz.*, **17**:77–102.
1905 L'embriologia dell'*Aplysia* e i problemi fondamentali dell'embriologia comparata. *Arch. Ital. Anat. Embriol.*, **4**:231–305 and **4**:459–504.

Carefoot, T. H.
1967a Growth and nutrition of three species of opisthobranch molluscs. *Comp. Biochem. Physiol.*, **2**:627–652.

1967b Studies on a sublittoral population of *Aplysia punctata. J. Mar. Biol. Assoc. U.K.,* **47**: 335–350.

1967c Growth and nutrition of *Aplysia punctata* feeding on a variety of marine algae. *J. Mar. Biol. Assoc. U.K.,* **47**:565–589.

1970 A comparison of absorption and utilization of food energy in two species of tropical *Aplysia. J. Exp. Mar. Biol. Ecol.,* **5**:47–62.

Carew, T. J., V. Castellucci, J. Byrne, and E. R. Kandel.

1976 Quantitative analysis of the contribution of central and peripheral nervous systems to the gill-withdrawal reflex in *Aplysia californica. Neuroscience Abstracts,* **2**:449.

Carew, T. J., and E. R. Kandel.

1973 Acquisition and retention of long-term habituation in *Aplysia:* Correlation of behavioral and cellular processes. *Science (Wash., D.C.)* **182**:1158–1160.

1975 Two functional consequences of decreased conductance EPSP's: Increased electrotonic coupling and amplification. *Proc. Soc. Neurosci.,* **1**:587–589.

1977a Inking in *Aplysia californica.* I: Neural circuit of an all-or-none behavioral response. *J. Neurophysiol.,* 40:692–707.

1977b Inking in *Aplysia californica.* II: Central program for inking. *J. Neurophysiol.,* 40: 708–720.

1977c Inking in *Aplysia californica.* III: Two different synaptic conductance mechanisms for triggering central program for inking. *J. Neurophysiol.,* 40:721–734.

Carew, T. J., and I. Kupfermann.

1974 The influence of different natural environments on habituation in *Aplysia californica. Behav. Biol.,* **12**:339–345.

Carew, T. J., H. M. Pinsker, and E. R. Kandel.

1972 Long-term habituation of a defensive withdrawal reflex in *Aplysia. Science (Wash., D.C.),* **175**:451–454.

Carew, T. J., H. Pinsker, K. Rubinson, and E. R. Kandel.

1974 Physiological and biochemical properties of neuromuscular transmission between identified motoneurons and gill muscle in *Aplysia. J. Neurophysiol.,* **37**:1020–1040.

Carlson, A. J.

1905 The rhythm produced in the resting heart of molluscs by the stimulation of the cardio-accelerator nerves. *Am. J. Physiol.,* **12**:55–66.

Castellucci, V., T. J. Carew, and E. R. Kandel.

1978 Cellular analysis of long-term habituation of the gill-withdrawal reflex of *Aplysia californica. Science,* in press.

Castellucci, V. F., and E. R. Kandel.

1974 A quantal analysis of the synaptic depression underlying habituation of the gill-withdrawal reflex in *Aplysia. Proc. Nat. Acad. Sci. (U.S.A.),* **71**:5004–5008.

1976 Presynaptic facilitation as a mechanism for behavioral sensitization in *Aplysia. Science (Wash., D.C.),* **194**:1176–1178.

Castellucci, V., H. Pinsker, I. Kupfermann, and E. R. Kandel.

1970 Neuronal mechanisms of habituation and dishabituation of the gill-withdrawal reflex in *Aplysia. Science (Wash., D.C.),* **167**:1745–1748.

Cather, J. N.

1971 Cellular interactions in the regulation of development in annelids and molluscs. *Adv. Morphog.* **9**:67–125.

Cedar, H., E. R. Kandel, and J. H. Schwartz.

1972 Cyclic adenosine monophosphate in the nervous system of *Aplysia californica.* I: Increased synthesis in response to synaptic stimulation. *J. Gen. Physiol.,* **60**:558–569.

Cedar, H., and J. H. Schwartz.

1972 Cyclic adenosine monophosphate in the nervous system of *Aplysia californica.* II: Effect of serotonin and dopamine. *J. Gen. Physiol.,* **60**:570–587.

Chace, L.
1952 The aerial mating of the great slug. *Discovery,* Vol. 13, No. 11.

Chiaje, S., delle.
1828 Descrizione ed anatomia delle Aplisie. *Atti R. Ist. Incoragg. Sci. Nat. Napoli,* **4**:25–76.

Civil, G. W., and T. E. Thompson.
1972 Experiments with the isolated heart of the gastropod *Helix pomatia* in an artificial pericardium. *J. Exp. Biol.,* **56**:239–247.

Clarke, A. H., Jr., and R. J. Menzies.
1959 *Neopilina (Vema) ewingi,* a second living species of the paleozoic class Monoplacophora. *Science (Wash., D.C.),* **129**:1026–1027.

Clarke, B.
1975 The causes of biological diversity. *Sci. Am.,* **223**(2):50–60.

Clement, A. C.
1956 Experimental studies on germinal localization in *Ilyanassa.* II: The development of isolated blastomeres. *J. Exp. Zool,* **132**:427–445.
1968 Development of the vegetal half of the *Ilyanassa* egg after removal of most of the yolk by centrifugal force, compared with the development of animal halves of similar visible composition. *Dev. Biol.,* **17**:165–186.

Coggeshall, R. E.
1967 A light and electron microscope study of the abdominal ganglion of *Aplysia california. J. Neurophysiol.,* **30**:1263–1287.
1969 A fine structural analysis of the statocyst in *Aplysia californica. J. Morphol.,* **127**:113–131.
1970 A cytologic analysis of the bag cell control of egg laying in *Aplysia. J. Morphol.,* **132**: 461–485.
1971 A possible sensory-motor neuron in *Aplysia californica. Tissue and Cell,* **3**:637–647.
1972 The structure of the accessory genital mass in *Aplysia californica. Tissue and Cell,* **4**:105–127.

Coggeshall, R. E., B. A. Yaksta, and F. J. Swartz.
1970 A cytophotometric analysis of the DNA in the nucleus of the giant cell, R-2, in *Aplysia. Chromosoma (Berlin),* **32**:205–212.

Cohen, J., K. Weiss, and I. Kupfermann.
1974 Physiology of the neuromuscular system of buccal muscle of *Aplysia. The Physiologist,* **17**:198 (Abstract)
1978 Motor control of buccal muscles in *Aplysia. J. Neurophysiol.,* **41**:157–180.

Conklin, E. G.
1897 The embryology of *Crepidula,* a contribution to the cell lineage and early development of some marine gasteropods. *J. Morphol.,* **13**:1–226.
1907 The embryology of *Fulgur:* A study of the influence of yolk on development. *Proc. Acad. Sci. Phila.,* **59**:320–359.

Cooper, J. G.
1863 On new or rare mollusca inhabiting the Coast of California—No. II. *Proc. Calif. Acad. Nat. Sci.,* **3**:56–60.

Copeland, M.
1918 The olfactory reactions and organs of the marine snails *Alectrion obsoleta* (Say) and *Busycon canaliculatum* (Linn.). *J. Exp. Zool.,* **25**:177–227.

Cottrell, G. A., and N. N. Osborne.
1970 Subcellular localization of serotonin in an identified serotonin-containing neurone. *Nature (London),* **225**:470–472.

Crampton, H. E., Jr.
1896 Experimental studies on gasteropod development. *Wilhelm Roux' Arch Entwickelungsmech. Org.* **3**:1–19.

Creek, G. A.
  1951 The reproductive system and embryology of the snail *Pomatias elegans* (Müller). *Proc. Zool. Soc. London,* **121:**599–640.
Crick, F. H. C.
  1958 On protein synthesis. *Symp. Soc. Exp. Biol.,* **12:**138–163.
Crofts, D. R.
  1929 Haliotis. *Liverpool Marine Biology Committee Memoir* No. 29 *Proc. Trans. Liverpool Biol. Soc.,* vol. 43 (174 pp).
  1937 The development of *Haliotis tuberculata,* with special reference to organogenesis during torsion. *Philos. Trans. R. Soc. London (B),***228:**219–268.
  1955 Muscle morphogenesis in primitive gastropods and its relation to torsion. *Proc. Zool. Soc. London,* **125:**711–750.
Crow, T., and D. Alkon.
  1977 Acquisition and retention of a long-term behavioral change in *Hermissenda Crassicornis. Neuroscience Abstracts,* **3:**533.
Crozier, W. J., and L. B. Arey.
  1919 Sensory reactions of *Chromodoris zebra. J. Exp. Zool.,* **29:**261–310.
Cuvier, G.
  1803 Mémoire sur le genre *Laplysia,* vulgairement nommé *Lièvre marin;* sur son anatomie, et sur quelques-unes de ses espèces. *Ann. Mus. Nat. Hist. Paris,* **2:**287–314.
  1817 *Mémoires pour Servir a l'Histoire et a l'Anatomie des Mollusques.* Paris: Chez Deterville, Libraire.
Dainton, B. H.
  1954 The activity of slugs. I. The induction of activity by changing temperatures. *J. Exp. Biol.,* **31:**165–187.
Davidson, E. H.
  1976 *Gene Activity in Early Development,* 2nd ed. New York: Academic.
Davis, W. J., and G. J. Mpitsos.
  1971 Behavioral choice and habituation in the marine mollusk *Pleurobranchaea californica* MacFarland (Gastropoda, Opisthobranchia). *Z. vgl. Physiol.,* **75:**207–232.
Davis, W. J., G. J. Mpitsos, and J. M. Pinneo.
  1974a The behavioral hierarchy of the mollusk *Pleurobranchaea.* I: The dominant position of the feeding behavior. *J. Comp. Physiol.,* **90:**207–224.
  1974b The behavioral hierarchy of the mollusk *Pleurobranchaea.* II: Hormonal suppression of feeding associated with egg-laying. *J. Comp. Physiol.,* **90:**225–243.
Davis, W. J., G. J. Mpitsos, J. M. Pinneo, and J. L. Ram.
  1977 Modification of the behavioral hierarchy of *Pleurobranchaea.* I: Satiation and feeding motivation. *J. Comp. Physiol. (A),* **117:**99–125.
Davis, W. J., M. V. S. Siegler, and G. J. Mpitsos.
  1973 Distributed neuronal oscillators and efference copy in the feeding system of *Pleurobranchaea. J. Neurophysiol.,* **36:**258–274.
Dawson, E. Y.
  1956 *How to Know the Seaweeds.* Dubuque, Iowa: W. C. Brown.
Dethier, V. G.
  1964 Microscopic brains. *Science (Wash., D.C.),* **143:**1138–1145.
  1966 Insects and the concept of motivation. *Nebr. Symp. Motiv.,* pp. 105–136.
  1976 *The Hungry Fly: A Physiological Study of the Behavior Associated with Feeding.* Harvard University Press.
Detwiler, P. B., and D. L. Alkon.
  1973 Hair-cell interactions in the statocyst of *Hermissenda. J. Gen. Physiol.,* **62:**618–642.
Dieringer, N., J. Koester, and K. R. Weiss.
  1978 Adaptive changes in heart rate of *Aplysia californica. J. Comp. Physiol.,* **123:**11–21.

Dijkgraaf, S., and H. G. A. Hessels.
    1969 Über Bau und Funktion der Statocyste bei der Schnecke *Aplysia limacina. Z. vgl. Physiol.,* **62:**38–60.

Dobzhansky, T. G.
    1941 *Genetics and the Origin of Species,* 2nd ed. Columbia University Press.

Dogiel, J.
    1877 Die Muskeln und Nerven des Herzens bei einigen Mullusken. *Arch. Mikrosk. Anat.,* **14:**59–65.

Dorsett, D. A.
    1968 The pedal neurons of *Aplysia* punctata. *J. Exp. Biol.,* **48:**127–140.
    1974 Neuronal homologies and the control of branchial tuft movements in two species of *Tritonia. J. Exp. Biol.,* **61:**639–654.

Doty, R. W.
    1968 Neural organization of deglutition. In *Handbook of Physiology,* Section 6: *Alimentary Canal,* edited by C. F. Code. Washington, D.C.: American Physiological Society. **4:** 1861–1902.

Downey, P., and B. Jahan-Parwar.
    1972 Cooling as reinforcing stimulus in *Aplysia. Am. Zool.,* **12:**507–512.

Drew, G. A.
    1899 The anatomy, habits, and embryology of *Yoldia limatula,* Say. *Mem. Biol. Lab. Johns Hopkins Univ.,* **4:**3.

Dudek, F. E., and J. E. Blankenship.
    1976 Neuroendocrine (bag) cells of *Aplysia:* Spike blockade and a mechanism for potentiation. *Science (Wash., D.C.),* **192:**1009–1010.
    1977a Neuroendocrine cells of *Aplysia brasiliana.* I: Bag-cell action potentials and afterdischarge. *J. Neurophysiol.,* **40:**1301–1311.
    1977b Neuroendocrine cells of *Aplysia. brasiliana.* II: Bag-cell prepotentials and potentiation. *J. Neurophysiol.,* **40:**1312–1324.

Duffy, E.
    1941 The conceptual categories of psychology: A suggestion for revision. *Psychol. Rev.,* **48:**177–203.
    1957 The psychological significance of the concept of "arousal" or "activation." *Psychol. Rev.,* **64:**265–275.

Eales, N. B.
    1921 *Aplysia. Liverpool Marine Biology Committee, Proc. Trans. Liverpool Biol. Soc.,* L.M.B.C. Mem. vol. 35, **24:**183–266.
    1944 *Aplysiids* from the Indian Ocean, with a review of the family *Aplysiidae. Proc. Malacol. Soc. London,* **26:**1–22.
    1950a Torsion in Gastropoda. *Proc. Malacol. Soc. London,* **28:**53–61.
    1950b Secondary symmetry in Gastropods. *Proc. Malacol. Soc. London,* **28:**185–196.
    1960 Revision of the world species of *Aplysia* (Gastropoda, Opisthobranchia). *Bull. Br. Mus. (Nat. Hist.) Zool.,* **5:**276–404.

Eisenstadt, M., J. E. Goldman, E. R. Kandel, H. Koike, J. Koester, and J. H. Schwartz.
    1973 Intrasomatic injection of radioactive precursors for studying transmitter synthesis in identified neurons of *Aplysia californica. Proc. Nat. Acad. Sci. U.S.A.,* **70:**3371–3375.

Emery, D. G.
    1976 Taste receptors in the lip of *Aplysia. Am. Zool.,* **16:**241, Abstract 349.

Erlanger, R. V.
    1891a Zur Entwicklung von *Paludina vivipara. Morphol. Jahrb.,* **17:**337–379.
    1891b Zur Entwicklung von *Paludina vivipara.* II. Theil. *Morphol. Jahrb.,* **17:**636–680.
    1892 Beitrage zur Entwicklungsgeschichte der Gasteropoden. *Mitt. Zool. Stn. Neapel.,* **10:** 376–407.

Eskin, A.
　　1971 Properties of the *Aplysia* visual system: *In vitro* entrainment of the circadian rhythm and centrifugal regulation of the eye. *Z. vgl. Physiol.,* **74:**353–371.

Farmer, W. M.
　　1970 Swimming gastropods (Opisthobranchia and Prosobranchia). *Veliger,* **13:**73–89.

Feder, H., and A. M. Christensen.
　　1966 Aspects of asteroid biology. In *Physiology of Echinodermata,* ed. by R. A. Boolootian. New York: Interscience, pp. 87–127.

Feinstein, R., R. Pinsker, M. Schmale, and B. A. Gooder.
　　1977 Bradycardial response in *Aplysia* exposed to air. *J. Comp. Physiol (B),* **122:**311 324.

Filatova, Z. A., M. N. Sokolova, and R. Y. Levenstein.
　　1968 Mollusc of the Cambro-Devonian class Monoplacophora found in the Northern Pacific. *Nature (London),* **220:**1114–1115.

Florkin, M.
　　1966 Nitrogen metabolism. In *Physiology of Mollusca,* vol. 2, edited by K. M. Wilbur and C. M. Yonge. New York: Academic, pp. 309–351.

Flury, F.
　　1915 Über das Aplysiengift. *Nauym Schiedebergs Archiv für experimentelle Pathologie und Pharmakologie,* **79:**250–263.

Forrest, J. E.
　　1953 On the feeding habits and the morphology and mode of functioning of the alimentary canal in some littoral dorid nudibranchiate Mollusca. *Proc. Linn. Soc. London,* **164:** 225–235.

Frazier, W. T., E. R. Kandel, I. Kupfermann, R. Waziri, and R. E. Coggeshall.
　　1967 Morphological and functional properties of identified neurons in the abdominal ganglion of *Aplysia californica. J. Neurophysiol.,* **30:**1288–1351.

Fredman, S. M., and B. Jahan-Parwar.
　　1977 Identifiable cerebral neurons mediating an anterior tentacular withdrawal reflex in *Aplysia. J. Neurophysiol.,* **40:**608–615.

French, J. D., and H. W. Magoun.
　　1952 Effects of chronic lesions in central cephalic brain stem of monkeys. *Arch. Neurol. and Psychiat.,* **68:**591–604.

Fretter, V.
　　1969 Aspects of metamorphosis in prosobranch gastropods (Presidential address). *Proc. Malac. Soc. Lond.,* **38:**375–386.

Fretter, V., and A. Graham.
　　1962 *British Prosobranch Molluscs: Their Functional Anatomy and Ecology.* London: Ray Society.
　　1964 Reproduction. In *Physiology of Mollusca,* vol. 1, edited by K. M. Wilbur and C. M. Yonge. New York: Academic, pp. 127–164.
　　1976 *A Functional Anatomy of Invertebrates.* New York: Academic.

Fretter, V. and J. Peake.
　　1975 *Pulmonates,* Vol. I: *Functional Anatomy and Physiology.* New York: Academic.

Friedrich, H.
　　1932 Studien uber die Gleichgewichtserhaltung und Bewegungsphysiologie bei Pterotrachea. *Z. vgl. Physiol.,* **16:**345–361.

Frings, H., and C. Frings.
　　1965 Chemosensory bases of food finding and feeding in *Aplysia juliana* (Mollusca, Opisthobranchia). *Biol. Bull. (Woods Hole),* **128:**211–217.

Fröhlich, F. W.
　　1910a Experimentelle Studien am Nervensystem der Mollusken. 10: Die Fortpflanzungsgeschwindigkeit der Errengung in den Flügelnerven von *Aplysia limacina. Z. allg. Physiol.,* **11:**141–144.

1910b Experimentelle Studien am Nervensystem der Mollusken. 11: Die Wirkung von Karbolsäure und Strychnin auf das Nervensystem von *Aplysia limacina. Z. allg. Physiol.,* **11:**269–274.

1910c Experimentelle Studien am Nervensystem der Mollusken. 13: Über die durch das Pedalganglion von *Aplysia limacina* vermittelte "Reflexverkettung." *Z. allg. Physiol.,* **11:**351–370.

Furshpan, E. J., and D. D. Potter.

1968 Low-resistance junctions between cells in embryos and tissue culture. *Curr. Top. Dev. Biol.,* **3:**95–127.

Gallin, E. K., and M. L. Wiederhold.

1977 Response of *Aplysia* statocyst receptor cells to physiologic stimulation. *J. Physiol. (London),* **266:**123–137.

Garcia, J., W. G. Hankins, and K. W. Rusiniak.

1974 Behavioral regulation of the milieu interne in man and rat. *Science (Wash., D.C.),* **185:**824–831.

Garcia, J., B. K. McGowan, and K. F. Green.

1972 Biological constraints on conditioning. In *Classical Conditioning, II: Current Research and Theory,* edited by A. H. Black and W. F. Prokasy. New York: Appleton-Century-Crofts, pp. 3–27.

Gardner, D.

1969 Symmetry and redundancy of interneuronal connections in the buccal ganglion of *Aplysia. Physiologist,* **12:**232.

1971 Bilateral symmetry and interneuronal organization in the buccal ganglia of *Aplysia. Science (Wash., D.C.),* **173:**550–553.

Gardner, D., and E. R. Kandel.

1972 Diphasic postsynaptic potential: A chemical synapse capable of mediating conjoint excitation and inhibition. *Science (Wash., D.C.),* **176:**675–678.

Garstang, W.

1890 A complete list of the opisthobranchiate mollusca found at Plymouth; with further observations on their morphology, colours, and natural history. *J. Mar. Biol. Assoc. U.K., N.S.,* **1:**399–457.

1928 The origin and evolution of larval forms. *Rept. Br. Assoc. Adv. Sci.,* pp. 77–98.

Gascoigne, T.

1956 Feeding and reproduction in the Limapontiidae. *Trans. R. Soc. Edinb.,* **63:**129–151.

Gelperin, A.

1967 Stretch receptors in the foregut of the blowfly. *Science,* **157:**208–210.

1975a Rapid food-aversion learning by a terrestrial mollusk. *Science (Wash, D.C.),* **189:** 567–570.

1975b An identified serotonergic input has reciprocal effects on two electrically coupled motoneurons in the terrestrial slug, *Limax maximus. Biol. Bull. (Woods Hole),* **149:**426–427.

Gelperin, A., J. J. Chang, and S. C. Reingold.

1978 Feeding motor program in *Limax.* I: Neuromuscular correlates and control by chemosensory input. *J. Neurobiol.,* **9:** 285–300.

Gelperin, A., and D. Forsythe.

1976 Neuroethological studies of learning in mollusks. In *Simpler Networks and Behavior,* edited by J. C. Fentress. Sunderland, Mass.: Sinauer, pp. 239–246.

Gerhardt, U.

1934 Zur Biologie der Kopulation der Limaciden. II. Mitteilung. *Z. Morphol. Ökol. Tiere,* **28:**229–258.

Gerschenfeld, H. M., P. Ascher, and L. Tauc.

1967 Two different excitatory transmitters acting on a single molluscan neurone. *Nature (London),* **213:**358–359.

Getting, P. A.

1975 *Tritonia* swimming: Triggering of a fixed action pattern. *Brain Res.,* **96:**128–133.

1976 Afferent neurons mediating escape swimming of the marine mollusc, *Tritonia. J. Comp. Physiol. (A),* **110:**271–286.

1977 Neuronal organization of escape swimming in *Tritonia. J. Comp. Physiol. (A),* **121:** 325–342.

Ghiretti, F.

1966 Molluscan hemocyanins. In *Physiology of Mollusca,* vol. 2, edited by K. M. Wilbur and C. M. Yonge. New York: Academic, pp. 233–248.

Ghiselin, M. T.

1965 Reproductive function and the phylogeny of opisthobranch gastropods. *Malacologia,* **3:**327–378.

Giller, E., Jr., and J. H. Schwartz.

1971a Acetylcholinesterase in identified neurons of abdominal ganglion of *Aplysia californica. J. Neurophysiol.,* **34:**108–115.

1971b Choline acetyltransferase in identified neurons of abdominal ganglion of *Aplysia californica. J. Neurophysiol.,* **34:**93–107.

Gillette, R., and W. J. Davis.

1977 The role of the metacerebral giant neuron in the feeding behavior of *Pleurobranchaea. J. Comp. Physiol. (A),* **116:**129–159.

Gillette, R., M. P. Kovac, and W. J. Davis.

1978 Command neurons receive synaptic feedback from the motor network they excite. *Science (Wash., D.C.),* **199:**798–801.

Gilula, N. B., and P. Satir.

1971 Septate and gap junctions in molluscan gill epithelium. *J. Cell Biol.,* **51:**869–872.

Globus, A., H. D. Lux, and P. Schubert.

1968 Somadendritic spread of intracellularly injected tritiated glycine in cat spinal motoneurons. *Brain Res.,* **11:**440–445.

1973 Transfer of amino acids between neuroglia cells and neurons in the leech ganglion. *Exp. Neurol.,* **40:**104–113.

Goldman, J. E., R. T. Ambron, and J. H. Schwartz.

1974 Axonal transport of serotonin and membrane glycoproteins in metacerebral neurons of *Aplysia californica.* Society for Neuroscience, Fourth Annual Meeting, Abstract No. 249.

Goodrich, E. S.

1946 The study of nephridia and genital ducts since 1895. *Q. J. Microsc. Sci.,* **86:**113–392.

Graham, A.

1949 The molluscan stomach. *Trans. R. Soc. Edinb.,* **61:**737–778.

1953 Form and function in the molluscs. *Proc. Linn. Soc. London,* **164:**213–217.

Granzow, B., and S. R. Kater.

1977 Identified higher-order neurons controlling the feeding motor program of *Helisoma. Neuroscience,* **2:**1049–1063.

Grasse, P. P.

1968 *Traite de Zoologie. Anatomie, Systematique, Biologie. Tome V. Mollusques, Gasteropodes et Schaphopodes.* (Fascicule III). Paris: Masson et Cie.

Graubard, K.

1973 Morphological and electrotonic properties of identified neurons of the mollusc, *Aplysia californica.* Ph. D. dissertation, University of Washington.

Gray, J.

1968 *Animal Locomotion.* New York: Norton.

Grether, W. F.

1938 Pseudo-conditioning without paired stimulation encountered in attempted backward conditioning. *J. Comp. Psychol.,* **25:**91–96.

Grillner, S.
  1975 Locomotion in vertebrates: Central mechanisms and reflex interaction. *Physiol. Rev.,* **55:**247–304.

Groves, P. M., and R. F. Thompson.
  1970 Habituation: A dual-process theory. *Psychol. Rev.,* **77:**419–450.

Guiart, J.
  1901 Contribution a l'étude des gastéropodes opisthobranches et en particulier des céphalaspides. *Mém. Soc. Zool. Fr.,* **14:**5–219.

Haas, O., and G. G. Simpson.
  1946 Analysis of some phylogenetic terms, with attempts at redefinition. *Proc. Am. Philos. Soc.,* **90:**319–349.

Hadfield, M. G., and R. H. Karlson.
  1969 Externally induced metamorphosis in a marine gastropod. *Am. Zool.,* **9**(4):1122. Abstract 317.
  1977 Interactions of larval settling of a marine gastropod. In NATO *Conference on Marine Natural Products,* D. J. Faulkner and W. H. Fenical, eds., New York: Plenum, pp. 403–413.

Haeckel, W.
  1913 Beiträge zur Anatomie der Gattung Ghilina. *Zool. Jahrb. Suppl.,* **13:**89–136.

Hamilton, P. V., and H. W. Ambrose III.
  1975 Swimming and orientation in *Aplysia brasiliana* (Mollusca: Gastropoda). *Mar. Behav. Physiol.,* **3:**131–143.

Hardy, A.
  1959 *The Open Sea: Its Natural History, Part II: Fish and Fisheries.* London: Collins.

Harris, A. J., S. W. Kuffler, and M. J. Dennis.
  1971 Differential chemosensitivity of synaptic and extra-synaptic areas on the neuronal surface membrane in parasympathetic neurons of the frog, tested by microapplication of acetylcholine. *Proc. R. Soc. Lond. (B),* **177:**541–553.

Harris, L. G.
  1975 Studies on the life history of two coral-eating nudibranchs of the genus *Phestilla. Biol. Bull. (Woods Hole),* **149:**539–550.

Hawkins, R. D., V. Castellucci, and E. R. Kandel.
  1976 Identification of individual neurons mediating the heterosynaptic facilitation underlying behavioral sensitization in *Aplysia. Neurosci. Abstr.,* **1:**325.

Hebb, D. O.
  1955 Drives and the C.N.S. (conceptual nervous system). *Psychol. Rev.,* **62:**243–254.
  1966 *A Textbook of Psychology,* 2nd ed. Philadelphia: Saunders.

Hegner, R. W., and J. G. Engemann.
  1968 *Invertebrate Zoology,* 2nd ed. New York: Macmillan.

Heinroth, O.
  1911 Beiträge zur Biologie, namentlich Ethologie und Psychologie der Anatiden. *Verh. V. Int. Ornithol. Kongr. (Berlin),* **1910:**589–702.

Hening, W., T. Carew, and E. R. Kandel.
  1976 Interganglionic integration of different behavioral components of a centrally commanded behavior. *Neurosci. Abstracts,* **2:**485.
  1977 Motor program for locomotion in *Aplysia* does not require peripheral feedback. *Neurosci. Abstracts,* **3:**1219.

Henkart, M.
  1975 Light-induced changes in the structure of pigmented granules in *Aplysia* neurons. *Science (Wash., D.C.),* **188:**155–157.

Herter, K.
  1931 Der Jordansche "Halbtierversuch." *Z. vergl. Physiol.,* **15:**261–308.

Heyer, C. B., S. B. Kater, and U. L. Karlsson.
  1973 Neuromuscular systems in molluscs. *Am. Zool.,* **13:**247–270.
Hill, R. B. and J. H. Welsh.
  1966 Heart, circulation, and blood cells. In *Physiology of Mollusca,* vol. 2, edited by K. M. Wilbur and C. M. Yonge. New York: Academic, pp. 125–174.
Hinde, R. A.
  1970 *Animal Behaviour: A Synthesis of Ethology and Comparative Psychology,* 2nd. ed. New York: McGraw-Hill.
Hodos, W.
  1970 Evolutionary interpretation of neural and behavioral studies of living vertebrates. In *The Neurosciences: Second Study Program,* edited by F. O. Schmitt, G. C. Quarton, T. Melnechuk, and G. Adelman. New York: Rockefeller University Press, pp. 26–39.
  1974 The comparative study of brain-behavior relationships. In *Birds: Brain and Behavior,* edited by I. J. Goodman and M. W. Schein. New York: Academic, pp. 15–25.
Hodos, W., and C. B. G. Campbell.
  1969 *Scala Naturae:* Why there is no theory in comparative psychology? *Psychol. Rev.,* **76:**337–350.
Hoffmann, F. B.
  1910 Gibt es in der Muskulatur der Mollusken periphere, kontinuierlich leitende Nervennetze bei Abwesenheit von Ganglienzellen? II. Mitteilung: Weitere Untersuchungen an den Chromatophoren der Kephalopoden. Innervation der Mantellappen von *Aplysia. Pflügers Archiv gesamte Physiol. Menschen Tiere,* **132:**43–81.
Hoffmann, H.
  1939 Opisthobranchia, Teil 1. In *Klassen und Ordnungen des Tierreichs, Vol. 3: Mollusca, Abteilung 2: Gastropoda, Buch 3: Opisthobranchia,* H. G. Bronn, ed., Leipzig: Akademische Verlagsgesellschaft M.B.H.
Howells, H. H.
  1942 The structure and function of the alimentary canal of *Aplysia punctata. Q. J. Microsc. Sci.,* **83:**357–397.
Hoy, R. R., G. D. Bittner, and D. Kennedy.
  1967 Regeneration in crustacean motoneurons: Evidence for axonal fusion. *Science (Wash., D.C.),* **156:**251–252.
Huber, F.
  1962 Central nervous control of sound productions in crickets and some speculations on its evolution. *Evolution,* **16:**429–442.
Hughes, G. M.
  1965 Neuronal pathways in the insect central nervous system. In *The Physiology of the Insect Central Nervous System,* edited by J. E. Treherne and J. W. L. Beament. New York: Academic, pp. 79–112.
Hughes, G. M., and W. D. Chapple.
  1967 The organization of nervous systems. In *Invertebrate Nervous Systems. Their Significance for Mammalian Neurophysiology,* edited by C. A. G. Wiersma. University of Chicago Press, pp. 177–195.
Hughes, G. M., and L. Tauc.
  1963 An electrophysiological study of the anatomical relations of two giant nerve cells in *Aplysia depilans. J. Exp. Biol.,* **40:**469–486.
Hughes, H. P. I.
  1970 A light and electron microscope study of some opisthobranch eyes. *Z. Zellforsch. Mikrosk. Anat.,* **106:**79–98.
Hurst, A.
  1965 Studies on the structure and function of the feeding apparatus of *Philine aperta* with a comparative consideration of some other opisthobranchs. *Malacologia,* **2:**281–347.

Hyman, L. H.
   1967 *The Invertebrates, Vol. 6: Mollusca I.* New York: McGraw-Hill.
Iles, J. F., and B. Mulloney.
   1971 Procion Yellow staining of cockroach motor neurones without the use of micro-electrodes. *Brain Res.,* **30:**397–400.
Inaba, A.
   1959 Cytological studies in molluscs, III: A chromosome survey in the opisthobranchiate gastropoda. *Annot. Zool. Jap.,* **32:**81–88.
Jacklet, J. W.
   1969a Circadian rhythm of optic nerve impulses recorded in darkness from isolated eye of *Aplysia. Science (Wash., D.C.),* **164:**562–563.
   1969b Electrophysiological organization of the eye of *Aplysia. J. Gen. Physiol.,* **53:**21–42.
   1971 A circadian rhythm in optic nerve impulses from an isolated eye in darkness. In *Bio-chronometry,* M. Menaker, ed. Washington, D.C.: National Academy of Sciences, pp. 351–362.
   1972 Circadian locomotor activity in *Aplysia. J. Comp. Physiol.,* **79:**325–341.
   1973a Model for the circadian neuronal activity of the eye of *Aplysia. Physiologist,* **16:**352.
   1973b Neuronal population interactions in a circadian rhythm in *Aplysia.* In: *Neurobiology of Invertebrates,* J. Salánki, ed. Budapest: Hungarian Academy of Sciences, pp. 363–380.
   1976 Dye marking neurons in the eye of *Aplysia. Comp. Biochem. Physiol. (A),* **55:**373–377.
Jacklet, J. W., R. Alvarez, and B. Bernstein.
   1972 Ultrastructure of the eye of *Aplysia. J. Ultrastruct. Res.,* **38:**246–261.
Jacklet, J. W., and J. Geronimo.
   1971 Circadian rhythm: Population of interacting neurons. *Science (Wash., D.C.),* **174:**299–302.
Jacklet, J. W., and K. Lukowiak.
   1974 Neural processes in habituation and sensitization in model systems. *Progr. Neurobiol. (Oxf.),* **4:**1–56.
Jacklet, J. W., and J. Rine.
   1977 Facilitation at neuromuscular junctions: Contribution to habituation and dishabituation of the *Aplysia* gill withdrawal reflex. *Proc. Nat. Acad. Sci. U.S.A.,* **74:**1267–1271.
Jacobson, M.
   1978 *Developmental Neurobiology,* 2nd ed. New York: Plenum Press.
Jahan-Parwar, B.
   1972a Behavioral and electrophysiological studies on chemoreception in *Aplysia. Am. Zool.,* **12:**525–537.
   1972b Central projection of chemosensory pathways of *Aplysia. Physiologist,* **15:**180.
   1976 Proprioceptive reflexes in the buccal mass of *Aplysia. Neuroscience Abstracts,* **2:**347.
Jahan-Parwar, B., and S. M. Fredman.
   1978a Control of pedal and parapodial movements in *Aplysia.* I: Proprioceptive and tactile reflexes. *J. Neurophysiol.,* **41:** 600–608.
   1978b Control of pedal and parapodial movements in Aplysia. II: Cerebral ganglion neurons. *J. Neurophysiol.,* **41:**609–620.
   1978c Pedal locomotion in *Aplysia.* I: Sensory and motor functions of foot neurons. *Comp. Physiol. Biochem.,* in press.
Jahan-Parwar, B., M. Smith, and R. von Baumgarten.
   1969 Activation of neurosecretory cells in *Aplysia* by osphradial stimulation. *Am. J. Physiol.,* **216:**1246–1257.
Jennings, H. S.
   1906 *Behavior of the Lower Organisms.* New York: Columbia University Press.
Jones, H. D.
   1970 Hydrostatic pressures within the heart and pericardium of *Patella vulgata* L. *Comp. Biochem. Physiol.,* **34:**263–272.

1973 The mechanism of locomotion of *Agriolimax reticulatus* (Mollusca: Gastropoda), *J. Zool. (London)*, **171**:489–498.

Jordan, H.

1901 Die Physiologie der Locomotion bei *Aplysia limacina. Z. Biol.*, **41**:196–238.

1917 Das Wahrnehmen der Nahrung bei *Aplysia limacina* und *Aplysia depilans. Biol. Zentralbl.*, **37**:2–9.

1929 *Allgemeine vergleichende Physiologie der Tiere.* Berlin and Leipzig: W. de Gruytert.

Jourdan, F., and G. Nicaise.

1971 L'Ultrastructure des synapses dans le ganglion pleural de l'*Aplysie. J. Microsc. (Paris)*, **11**:69–70.

Jouvet, M.

1972 The role of monoamines and acetylcholine containing neurons in the regulation of the sleep-walking cycle. *Ergeb. Physiol. Biol. Chem. Exp. Pharmakol.*, **64**:166–307.

Kandel, E. R.

1976 *Cellular Basis of Behavior: An Introduction to Behavioral Neurobiology:* San Francisco: W. H. Freeman.

Kandel, E. R., M. Brunelli, J. Byrne, and V. Castellucci.

1976 A common presynaptic locus for the synaptic changes underlying short-term habituation and sensitization of the gill-withdrawal reflex in *Aplysia.* In *Cold Spring Harbor. Symp. Quant. Biol.*, **40**:465–482.

Kandel, E. R., and W. A. Spencer.

1968 Cellular neurophysiological approaches in the study of learning. *Physiol. Rev.*, **48**:65–134.

Kandel, E. R., and L. Tauc.

1965a Heterosynaptic facilitation in neurones of the abdominal ganglion of *Aplysia depilans. J. Physiol. (London)*, **181**:1–27.

1965b Mechanism of heterosynaptic facilitation in the giant cell of the abdominal ganglion of *Aplysia depilans. J. Physiol. (London)*, **181**:28–47.

Kaneko, C. R. S., S. B. Kater, and M. Merickel.

1978 Centrally programmed feeding in *Helisoma* controlled by electrically coupled network. I: Premotor neuron identification and characteristics. *Brain Res.*, **146**:1–21.

Karsznia, R., W. Spielmann and J. Trinkhaus.

1969 Zum Nachweis der Blutgruppenantigene A und B an menschlichen Spermien. *Gynaecologia*, **167**:14–22.

Kater, S. B.

1974 Feeding in *Helisoma trivolvis:* The morphological and physiological bases of a fixed action pattern. *Am. Zool.*, **14**:1017–1036.

Kater, S. B., and C. H. Fraser Rowell.

1973 Integration of sensory and centrally programmed components in generation of cyclical feeding activity of *Helisoma trivolvis. J. Neurophysiol.*, **36**:142–155.

Kater, S. B., C. Heyer, and J. P. Hegmann.

1971 Neuromuscular transmission in the gastropod mollusc *Helisoma trivolvis:* Identification of motoneurons. *Z. vgl. Physiol.*, **74**:127–139.

Kater, S. B., and C. Nicholson, eds.

1973 *Intracellular Staining in Neurobiology.* New York: Springer Verlag.

Kay, E. A.

1964 The Aplysiidae of the Hawaiian Islands. *Proc. Malacol. Soc. Lond.*, **36**:173–190.

Kennedy, D., and W. J. Davis.

1977 The organization of invertebrate motor systems. in "Cellular Biology of Neurons" (vol. 1, sec. 1, *Handbook of Physiology, The Nervous System*), edited by E. R. Kandel. Baltimore: Williams and Wilkins, pp. 1023–1087.

King, D. G.

1975 Organization of Crustacean neuropil: Identified synaptic processes in Stomatogastric ganglion. *Neurosci. Abstracts,* **1:**879.

1976a Organization of crustacean neuropil. I: Patterns of synaptic connections in lobster stomatogastric ganglion. *J. Neurocytol.,* **5:**207–237.

1976b Organization of crustacean neuropil. II: Distribution of synaptic contacts on identified motor neurons in lobster stomatogastric ganglion. *J. Neurocytol.,* **5:**239–266.

Kirschner, L. B.

1967 Comparative physiology: Invertebrate excretory organs. *Annu. Rev. Physiol.,* **29:** 169–196.

Klein, M., and E. R. Kandel.

1978 Presynaptic modulation of voltage-dependent Ca^{2+} current: Mechanism for behavioral sensitization in *Aplysia californica. PNAS,* **75:**3512–3516.

Knight, J. B.

1952 Primitive fossil gastropods and their bearing on gastropod classification. *Smithson. Misc. Collect.,* **117(13):**1–56.

Knight, J. B., and E. L. Yochelson.

1958 A reconsideration of the relationships of the Monoplacophora and the primitive Gastropoda. *Proc. Malacol. Soc. London,* **33:**37–48.

Koester, J., and E. R. Kandel.

1977 Further identification of neurons in the abdominal ganglion of *Aplysia* using behavioral criteria. *Brain Res.,* **121:**1–20.

Koester, J., E. Mayeri, G. Liebeswar, and E. R. Kandel.

1974 Neural control of circulation in *Aplysia.* II. Interneurons. *J. Neurophysiol.,* **37:**476–496.

Kohn, A. J.

1961 Chemoreception in gastropod molluscs. *Am. Zool.,* **1:**291–308.

Koningsor, R. L., Jr., N. McLean, and D. Hunsaker II.

1972 Radiographic evidence for a digestive cellulase in the sea hare, *Aplysia vaccaria* (Gastropoda: Opisthobranchia). *Comp. Biochem. Physiol. (B),* **43:**237–240.

Kovac, M. P., and W. J. Davis.

1977 Behavioral choice: Neural mechanism in *Pleurobranchaea. Science (Wash., D.C.),* **198:**632–634.

Krasne, F. B., and Ch. A. Stirling.

1972 Synapses of crayfish abdominal ganglia with special attention to afferent and efferent connections of the lateral giant fibers. *Z. Zellforsch. Mikrosk. Anat.,* **127:**526–544.

Krauhs, J. M., L. A. Sordahl, and A. M. Brown.

1977 Isolation of pigmented granules involved in extra-retinal photoreception in *Aplysia californica* neurons. *Biochimica et Biophysica Acta,* **471:**25–31.

Kriegstein, A. R.

1977a Development of the nervous system of *Aplysia californica. Proc. Natl. Acad. Sci. U.S.A.,* **74:**375–378.

1977b Stages in the post-hatching development of *Aplysia californica. J. Exp. Zool.,* **199:** 275–288.

Kriegstein, A. R., V. Castellucci, and E. R. Kandel.

1974 Metamorphosis of *Aplysia californica* in laboratory culture. *Proc. Nat. Acad. Sci. U.S.A.,* **71:**3654–3658.

Krijgsman, B. J., and G. A. Divaris.

1955 Contractile and pacemaker mechanisms of the heart of molluscs. *Biol. Rev. Camb. Philos. Soc.,* **30:**1–39.

Krull, H.

1934 Die Aufhebung der Chiastoneurie bei den Pulmonaten. *Zool. Anz.,* **105:**173–182.

Kuffler, S. W., and J. G. Nicholls.

1966 The physiology of neuroglial cells. *Ergeb. Physiol. Biol. Chem. Exp. Pharmakol.,* **57:**1–90.

1976 *From Neuron to Brain: A Cellular Approach to the Function of the Nervous System.* Sunderland, Mass.: Sinauer.

Kumé, M., and K. Dan, eds.

1957 *Invertebrate Embryology,* Translated by J. C. Dan. Published for the National Library of Medicine (Public Health Service, U.S. Dept. of Health, Education, and Welfare) and the National Science Foundation, Washington, D.C., by NOLIT Publishing House, Belgrade, Yugoslavia, 1968. (Distributed by Clearinghouse for Federal Scientific and Technical Information, U.S. Dept. of Commerce, National Bureau of Standards, Institute for Applied Technology, Springfield, Virginia.)

Kunze, H.

1918 Über das ständige Auftreten bestimmter Zellelemente im Centralnervensystem von *Helix pomatia* L. *Zool. Anz.,* **49:**123–137.

Kupfermann, I.

1965 Locomotor activity patterns in *Aplysia californica. Physiologist,* **8:**214.

1967 Stimulation of egg laying: Possible neuroendocrine function of bag cells of abdominal ganglion of *Aplysia californica. Nature (London),* **216:**814–815.

1968 A circadian locomotor rhythm in *Aplysia californica. Physiol. Behav.,* **3:**179–181.

1970 Stimulation of egg-laying by extracts of neuroendocrine cells (bag cells) of abdominal ganglion of *Aplysia. J. Neurophysiol.,* **33:**877–881.

1972 Studies on the neurosecretory control of egg laying in *Aplysia. Am. Zool.,* **12:**513–519.

1974a Feeding behavior in *Aplysia:* A simple system for the study of motivation. *Behav. Biol.,* **10:**1–26.

1974b Dissociation of the appetitive and consummatory phases of feeding behavior in *Aplysia:* A lesion study. *Behav. Biol.,* **10:**89–97.

Kupfermann, I., and T. J. Carew.

1974 Behavior patterns of *Aplysia californica* in its natural environment. *Behav. Biol.,* **12:** 317–337.

Kupfermann, I., T. J. Carew, and E. R. Kandel.

1974 Local, reflex, and central commands controlling gill and siphon movements in *Aplysia. J. Neurophysiol.,* **37:**996–1019.

Kupfermann, I., and J. Cohen.

1971 The control of feeding by identified neurons in the buccal ganglion of *Aplysia. Am. Zool.,* **11:**667, Abstract #243.

Kupfermann, I., and E. R. Kandel.

1969 Neuronal controls of a behavioral response mediated by the abdominal ganglion of *Aplysia. Science (Wash., D.C.),* **164:**847–850.

1970 Electrophysiological properties and functional interconnections of two symmetrical neurosecretory clusters (bag cells) in abdominal ganglion of *Aplysia. J. Neurophysiol.,* **33:**865–876.

Kupfermann, I., and M. Kesselman.

UNPUBL. Withdrawal reflexes in intact and "deganglionated" *Aplysia californica.*

Kupfermann, I., and H. Pinsker.

1968 A behavioral modification of the feeding reflex in *Aplysia californica. Commun. Behav. Biol. Part A,* **2:**13–17.

Kupfermann, I., H. Pinsker, V. Castellucci, and E. R. Kandel.

1971 Central and peripheral control of gill movements in *Aplysia. Science (Wash., D.C.),* **174:**1252–1256.

Kupfermann, I., and K. R. Weiss.

1974 Functional studies on the metacerebral cells in *Aplysia. Abst. and Proc. Soc. Neurosci.,* **3:**375.

1976 Water regulation by a presumptive hormone contained in identified neurosecretory cell R15 of *Aplysia. J. Gen. Physiol.,* **67:**113–123.

De Lacaze-Duthiers, H.
1898 Les ganglions dits Palléaux et le stomato-gastrique de quelques Gastéropodes. *Arch. Zool. Exp. Gen., 3rd Serie,* **6:**331–428.

Lang, A.
1900 *Lehrbuch der vergleichenden Anatomie der wirbellosen Thiere, Part I: Mollusca,* revised by K. Hescheler. Jena: Gustav Fischer.

Lang, F., and H. L. Atwood.
1973 Crustacean neuromuscular mechanisms: Functional morphology of nerve terminals and the mechanism of facilitation. *Am. Zool.,* **13:**337–355.

Lankester, E. R.
1883 Mollusca. In *Encyclopaedia Britannica,* vol. 16., 9th ed. Chicago: Werner, pp. 632–695.

Lasek, R. J., and W. J. Dower.
1971 *Aplysia californica:* Analysis of nuclear DNA in individual nuclei of giant neurons. *Science (Wash., D.C.)* **172:**278–280.

Lasek, R. J., C. K. Lee, and R. J. Przybylski.
1972 Granular extensions of the nucleoli in giant neurons of *Aplysia californica. J. Cell Biol.,* **55:**237–242.

Laverack, M. S.
1970 Responses of a receptor associated with the buccal mass of *Aplysia dactylomela. Comp. Biochem. Physiol.,* **33:**471–473.

Lee, R. M.
1969 *Aplysia* behavior: Effects of contingent water-level variation. *Communic. Behav. Biol. Part A,* **4:**157–164.
1970 *Aplysia* behavior: Operant-response differentiation. *Proc. Annu. Conv. Am. Psychol. Assoc.,* **78:**249–250.

Lee, R. M., and R. J. Liegeois.
1974 Motor and sensory mechanisms of feeding in *Pleurobranchaea. J. Neurobiol.,* **5:**545–564.

Lee, R. M., and R. A. Palovcik.
1976 Behavioral states and feeding in the gastropod *Pleurobranchaea. Behav. Biol.,* **16:** 251–266.

Lee, R. M., M. R. Robbins, and R. Palovcik.
1974 *Pleurobranchaea* behavior: Food finding and other aspects of feeding, *Behav. Biol.,* **12:**297–315.

Lehrman, D. S.
1961 Hormonal regulation of parental behavior in birds and infrahuman mammals. In *Sex and Internal Secretions,* 3rd ed., vol. 2, edited by W. C. Young. Baltimore: Williams and Wilkins, pp. 1268–1382.

Lemche, H.
1957 A new living deep-sea mollusc of the Cambro-Devonian class Monoplacophora. *Nature (London),* **179:**413–416.
1958 Molluscan phylogeny in the light of *Neopilina. Proc. Int. Congr. Zool., Lond.,***15:**380–381.

Lemche, H. and K. G. Wingstrand.
1959 The anatomy of *Neopilina galathea* Lemche. 1957 (Mollusca. Tryblidiacea). *Galathea Rep.,* **3:**9–71.

Lent, C. M.
1973 Retzius cells: Neuroeffectors controlling mucus release by the leech. *Science (Wash., D.C.).* **179:**693–696.

Levi-Montalcini, R.
1975 NGF: An uncharted route. In *The Neurosciences: Paths of Discovery,* edited by F. G. Worden, J. P. Swazey, and G. Adelman. Cambridge, Mass.: MIT Press, pp. 245–265.

Levitan, I. B., and S. H. Barondes.
  1974 Octopamine- and serotonin-stimulated phosphorylation of specific protein in the ab-
       dominal ganglion of *Aplysia californica. Proc. Nat. Acad. Sci. U.S.A.,* **71**:1145–1148.
Levitan, H., L. Tauc, and J. P. Segundo.
  1970 Electrical transmission among neurons in the buccal ganglion of a mollusc, *Navanax
       inermis. J. Gen. Physiol.,* **55**:484–496.
Lewis, E. R., T. E. Everhart, and Y. Y. Zeevi.
  1969 Studying neural organization in *Aplysia* with the scanning electron microscope. *Science
       (Wash., D.C.),* **165**:1140–1143.
Lewis, J. R.
  1964 *The Ecology of Rocky Shores.* London: The English Universities Press Ltd.
Lewis, P. R.
  1952 The free amino-acids of invertebrate nerve. *Biochem. J.,* **52**:330–338.
Lewis, R. D.
  1969 Studies on the locomotor activity of the slug *Arion ater* (Linnaeus). II: Locomotor
       activity rhythms. *Malacologia,* **7**:307–312.
Lickey, M. E.
  1968 Learned behavior in *Aplysia vaccaria. J. Comp. Physiol. Psychol.,* **66**:712–718.
Lickey, M. E., and R. W. Berry.
  1966 Learned behavioral discrimination of food objects by *Aplysia californica. Physiologist,*
       9:230.
Lickey, M. E., D. Block, D. J. Hudson, and J. T. Smith.
  1976 Circadian oscillators and photoreceptors in the gastropod, *Aplysia. Photochem. Photo-
       biol.,* **23**:253–273.
Liebeswar, G., J. E. Goldman, J. Koester, and E. Mayeri.
  1975 Neural control of circulation in *Aplysia.* III: Neurotransmitters. *J. Neurophysiol.,*
       38:767–779.
Linneaus, C.
  1756 *Systema Naturae.* 9th ed.
  1758–1759 *Systema Naturae.* 10th ed. rev. Holmiae, imprensis direct. L. Salvii., 2v.
  1766–1768 *Systema Naturae.* 12th ed. rev. Holmiae, imprensis direct. L. Salvii. 3 v in 4.
Lissmann, H. W.
  1945a The mechanism of locomotion in gastropod molluscs. I: Kinematics. *J. Exp. Biol.,*
        21:58–69.
  1945b The mechanism of locomotion in gastropod molluscs. II: Kinetics. *J. Exp. Biol.,* **22**:
        37–50.
Little, C.
  1965 The formation of urine by the prosobranch gastropod mollusc *Viviparus viviparus* Linn.
       J. Exp. Biol., **43**:39–54.
Lloyd, D. P. C.
  1957 Input-output relation in a flexion reflex. *J. Gen. Physiol.,* **41**:297–306.
Lockard, R. B.
  1971 Reflections on the fall of comparative psychology: Is there a message for us all? *Am.
       Psychol.,* **26**:168–79.
Loeb, J.
  1918 *Forced Movements, Tropisms, and Animal Conduct.* Philadelphia: Lippincott.
Loh, Y. P., Y. Sarne, and H. Gainer.
  1975 Heterogeneity of proteins synthesized, stored and released by the bag cells of *Aplysia
       californica. J. Comp. Physiol. (B),* **100**:283–295.
Luborsky-Moore, J. L.
  1975 Ultrastructure and fluorescence histochemistry of secondary cells in the eye of *Aplysia
       californica. Proc. 33rd Annu. Meet. Electron Microsc. Soc. Am.,* pp. 468–469.

Luborsky-Moore, J. L., and J. W. Jacklet.

1976 Localization of catecholamines in the eyes and other tissues of *Aplysia. J. Histochem. and Cytochem.*, **24:**1150–1158.

Lukowiak, K., and J. W. Jacklet.

1972 Habituation and dishabituation: Interactions between peripheral and central nervous systems in *Aplysia. Science (Wash., D.C.),* **178:**1306–1308.

Lupo di Prisco, C., and F. Dessi' Fulgheri.

1975 Alternative pathways of steroid biosynthesis in gonads and hepatopancreas of *Aplysia depilans. Comp. Biochem. Physiol. (B),* **50:**191–195.

Lupo di Prisco, C., F. Dessi' Fulgheri, and M. Tomasucci.

1973 Identification and biosynthesis of steroids in the marine mollusc *Aplysia depilans. Comp. Biochem. Physiol. (B),* **45:**303–310.

MacFarland, F. M.

1909 *The Opisthobranchiate Mollusca of the Branner–Agassiz Expedition to Brazil.* Leland Stanford Junior University Publications, University Series No. 2.

1918 Reports on the scientific results of the expedition to the tropical Pacific, in charge of Alexander Agassiz, by the U.S. Fish Commission Steamer "Albatross", from August, 1899, to March, 1900, Commander Jefferson F. Moser, U.S.N., Commanding XIX. The Dolabellinae. *Mem. Mus. Comp. Zool. Harv. Univ.,* **35:**297–348.

MacGinitie, G. E.

1934 The egg-laying activities of the sea hare, *Tethys californicus* (Cooper). *Biol. Bull. (Woods Hole),* **67:**300–303.

1935 Ecological aspects of a California marine estuary. *Am. Midl. Nat.,* **16:**629–765.

MacGinitie, G. E., and N. MacGinitie.

1968 *Natural History of Marine Animals,* 2nd ed. New York: McGraw-Hill.

Magoun, H. W.

1954 The ascending reticular system and wakefulness. In *Brain Mechanisms and Consciousness,* edited by J. F. Delafresnaye, Blackwell, Oxford, pp. 1–15.

Malmo, R. B.

1959 Activation: A neuropsychological dimension. *Psychol. Rev.,* **66:**367–386.

Manning, A.

1965 *Drosophila* and the evolution of behaviour. In *Viewpoints in Biology,* vol. 4, edited by J. D. Carthy and C. L. Duddington. London: Butterworths, pp. 125–169.

1972 *An Introduction to Animal Behavior,* 2nd ed. Reading, Mass.: Addison-Wesley.

1976 Animal learning: Ethological approaches. In *Neural Mechanisms of Learning and Memory,* edited by M. R. Rosenzweig and E. L. Bennett. Cambridge, Mass.: MIT Press, pp. 147–158.

Manwell, C.

1960 Comparative physiology: Blood pigments. *Annu. Rev. Physiol.,* **22:**191–244.

Marcus, E.

1953 Three Brazilian sand-opisthobranchia. *Bol. Fac. Filos. Cienc. Let. Univ. Sao Paulo Ser. Zool.,* #18:165–203.

1972 On the Anaspidea (Gastropods: Opisthobranchia) of the warm waters of the western Atlantic. *Bull. Mar. Sci.,* **22:**(4):841–874.

Margoliash, E.

1972 The molecular variations of cytochrome *C* as a function of the evolution of species. *The Harvey Lectures, 1970–1971,* Series 66, New York: Academic Press, pp. 177–247.

Marler, P., and W. J. Hamilton III.

1966 *Mechanisms of Animal Behavior.* New York: Wiley.

Martin, A. W., and F. M. Harrison.

1966 Excretion. In *Physiology of Mollusca,* edited by K. M. Wilbur and C. M. Yonge, Vol. II. New York: Academic Press, pp. 353–386.

Martin, A. W., D. M. Stewart, and F. M. Harrison.
    1965 Urine formation in the pulmonate land snail, *Achatina fulica. J. Exp. Biol.,* **42:**99–123.
Martin, R.
    1966 On the swimming behaviour and biology of *Notarchus punctatus* Philippi (Gastropoda, Opisthobranchia). *Pubbl. Stn. Zool. Napoli.,* **35:**61–75.
Mayeri, E., J. Koester, I. Kupfermann, G. Liebeswar, and E. R. Kandel.
    1974 Neural control of circulation in *Aplysia.* I: Motoneurons. *J. Neurophysiol.,* **37:**458–475.
Mayeri, E., and S. Simon.
    1975 Modulation of synaptic transmission and burster neuron activity after release of a neurohormone. *Neuroscience Abstracts,* **1:**912.
Mayr, E.
    1942 *Systematics and the Origin of Species from the Viewpoint of a Zoologist.* Columbia University Press. (Corrected edition, with new preface, published 1964, New York: Dover.)
    1963 *Animal Species and Evolution.* Harvard University Press.
    1974 Behavior programs and evolutionary strategies. *Amer. Scientist,* **62:**650–659.
Mazzarelli, G.
    1893 Monografia delle *Aplysiidae* del Golfo de Napoli. *Mem. Mat. Fis. Soc. Ital. Sci. (Rome),* 3rd serie, **9:**1–222.
McCaman, R. E., and S. A. Dewhurst.
    1970 Choline acetyltransferase in individual neurons of *Aplysia californica. J. Neurochem.,* **17:**1421–1426.
McKee, A. E., and M. L. Wiederhold.
    1974 *Aplysia* statocyst receptor cells: Fine structure. *Brain Res.,* **81:**310–313.
Meglitsch, P. A.
    1972 *Invertebrate Zoology,* 2nd ed. London: Oxford University Press.
Merickel, M. B., E. D. Eyman, and S. B. Kater.
    1977 Analysis of a network of electrically coupled neurons producing rhythmic activity in the snail *Helisoma trivolvis. I.E.E.E. Trans. Biomed. Eng.,* **24:**277–287.
Merton, H.
    1920 Untersuchungen über die Hautsinnesorgane der Mollusken. I: Opisthobranchia. *Abh. Senckenb. Naturforsch. Ges.,* **36:**447–473.
Miller, M. C.
    1960 A note on the life history of *Aplysia punctata* Cuvier in Manx waters. *Proc. Malacol. Soc. Lond.,* **34:**165–167.
    1962 Annual cycles of some Manx nudibranchs, with a discussion of the problem of migration. *J. Anim. Ecol.,* **31:**545–569.
Miller, N. E.
    1967 Certain facts of learning relevant to the search for its physical basis. In *The Neurosciences: A Study Program,* edited by G. C. Quarton, T. Melnechuk, and F. O. Schmitt. New York: Rockefeller University Press, pp. 643–652.
Milne-Edwards, H.
    1849 Observations et expériences sur la circulation chez les Mollusques. *Mem. Acad. Sci. Inst. France.,* **20:**443–484.
Morgan, T. H.
    1927 *Experimental Embryology.* Columbia University Press.
Moritz, C. E.
    1939 Organogenesis in the gasteropod *Crepidula adunca* Sowerby. *Univ. Calif. Publ. Zool.,* **43:**217–248.
Morris, D.
    1958 The reproductive behaviour of the ten-spined stickleback (Pygosteus Pungitius L.). *Behaviour,* Suppl. VI, Leiden: Brill.

Morris, M. C.
   1950 Dilation of the foot in *Uber (Polinices) strangei* (Mollusca, class Gastropoda). *Proc. Linn. Soc. N.S.W.,* **75:**70–80.
Morton, J. E.
   1958a *Molluscs.* London: Hutchinson University Library.
   1958b Torsion and the adult snail: A re-evaluation. *Proc. Malacol. Soc. Lond.,* **33:**2–10.
   1963 The molluscan pattern: Evolutionary trends in a modern classification. *Proc. Linn. Soc. Lond.,* **174:**53–72.
   1964 Locomotion. In *Physiology of Mollusca,* vol. 1, edited by K. M. Wilbur and C. M. Yonge. New York: Academic, pp. 383–423.
Morton, J. E., and C. M. Yonge.
   1964 Classification and structure of the mollusca. In *Physiology of Mollusca,* vol. 1, edited by K. M. Wilbur and C. M. Yonge. New York: Academic. pp. 1–58.
Moruzzi, G., and H. W. Magoun.
   1949 Brain stem reticular formation and activation of the EEG. *EEG Clin. Neurophysiol.,* **1:**455–473.
Mpitsos, G. J., and S. D. Collins.
   1975 Learning: Rapid aversive conditioning in the gastropod mollusk *Pleurobranchaea. Science (Wash., D.C.).* **188:**954–957.
Mpitsos, G. J., and W. J. Davis.
   1973 Learning: Classical and avoidance conditioning in the mollusk *Pleurobranchaea. Science (Wash., D.C.),* **180:**317–320.
Muller, K. J., and U. J. McMahan.
   1975 The arrangement and structure of synapses formed by specific sensory and motor neurons in segmental ganglia of the leech. *American Association of Anatomists, Eighty Eighth Annual Session, Abstracts.* p. 432.
   1976 The shapes of sensory and motor neurones and the distribution of their synapses in ganglia of the leech: A study using intracellular injection of horseradish peroxidase. *Proc. R. Soc. London (B),* **194:**481–499.
Murray, M. J.
   1971 *The Biology of a Carnivorous Mollusc: Anatomical, Behavioral, Electrophysiological Observations on* Navanax inermis. Ph.D. dissertation, University of California, Berkeley.
Naef, A.
   1911 Studien zur generellen Morphologie der Mollusken. I: Teil: Über Torsion und Asymmetrie der Gastropoden. *Ergeb. Fortschr. Zool.,* **3:**73–164.
   1926 Studien zur generellen Morphologie der Mollusken. III: Teil: Die typischen Beziehungen der Weichtierklassen untereinander und das Verhältnis ihrer Urformen zu anderen Cölomaten. *Ergeb. Fortschr. Zool.,* **6:**27–124.
Neu, W.
   1932 Wie schwimmt *Aplysia depilans* L.? *Z. vgl. Physiol.,* **18:**244–254.
Newby, N. A.
   1972 *Habituation to Light and Spontaneous Activity in the Isolated Siphon of Aplysia: The Effects of Synaptically Active Pharmacological Agents.* Ph.D. thesis. Case Western Reserve University, Cleveland.
Newell, R. C.
   1970 *Biology of Intertidal Animals.* New York: American Elsevier.
Nordlander, R. H., and J. S. Edwards.
   1968 Morphological cell death in the post-embryonic development of the insect optic lobes. *Nature (London),* **218:**780–781.
Novikoff, A. B., and E. Holtzman.
   1976 *Cells and Organelles.* 2nd ed. New York: Holt, Rinehart, and Winston.

Odhner, N. H.

1932 Beiträge zur Malakozoologie der Kanarischen Inseln. Lamellibranchien, Cephalopoden, Gastropoden. *Ark. Zool. (A),* **23**(14):1-116.

1939 Opisthobranchiate Mollusca from the Western and Northern coasts of Norway. *K. Nor. Vidensk. Selsk. Skr.,* No. 1, pp. 1-93.

Orkand, P. M., and R. K. Orkand.

1975 Neuromuscular junctions in the buccal mass of *Aplysia:* Fine structure and electrophysiology of excitatory transmission. *J. Neurobiol.,* **6**:531-548.

Orkand, R. K.

1977 Glial cells. In "Cellular Biology of Neurons" (Vol. 1, Sect. 1, *Handbook of Physiology, The Nervous System*), edited by E. R. Kandel. Baltimore: Williams and Wilkins, pp. 625-650.

Ortmann, R.

1960 Neurosecretion. In "Neurophysiology" (Vol. 2, Sect. 1, *Handbook of Physiology,* edited by. J. Field, H. W. Magoun, and V. E. Hall. Washington, D.C.: Am. Physiol. Soc., pp. 1039-1065.

Owen, G.

1966a Digestion. In *Physiology of Mollusca,* vol. 2, edited by K. M. Wilbur and C. M. Yonge. New York: Academic, pp. 53-96.

1966b Feeding. In *Ibid.,* pp. 1-51.

Paine, R. T.

1963a Trophic relationships of eight sympatric predatory gastropods. *Ecology,* **44**:63-73.

1963b Food recognition and predation on opisthobranchs by *Navanax inermis* (Gastropoda: Opisthobranchia). *Veliger,* **6**:1-9.

1965 Natural history, limiting factors, and energetics of the opisthobranch *Navanax inermis. Ecology,* **46**:603-619.

Palay, S. L.

1960 The fine structure of secretory neurons in the preoptic nucleus of the goldfish (*Carassius auratus*). *Anat. Rec.,* **138**:417-443.

Palay, S. L., and V. Chan-Palay.

1977 General morphology of neurons and neuroglia. In "Cellular Biology of Neurons" (Vol. 1, Sec. 1, *Handbook of Physiology, The Nervous System*), edited by E. R. Kandel. Baltimore: Williams and Wilkins, pp. 5-37.

Parker, G. H.

1917 The pedal locomotion of the sea-hare *Aplysia californica. J. Exp. Zool.,* **24**:139-145.

1919 *The Elementary Nervous System.* Philadelphia: Lippincott.

Pavlov, I. P.

1906 The scientific investigation of the psychical faculties or processes in the higher animals. *Science (Wash., D.C.),* **24**:613-619.

1927 *Conditioned Reflexes: An Investigation of the Physiological Activity of the Cerebral Cortex,* translated and edited by G. V. Anrep. London: Oxford University Press.

1928 *Lectures on Conditioned Reflexes,* translated and edited by W. H. Gantt. New York: International Publishers.

Pearson, K. G.

1972 Central programming and reflex control of walking in the cockroach. *J. Exp. Biol.,* **56**:173-193.

Pelseneer, P.

1888 Sur la valeur morphologique des bras et la composition du système nerveux central des Céphalopodes. *Arch. Biol.,* **8**:723-756.

1894 Recherches sur divers Opisthobranches. *Mém. couronnes Mem. Savants étrangers Acad. R. Sci. Belg.,* **53**:1-157.

1906 *Mollusca.* Part V, *A Treatise on Zoology,* edited by E. R. Lankester. London: Adam & Charles Black.

1912 Recherches sur l'embryologie des Gastropodes. *Acad. R. Belg. Cl. Sci. Mém., 2nd Ser.,* **3**(6):1–167.

Peltrera, A.

1940 Le capacità regolative dell'uovo di *Aplysia limacina* L. studiate con la centrifugazione e con le reazioni vitali. *Pubbl. Stn. Zool. Napoli,* **18**:20–49.

Pentreath, V. W., and M. S. Berry.

1975 Ultrastructure of the terminals of an identified dopamine-containing neurone marked by intracellular injection of radioactive dopamine. *J. Neurocytol.,* **4**(3):249–260.

Perdeck, A. C.

1958 The isolating value of specific song patterns in two sibling species of grasshoppers (*Chorthippus brunneus* Thunb. and *C. biguttulus* L.) *Behaviour,* **12**:1–75.

Peretz, B.

1969 Central neuron initiation of periodic gill movements. *Science (Wash., D.C.),* **166**:1167–1172.

1970 Habituation and dishabituation in the absence of a central nervous system. *Science (Wash., D.C.),* **169**:379–381.

Peretz, B., and D. B. Howieson.

1973 Central influence on peripherally mediated habituation of an *Aplysia* gill-withdrawal response. *J. comp. Physiol.,* **84**:1–18.

Peretz, B., J. W. Jacklet, and K. Lukowiak.

1976 Habituation of reflexes in *Aplysia:* Contribution of the peripheral and central nervous systems. *Science (Wash., D.C.),* **191**:396–399.

Peretz, B., and K. D. Lukowiak.

1975 Age-dependent CNS control of the habituating gill withdrawal reflex and of correlated activity in identified neurons in *Aplysia. J. comp. Physiol. (A),* **103**:1–17.

Peretz, B., and R. Moller.

1974 Control of habituation of the withdrawal reflex by the gill ganglion in *Aplysia. J. Neurobiol.,* **5**:191–212.

Perlman, A. J.

1975 *Central and Peripheral Control of the Siphon-Withdrawal Reflex in* Aplysia. Ph.D. dissertation, New York University School of Medicine.

Peters, A., S. L. Palay, and H. de F. Webster.

1976 *The Fine Structure of the Nervous System: The Neurons and Supporting Cells.* Philadelphia: W. B. Saunders.

Phillips, D. W.

1975 Localization and electrical activity of the distance chemoreceptors that mediate predator avoidance behaviour in *Acmaea limatula* and *Acmaea scutum* (Gastropoda, Prosobranchia). *J. Exp. Biol.,* **63**:403–412.

Pinsker, H. M., W. A. Hening, T. J. Carew, and E. R. Kandel.

1973 Long-term sensitization of a defensive withdrawal reflex in *Aplysia. Science (Wash., D.C.),* **182**:1039–1042.

Pinsker, H., I., Kupfermann, V. Castellucci, and E. Kandel.

1970 Habituation and dishabituation of the gill-withdrawal reflex in *Aplysia. Science (Wash., D.C.),* **167**:1740–1742.

Pitman, R. M., C. D. Tweedle, and M. J. Cohen.

1972 Branching of central neurons: Intracellular cobalt injection for light and electron microscopy. *Science (Wash., D.C.),* **176**:412–414.

Potts, W. T. W.

1967 Excretion in the molluscs. *Biol. Rev. Camb. Philos. Soc.,* **42**:1–41.

Power, M. E.

1943 The effect of reduction in numbers of ommatidia upon the brain of *Drosophila melanogaster*. *J. Exp. Zool.*, **94**:33–71.

Preston, R. J., and Lee, R. M.

1971 Food localization in *Aplysia californica:* Chemical cues and tactile stimulation. Paper read at the meeting of the APA in Washington, D.C.

1973 Feeding behavior in *Aplysia californica:* Role of chemical and tactile stimuli. *J. Comp. Physiol. Psychol.*, **82**:368–381.

Prior, D. J.

1972a Electrophysiological analysis of peripheral neurones and their possible role in the local reflexes of a mollusc. *J. Exp. Biol.*, **57**:133–145.

1972b A neural correlate of behavioural stimulus intensity discrimination in a mollusc. *Ibid.*, pp. 147–160.

Prior, D. J., and A. Gelperin.

1974 Behavioral and physiological studies on locomotion in the giant garden slug *Limax maximus*. *Malacol. Rev.*, **7**:50–51.

1977 Autoactive molluscan neuron: Reflex function and synaptic modulation during feeding in the terrestrial slug, *Limax maximus*. *J. Comp. Physiol. (A)*, **114**:217–232.

Prosser, C. L.

1946 The physiology of nervous systems of invertebrate animals. *Physiol. Rev.*, **26**:337–382.

Purchon, R. D.

1968 *The Biology of the Mollusca*. Oxford: Pergamon Press.

Purves, D., and U. J. McMahan.

1972 The distribution of synapses on a physiologically identified motor neuron in the central nervous system of the leech. An electron-microscope study after the injection of the flourescent dye Procion yellow. *J. Cell. Biol.*, **55**:205–220.

Ramsay, J. A.

1952 *A Physiological Approach to the Lower Animals*. Cambridge: The University Press.

Rang, S.

1828 Histoire naturelle des Aplysiens. Première famille de l'ordre des Tectibranches. In *Histoire Naturelle, Genérale et Particulière des Mollusques,* edited by Baron de Férussac. Paris.

Ranzi, S.

1928a Correlazioni tra organi di senso e centri nervosi in via di sviluppo. *Wilhelm Roux' Arch. Entwicklungsmech. Org.*, **114**:364–370.

1928b Suscettibilità differenziale nello sviluppo dei Cefalopodi. (Analisi sperimentale dell' embriogenesi). *Pubbl. Stn. Zool. Napoli*, **9**:81–159.

1932 Resultati di richerche di embriologia sperimentale sui Cefalopodi. *Arch. Zool. Ital.*, **16**:403–408.

Raven, C. P.

1942 The influence of lithium upon the development of the pond snail, *Limnaea stagnalis* L. *Proc. K. Ned. Akad. Wet.*, **45**:856–860.

1949 On the structure of cyclopic, synophthalmic, and anophthalmic embryos, obtained by the action of lithium in *Limnaea stagnalis*. *Arch. Neerl. Zool.*, **8**:323–352.

1952 Morphogenesis in *Limnaea stagnalis* and its disturbance by lithium. *J. Exp. Zool.*, **121**:1–77.

1966 *Morphogenesis: The Analysis of Molluscan Development,* 2nd ed. New York: Pergamon.

Raven, C. P., and A. M. T. Beenakkers.

1955 On the nature of head malformations obtained by centrifuging the eggs of *Limnaea stagnalis*. *J. Embryol. Exp. Morphol.*, **3**:286–303.

Raven, C. P., A. C. De Roon, and A. M. Stadhouders.

1955 Morphogenetic effects of a heat shock on the eggs of *Limnaea stagnalis*. *J. Embryol. Exp. Morphol.*, **3**:142–159.

Read, K. R. H.
  1966 Molluscan hemoglobin and myoglobin. In *Physiology of Mollusca,* vol. 2, edited by
        K. M. Wilbur and C. M. Yonge. New York: Academic, pp. 209–232.
Redi, F.
  1684 *Osservazioni intorno agli animali viventi che si trovano negli animali viventi.* Florence.
Riegel, J. A.
  1972 *Comparative Physiology of Renal Excretion.* New York: Hafner.
Ries, E., and M. Gersch.
  1936 Die Zelldifferenzierung und Zellspezialisierung während der Embryonalentwicklung
        von *Aplysia limacina* L. zugleich ein Beitrag zu Problemen der vitalen Färbung. *Pubbl.
        Stn. Zool. Napoli,* **15**:223–273.
Rokop, F. J.
  1972 A new species of monoplacophoran from the abyssal North Pacific. *Veliger,* **15**:91–95.
Rondelet, G.
  1554 *Libri de Piscibus Marinis, in quibus verae Piscium effigies expressae sunt.*
Rose, R. M.
  1971 Functional morphology of the buccal mass of the nudibranch *Archidoris pseudoargus.
        J. Zool. (London),* **165**:317–336.
  1972 Burst activity of the buccal ganglion of *Aplysia depilans. J. Exp. Biol.,* **56**:735–754.
Rosenbluth, J.
  1963 The visceral ganglion of *Aplysia californica. Z. Zellforsch. Mikrosk. Anat.,* **60**:213–236.
  1972 Obliquely striated muscle. In *The Structure and Function of Muscle,* 2nd ed., Vol. 1,
        ed. by G. H. Bourne. New York: Academic, pp. 389–420.
Rossi-Fanelli, A., and E. Antonini.
  1957 A new type of myoglobin isolated and crystallized from the muscles of *Aplysiae. Bio-
        khimiya,* **22**:336–344.
Rossner, K. L.
  1974 Central projections of the *Aplysia* visual system. *Comp. Biochem. Physiol. (A),* **48**:
        609–615.
Rudman, W. B.
  1971 Structure and functioning of the gut in the Bullomorpha (Opisthobranchia). Part I:
        Herbivores. *J. Nat. Hist.,* **5**:647–675.
  1972 Structure and functioning of the gut in the Bullomorpha (Opisthobranchia). Part 4:
        Aglajidae. *J. Nat. Histo.,* **6**:547–560.
Runham, N. W.
  1963 A study of the replacement mechanism of the pulmonate radula. *Q. J. Microsc. Sci.,*
        104:271–277.
Runham, N. W., and P. J. Hunter.
  1970 *Terrestrial Slugs.* London: Hutchinson.
Runnegar, B., and J. Pojeta, Jr.
  1974 Molluscan phylogeny: The paleontological viewpoint. *Science (Wash., D.C.),* **186**:
        311–317.
Runnström, J., and B. Markman.
  1966 Gene dependency of vegetalization in sea urchin embryos treated with lithium. *Biol.
        Bull. (Woods Hole),* **130**:402–414.
Russell-Hunter, W. D.
  1968 *A Biology of Lower Invertebrates.* New York: Macmillan.
Sakharov, D. A.
  1970 Cellular aspects of invertebrate neuropharmacology. *Annu. Rev. Pharmacol.,* **10**:335–352.
Sakharov, D. A., and I. Zs.-Nagy.
  1968 Localization of biogenic monoamines in cerebral ganglia of *Lymnaea stagnalis* L.
        Acta Biol. Acad. Sci. Hung., **19**:145–157.

Sarne, Y., E. A. Neale, and H. Gainer.
1976a Protein metabolism in transected peripheral nerves of the crayfish. *Brain Res.*, **110**(1): 73–89.

Sarne, Y., B. K. Schrier, and H. Gainer.
1976b Evidence for the local synthesis of a transmitter enzyme (glutamic acid decarboxylase) in the crayfish peripheral nerve. *Brain Res.*, **110**(1):91–97.

Saunders, A. M. C., and M. Poole.
1910 The development of *Aplysia punctata*. *Q. J. Microsc. Sci.*, N.S., **55**:497–539.

Saunders, J. W., Jr.
1970 *Patterns and Principles of Animal Development*. London: Macmillan.

Schacher, S., and E. Kandel.
1977 Development of neurons and synapses in the abdominal ganglion of *Aplysia californica*. *Neurosci. Abstracts*, **3**:359.

Scharrer, E., and B. Scharrer.
1963 *Neuroendocrinology*. New York: Columbia University Press.

Scheltema, R. S.
1974 Biological interactions determining larval settlement of marine invertebrates. *Thalassia Jugosl.*, **10**:263–296.

Schoenlein, K.
1894 Ueber das Herz von *Aplysia limacina*. *Z. Biol.*, **30**:187–220.

Schrader, K.
1938 Untersuchungen über die Normalentwicklung des Gehirns und Gehirntransplantationen bei der Mehlmotte *Ephestia kühniella Zeller* nebst einigen Bemerkungen über das Corpus allatum. *Biol. Zentralbl.*, **58**:52–90.

Schwartz, J. H.
1979 Axonal transport: Components, mechanisms, and specificity. In *Annual Review of Neuroscience,* Vol II, edited by W. M. Cowan, Z. W. Hall, and E. R. Kandel. Palo Alto: Annual Reviews Inc. (in press).

Schwartz, J. H., V. F. Castellucci, and E. R. Kandel.
1971 Functioning of identified neurons and synapses in abdominal ganglion of *Aplysia* in absence of protein synthesis. *J. Neurophysiol.*, **34**:939–953.

Seligman, M. E. P., and J. L. Hager, eds.
1972 *Biological Boundaries of Learning*. New York: Appleton-Century-Crofts.

Selverston, A. I.
1973 The use of intracellular dye injections in the study of small neural networks. In *Intracellular Staining in Neurobiology,* edited by S. B. Kater and C. Nicholson, New York: Springer-Verlag, pp. 255–280.

Sener, R.
1972 Site of circadian rhythm production in *Aplysia* eye. *Physiologist*, **15**:262.

Senseman, D., and A. Gelperin.
1974 Comparative aspects of the morphology and physiology of a single identifiable neuron in *Helix aspersa, Limax maximus,* and *Ariolimax californica. Malacol. Rev.*, **7**:51–52.

Sherrington, C. S.
1906 *The Integrative Action of the Nervous System*. Yale University Press.

Shimahara, T., and L. Tauc.
1972 Mécanisme de la facilitation hétérosynaptique chez l'aplysie. *J. Physiol. (Paris)*, **65**: 303A–304A.

1977 Cyclic AMP induced by serotonin modulates the activity of an identified synapse in *Aplysia* by facilitating the active permeability to calcium. *Brain Res.*, **127**:168–172.

Simpson. G. G.
1949 *The Meaning of Evolution. A Study of the History of Life and of Its Significance for Man.* Yale University Press.

Smith, F. G. W.
  1935 The development of *Patella vulgata. Philos. Trans. R. Soc. Lond. (B),* **225**:95–125.
Smith, J. E., G. Chapman, R. B. Clark, D. Nichols, and J. D. Carthy.
  1971 *The Invertebrate Panorama.* London: Weidenfeld and Nicolson.
Smith, S. T.
  1967 The development of *Retusa obtusa* (Montagu) (Gastropoda, Opisthobranchia). *Can. J. Zool.,* **45**:737–764.
Smith, S. T., and T. H. Carefoot.
  1967 Induced maturation of gonads in *Aplysia punctata* Cuvier. *Nature (London),* **215**: 652–653.
Smith, T. G., Jr., J. L. Barker, and H. Gainer.
  1975 Requirements for bursting pacemaker potential activity in molluscan neurones. *Nature (London),* **253**:450–452.
Sokolov, P. G., C. M. Beiswanger, D. J. Prior, and A. Gelperin.
  1977 A circadian rhythm in the locomotor behaviour of the giant garden slug *Limax maximus. J. Exp. Biol.,* **66**:47–64.
Sparrow, A. H., and A. F. Nauman.
  1976 Evolution of genome size by DNA doublings. *Science (Wash., D.C.),* **192**:524–529.
Spencer, W. A., R. F. Thompson, and D. R. Neilson, Jr.
  1966a Response decrement of the flexion reflex in the acute spinal cat and transient restoration by strong stimuli. *J. Neurophysiol.,* **29**:221–239.
  1966b Decrement of ventral root electrotonus and intracellularly recorded PSPs produced by iterated cutaneous afferent volleys. *J. Neurophysiol.,* **29**:253–274.
Spengel, J. W.
  1881 Die Geruchsorgane und das Nervensystem der Mollusken: Ein Beitrag zur Erkenntnis der Einheit des Molluskentypus. *Z. Wiss. Zool.,* **35**:333–383.
Spira, M. E., and M. V. L. Bennett.
  1972 Synaptic control of electrotonic coupling between neurons. *Brain Res.,* **37**:294–300.
Spitznas, M., and M. J. Hogan.
  1970 Outer segments of photoreceptors and the retinal pigment epithelium: Interrelationship in the human eye. *Arch. Opthalmol.,* **84**:810–819.
Spray, D. C., and M. V. L. Bennett.
  1975a Pharyngeal sensory neurons and feeding behavior in the opisthobranch mollusc *Navanax. Neuroscience Abstracts,* **1**:570.
  1975b Proprioceptive inputs to large buccal motor neurons controlling pharyngeal expansion in *Navanax. Fed. Proc.,* **34**:418.
Spray, D. C., M. E. Spira, and M. V. L. Bennett.
  1976 Sequential activity as a consequence of synaptic control of electrotonic coupling among neurons. *Neurosci. Abst.,* **1**:#357.
Starmühlner, F.
  1956 Beiträge zur Mikroanatomie und Histologie des Darmkanals einiger Opisthobranchier, I: *Sitzungsber. Österr. Akad. Wiss., Math-Naturwiss. Kl., Abt. I,* **165**:93–152.
Stasek, C. R.
  1972 The molluscan framework. *Chem. Zool.,* **7**:1–44.
Stempell, W.
  1898 Beiträge zur Kenntniss der Nuculiden. *Zool. Jahrb. Suppl.,* **4**:339–430.
Stephenson, T. A., and A. Stephenson.
  1949 The universal features of zonation between tide-marks on rocky coasts. *J. Ecol.,* **37**: 289–305.
  1972 *Life Between Tidemarks on Rocky Shores.* San Francisco: W. H. Freeman.
Stinnakre, J., and L. Tauc.
  1969 Central neuronal response to the activation of osmoreceptors in the osphradium of *Aplysia. J. Exp. Biol.,* **51**:347–361.

Stone, B. A., and J. E. Morton.
  1958 The distribution of cellulases and related enzymes in Mollusca. *Proc. Malacol. Soc. Lond.,* **33:**127–141.

Straub, W.
  1901 Zur Physiologie des Aplysienherzens. *Pflügers' Archiv gesamte Physiol. Menschen Tiere,* **86:**504–532.
  1904a Beiträge zur physiologischen Methodik mariner Thiere. 1: *Aplysia. Mitt. Zool. Sta. Neapel.,* **16:**458–468.
  1904b Fortgesetzte Studien am Aplysienherzen (Dynamik. Kreislauf und dessen Innervation) nebst Bemerkungen zur vergleichenden Muskelphysiologie. *Pflügers' Archiv gesamte Physiol. Menschen Tiere,* **103:**429–449.

Stretton, A. O. W., and E. A. Kravitz.
  1968 Neuronal geometry: Determination with a technique of intracellular dye injection. *Science (Wash., D.C.),* **162:**132–134.
  1973 Intracellular dye injection: The selection of Procion yellow and its application in preliminary studies of neuronal geometry in the lobster nervous system. In *Intracellular Staining in Neurobiology,* edited by S. B. Kater and C. Nicholson. New York: Springer Verlag, pp. 21–40.

Strumwasser, F.
  1965 The demonstration and manipulation of a circadian rhythm in a single neuron. In *Circadian Clocks,* edited by J. Aschoff. Amsterdam: North-Holland, pp. 442–462.
  1971 The cellular basis of behavior in *Aplysia. J. Psychiatr. Res.,* **8:**237–257.
  1974 Neuronal principles organizing periodic behaviors. In *The Neurosciences. Third Study Program,* edited by F. O. Schmitt and F. G. Worden, Cambridge, Mass.: M.I.T. Press, pp. 459–478.

Strumwasser, F., and R. Bahr.
  1966 Prolonged in vitro culture and autoradiographic studies of neurons in *Aplysia. Fed. Proc.,* **25:**512, Abstract 1815.

Strumwasser, F., J. W. Jacklet, and R. B. Alvarez.
  1969 A seasonal rhythm in the neural extract induction of behavioral egg-laying in *Aplysia. Comp. Biochem. Physiol.,* **29:**197–206.

Sulston, J. E.
  1976 Post-embryonic development in the ventral cord of *Caenorhabditis elegans. Phil. Trans. R. Soc. Lond. (B),* **275:**287–297.

Susswein, A. J.
  1975 *Studies on Satiation of Feeding in the Marine Mollusc* Aplysia californica. Ph.D. dissertation, New York University.

Susswein, A. J., and I. Kupfermann.
  1974 Effects of radiation on the biting reflex of *Aplysia. Abst. and Proc. Soc. of Neurosci.,* **3:**671.
  1975a Bulk as a stimulus for satiation in *Aplysia. Behav. Biol.,* **13:**203–209.
  1975b Localization of bulk stimuli underlying satiation in *Aplysia. J. Comp. Physiol. (A),* **101:**309–328.

Susswein, A. J., I. Kupfermann, and K. R. Weiss.
  1976a Arousal of feeding behavior of *Aplysia. Neurosci. Abst.,* 6th Ann. Mtg. Soc. Neurosci.
  1976b The stimulus control of biting in *Aplysia. J. Comp. Physiol. (A).,* **108:**75–96.

Susswein, A. J., K. R. Weiss, and I. Kupfermann.
  1978 The effect of food arousal on the latency of biting in *Aplysia. J. Comp. Physiol.,* **123:**31–41.

Switzer-Dunlap, M., and M. G. Hadfield.
  1977 Observations on development, larval growth, and metamorphosis of four species of Aplysiidae (Gastropoda, Opisthobranchia) in laboratory culture. *J. Exptl. Mar. Biol. Ecol.,* **29:**245–261.

Tardy, J.
1962 Observations et expériences sur la métamorphose et la croissance de *Capellinia exigua* (Ald. et H.) (Mollusque Nudibranche). *C. R. Hebd. Seances Acad. Sci.,* **254:**2242–2244.
1970 Contribution à l'étude des métamorphoses chez les Nudibranches. *Ann. Sci. Nat. Zool. Biol. Anim., (Paris),* 12 Serie, **12:**299–371.

Tauc, L., and H. M. Gerschenfeld.
1962 A cholinergic mechanism of inhibitory synaptic transmission in a molluscan nervous system. *J. Neurophysiol.,* **25:**236–262.

Ten Cate, J.
1928 Contribution à la physiologie du ganglion pédal d'*Aplysia limacina. Arch. Neerl. Physiol.,* Sèrie IIIc, **12:**529–537.
1931 Physiologie der Gangliensysteme der Wirbellosen. *Ergeb. Physiol.,* **33:**137–336.

Thiele, J.
1931 *Handbuch der systematischen Weichtierkunde, Vol. 1.* Jena: Fischer.
1935 *Handbuch der systematischen Weichtierkunde, Vol. 2.* Jena: Fischer.

Thompson, E. B., C. H. Bailey, V. F. Castellucci, and E. R. Kandel.
1976 Two different and compatible intracellular labels: A preliminary structural study of identified sensory and motor neurons which mediate the gill withdrawal reflex in *Aplysia californica.* Society for Neuroscience, Sixth Annual Meeting.

Thompson, E. B., J. H. Schwartz, and E. R. Kandel.
1976 A radioautographic analysis in the light and electron microscope of identified *Aplysia* neurons and their processes after intrasomatic injection of ^{3}H-L-fucose. *Brain Res.,* **112:**251–281.

Thompson, E. L.
1917 An analysis of the learning process in the snail, *Physa gyrina* Say. *Behav. Monogr.,* Vol. 3, No. 14 (97 pp).

Thompson, S. H.
1974 Pattern generator for rhythmic motor output in *Melibe. Society for Neuroscience,* **4:**450.

Thompson, T. E.
1958 The natural history, embryology, larval biology and post-larval development of *Adalaria proxima* (Alder and Hancock) (Gastropoda Opisthobranchia). *Philos. Trans. R. Soc. Lond. (B),* **242:**1–58.
1960a Defensive acid-secretion in marine gastropods. *J. Mar. Biol. Assoc. U.K.,* **39:**115–122.
1960b Defensive adaptations in opisthobranchs. *J. Mar. Biol. Assoc. U.K.,* **39:**123–134.
1961 The structure and mode of functioning of the reproductive organs of *Tritonia hombergi* (Gastropoda Opisthobranchia). *Q. J. Microsc. Sci.,* **102:**1–14.
1962 Studies on the ontogeny of *Tritonia hombergi* Cuvier (Gastropoda Opisthobranchia). *Philos. Trans. R. Soc. Lond. (B),* **245:**171–218.
1964 Grazing and the life cycles of British nudibranchs. In *Grazing in Terrestrial and Marine Environments.* Symposium of the British Ecological Society, 1962, edited by D. J. Crisp. Oxford: Blackwell Scientific Publications, pp. 275–297.
1967 Direct development in a nudibranch, *Cadlina laevis,* with a discussion of developmental processes in Opisthobranchia. *J. Mar. Biol. Assoc. U.K.,* **47:**1–22.
1976 *Biology of Opisthobranch Molluscs,* vol. I. London: Ray Society.

Thompson, T. E., and A. Bebbington.
1969 Structure and function of the reproductive organs of three species of *Aplysia* (Gastropoda: Opisthobranchia). *Malacologia,* **7:**347–380.

Thompson, T. E., and D. J. Slinn.
1959 On the biology of the opisthobranch *Pleurobranchus membranaceus. J. Mar. Biol. Assoc. U.K.,* **38:**507–524.

Thorndike, E. L.
1898 Animal intelligence: An experimental study of the associative processes in animals.

Psychol. Rev. Ser. Monogr., Suppl. 2, (4):1–109.

Thorpe. W. H.
1963 *Learning and Instinct in Animals.* Harvard University Press, 2nd ed.

Thorson, G.
1950 Reproductive and larval ecology of marine bottom invertebrates. *Biol. Rev. Camb. Philos. Soc.,* **25:**1–45.

Tiffany, W. J. III.
1974 Ultrastructural evidence for reabsorption in the nephridium of the bivalved mollusc *Mercenaria campechiensis. Trans. Am. Micros. Soc.,* **93:**23–28.

Tinbergen, N.
1951 *The Study of Instinct.* Oxford: The Clarendon Press.

Tobach, E., P. Gold, and A. Ziegler.
1965 Preliminary observations of the inking behavior of *Aplysia (varria). Veliger,* **8:**16–18.

Tobach, E., H. E. Adler, and L. L. Adler.
1973 Comparative psychology at issue. *Ann. N.Y. Acad. Sci.,* **223:**1–197.

Toevs, L. A. S., and R. W. Brackenbury.
1969 Bag cell-specific proteins and the humoral control of egg laying in *Aplysia californica. Comp. Biochem. Physiol.,* **29:**207–216.

Trappmann, W.
1916 Die Muskulatur von *Helix pomatia* L. *Z. Wiss. Zool.,* **115:**489–585.

Twarog, B. M.
1967 Excitation of *Mytilus* smooth muscle. *J. Physiol. (London),* **192:**857–868.

Ungerstedt, U.
1971 Stereotaxic mapping of the monoamine pathways in the rat brain. *Acta Physiol. Scand.* (Suppl. 367), **82:**1–48.

Van Lummel, L. E. Ae.
1930 Untersuchungen über einige Solenogastren. *Z. Morphol. Ökol. Tiere.,* **18:**347–383.

Verworn, M.
1889 *Psycho-physiologische Protisten-Studien. Experimentelle Untersuchungen.* Jena.

Villee, C. A., W. F. Walker, Jr., and F. E. Smith.
1963 *General Zoology,* 2nd ed. Philadelphia: Saunders.

Vles, F.
1907 Sur les ondes pédieuses des Mollusques reptateurs. *C. R. Hebd. Seances Acad. Sci.,* **145:**276–278.

Von Skramlik, E.
1941 Über den Kreislauf bei den Weichtieren. *Ergeb. Biol.,* **18:**88–286.

Vorwohl, G.
1961 Zur Funktion der Exkretionsorgane von *Helix pomatia* L. und *Archachatina ventricosa* Gould. *Z. vgl. Physiol.,* **45:**12–49.

Wachtel, H., and D. Impelman.
1973 A galloping escape behavior in *Aplysia californica. Fed. Proc.,* **32**(3):368, Abstract 845.

Wald, G., and E. B. Seldin.
1968 Spectral sensitivity of the common prawn, *Palaemontes vulgaris. J. Gen. Physiol.,* **51:** 694–700.

Watson, J. D.
1970 *Molecular Biology of the Gene,* 2nd ed. New York: Benjamin.

Weel, P. B. van.
1957 Observations on the osmoregulation in *Aplysia juliana* Pease (Aplysiidae, Mollusca). *Z. vgl. Physiol.,* **39:**492–506.

Weevers, R. de G.
1971 A preparation of *Aplysia fasciata* for intrasomatic recording and stimulation of single neurones during locomotor movements. *J. Exp. Biol.,* **54:**659–676.

Weinreich, D., S. A. Dewhurst, and R. E. McCaman.
1972 Metabolism of putative transmitters in individual neurons of *Aplysia californica*: Aromatic amino acid decarboxylase. *J. Neurochem.*, **19**:1125–1130.

Weiss, K. R., J. Cohen, and I. Kupfermann.
1975 Modulatory command function of the metacerebral cell on feeding behavior in *Aplysia. Fed. Proc.*, **34**:418, Abstract 1118.
1978 Modulatory control of buccal mass musculature by a serotonergic neuron (metacerebral cell) in *Aplysia. J. Neurophysiol.*, **41**:181–203.

Weiss, K. R., and I. Kupfermann.
1976 Homology of the giant serotonergic neurons (metacerebral cells) in *Aplysia* and pulmonate molluscs. *Brain Res.*, **117**:33–49.
1978 Serotinergic neuronal activity and arousal of feeding in *Aplysia californica. Soc. Neurosci. Symp.*, Vol. III. (in press).

Weiss, K. R., M. Schonberg, J. Cohen, D. Mandelbaum, and I. Kupfermann.
1976 Modulation of muscle contraction by a scrotonergic neuron: Possible role of cyclic AMP. *Neuroscience Abstracts*, **2**:338.

Wells, M. J.
1964 Hormonal control of sexual maturity in cephalopods. *Bull. Natl. Inst. Sci. India*, **27**:61–77.
1968 *Lower Animals*, New York: McGraw-Hill.

Wells, M. J., and J. Wells.
1959 Hormonal control of sexual maturity in *Octopus. Ibid.*, **36**:1–33.

Welsh, J. H.
1956 Neurohormones of invertebrates, I: Cardio-regulators of *Cyprina* and *Buccinum. J. Marine Biol. Assoc. U.K.*, **35**:193–201.

Wenz, W.
1938 Gastropoda 1-2. In *Handbuch der Paläozoologie, 6,* edited by O. H. Schindewolf. Berlin: Borntraeger.

Wheeler, W. M.
1893 A contribution to insect embryology. *J. Morphol.*, **8**:1–160.

Whitman, C. O.
1899 Animal Behavior. *Biological Lectures from the Marine Biological Laboratory, Woods Hole, Mass.,* **1898**:285–338.

Wiederhold, M. L.
1974 *Aplysia* statocyst receptor cells: intracellular responses to physiological stimuli. *Brain Res.,* **78**:490–494.
1977 Rectification in *Aplysia* statocyst receptor cells. *J. Physiol. (London)*, **266**:139–156.

Wilbur, K. M., and C. M. Yonge, eds.
1964 *Physiology of Mollusca, Vol. 1.* New York: Academic.
1966 *Physiology of Mollusca, Vol. 2.* New York: Academic.

Willigen, C. A. van der.
1920 *Onderzoekingen over den bouw van het zenuwstelsel der Lamellibranchiata.* Utrecht.

Willows, A. O. D.
1967 Behavioral acts elicited by stimulation of single, identifiable brain cells. *Science (Wash., D.C.)*, **157**:570–574.
1968 Behavioral acts elicited by stimulation of single identifiable nerve cells. In *Physiological and Biochemical Aspects of Nervous Integration,* edited by F. D. Carlson. Englewood Cliffs, N.J.: Prentice-Hall, pp. 217–243.
1971 Giant brain cells in mollusks. *Sci. Am.,* **224**(2):68–75.
1973 Learning in gastropod mollusks. In *Invertebrate Learning. Vol. 2, Arthropods and Gastropod Mollusks,* edited by W. C. Corning, J. A. Dyal, and A. O. D. Willows. New York: Plenum, pp. 187–273.

Willows, A. O. D., and G. Hoyle.
    1969 Neuronal network triggering a fixed action pattern. *Science (Wash., D.C.)*, **166**:1549–1551.

Wilmoth, J. H.
    1967 *Biology of Invertebrata*. Englewood Cliffs. N.J.: Prentice–Hall.

Wilson, E. B.
    1904 Experimental studies in germinal localization. II: Experiments on the cleavage-mosaic in *Patella* and *Dentalium*. *J. Exp. Zool.*, **1**:197–268.

Wilson, E. O.
    1975 *Sociobiology: The New Synthesis*. Harvard University Press.

Winckworth, R.
    1949 Shells used as horns. *Proc. Malacol. Soc. Lond.*, **28**:37–38.

Wine, J. J.
    1973 Invertebrate central neurons: Orthograde degeneration and retrograde changes after axonotomy. *Exp. Neurol.*, **38**:157–169.

Winkler, L. R.
    1955 A new species of *Aplysia* on the Southern California coast. *Bull. South. Calif. Acad. Sci.*, **54**:5–7.
    1957 *The Biology of California Sea Hares of the Genus Aplysia*. Ph.D. dissertation, University of Southern California.
    1959 A new species of sea hare from California waters. *Bull. South. Calif. Acad. Sci.*, **58**:8–10.

Winkler, L. R., and E. Y. Dawson.
    1963 Observations and experiments on the food habits of California sea hares of the genus *Aplysia*. *Pac. Sci.*, **17**:102–105.

Winkler, L. R., and B. E. Tilton.
    1962 Predation on the California sea hare, *Aplysia californica* Cooper, by the solitary great green sea anemone, *Anthopleura xanthogrammica* (Brandt), and the effect of sea hare toxin and acetylcholine on anemone muscle. *Pac. Sci.*, **16**:286–290.

Winlow, W., and E. R. Kandel.
    1976 The morphology of identified neurons in the abdominal ganglion of *Aplysia californica*. *Brain Res.*, **112**:221–249.

Wirén, A.
    1892 Studien über die Solenogastren. II: *Chaetoderma productum, Neomenia, Proneomenia acuminata*. *K. Sven. Ventensk.-Akad. Handl.*, N.F., **25**(6):1–100.

Wittenberg, B. A., J. B. Wittenberg, S. Stolzberg, and E. Valenstein.
    1965 A novel reaction of hemoglobin in invertebrate tissues. II: Observations on molluscan muscle. *Biochim. Biophys. Acta*, **109**:530–535.

Wittenberg, J. B.
    1970 Myoglobin-facilitated oxygen diffusion: Role of myoglobin in oxygen entry into muscle. *Physiol. Rev.*, **50**:559–636.

Wolff, H. G.
    1973a Multi-directional sensitivity of statocyst receptor cells of the opisthobranch gastropod *Aplysia limacina*. *Mar. Behav. Physiol.*, **1**:361–373.
    1973b Statische Orientierung bei Mollusken. *Fortschr. Zool.*, **21**:80–99.

Wölper, C.
    1950 Das Osphradium der *Paludina vivipara*. *Z. vgl. Physiol.*, **32**:272–286.

Wright, D. L.
    1960 *Cardiac Studies on Aplysia vaccaria*. Ph.D. dissertation, University of California, Los Angeles.

Yonge, C. M.
    1947 The pallial organs in the aspidobranch Gastropoda and their evolution throughout the Mollusca. *Phil. Trans. R. Soc. Lond. (B)*, **232**:443–518.

1949a *The Sea Shore.* London: Collins.

1949b On the structure and adaptations of the Tellinacea, deposit-feeding Eulamellibranchia. *Phil. Trans. R. Soc. Lond. (B),* **234**:29–76.

Young, J. Z.

1971 *The Anatomy of the Nervous System of Octopus vulgaris.* Oxford: Clarendon Press.

Zeevi, Y. Y.

1972 *Structural Functional Relationships in Single Neurons: Scanning Electron Microscopy and Theoretical Studies.* Ph.D. dissertation, University of California, Berkeley.

Zulch, K. J. Mihaud.

1954 Mangeldurchblutung an der Grenzzone zweier Gefässgebiete: Urasche bisher ungeklärter Ruckenmarkschädigungen. *Deutsche Z. Nervenheilkd.,* **172**:81–101.

1960 Study of the fiber of the neurinoma, its Schwannian origin and its neurectodermal nature. *Rev. Neurol.,* **103**:541–545.

Name Index

An *italic* page number refers to a figure or table;
a page number in parentheses refers to an entry in
the Selected Readings; a page number followed by
n refers to a footnote.

Abbott, R. T., 6
Abraham, F. D., 348
Advokat, C., 354, *355,* 356, 368
Alexander, R. D., 403
Alkon, D. L., 64, 374, 375, *376,* 377
Alvarez, R. B., 50, 51, 136, 324, 325
Ambron, R. T., 165
Ambrose, H. W., III, 290, 292
Anderson, J. A., 126
Andrew, R. J., 353
Andrews, E., 84
Andriaanse, A., 394
Anfinsen, C. B., 399
Antonini, E., 94, 95n, 96
Arch, S., 136, 324, 325
Arey, L B., 272n
Arvanitaki, A., 119, 163, 163n, 165
Ascher, P., 181
Aspey, W. P., 288, *289,* 290, 308
Atz, J. W., 2, 3, 394
Audesirk, T. E., 50, 304, 306, 307

Bahr, R., 120
Bailey, C. H., *160,* 171, 175, 179,
    187n, 199, 204
Bailey, D. F., 61
Bak, A., 64
Banks, F. W., 211
Barnes, R. D., *19*
Barondes, S. H., 337
Bauer, V., 74, 165
Baur, P. S., Jr., 163
Beach, F. A., 265, 266
Bebbington, A., 97, *102,* 103, 220,
    221, 322
Beeman, R. D., 29, 44n, 97, *101,*
    322, *327*

Beenakkers, A. M. T., 249
Beiswanger, C. M., 287
Belon, P., 27
Bennett, M. V. L., 166, 211, 313
Bentley, D. R., *401,* 403
Berg, C. J., Jr., 252n
Bergh, R., 119
Bernstein, B., 50, 51
Berrill, N. J., 224
Berry, M. S., 165, 171
Berry, R. W., 308, 377
Bethe, A., 86n, 188, 209, 272n
Bevelaqua, F. A., 83
Bidder, A. N., 89n
Bittner, G. D., 157
Blankenship, J. E., 288, *289,* 290,
    308, 324, 391, *392,* 395
Blochmann, F., 27, 29, 217n
Block, D., 53, 56, 284, 286, *286,* 287
Block, G. D., 286
Boettger, C. R., 31, 32n
Bohadsch, J. B., 27
Bolles, R. C., 359
Bonar, D. B., 260
Borradaile, L. A., *16*
Bottazzi, F., 29, 82, 120, 124, 127,
    130, 133, 136
Boutan, L., 18
Bouvier, E. L., *114, 122, 389*
Boyden, A., 3
Brace, R. C., 23, 24, 31, 32, 32n, *33,*
    34, 63n, 68, 112, 116n, 391
Brackenbury, R. W., 324
Bridges, C. B., 260
Bronn, H. G., (147)
Brower, M. E., 163, 165
Brown, A. C., 61

Brown, A. M., 163, 163n, *164,* 165
Brown, H. M., 163, 163n, *164*
Brown, J. S., 359
Brunelli, M., *193, 195, 339, 340,*
    343, 343, 358
Brunori, M., 94, 95n
Buchsbaum, R. M., *10, 110*
Bullock, T. H., 108n, *110, 111, 113,*
    115, 119, 123n, (147), 150, 187n,
    210, 214, (329)
Byrne, J., 65, 66, 136, 188n, 190, *191,*
    192, 193, *193, 195,* 203, 269,
    272n, 275, *276,* 276, *278, 339,*
    340, 343, 343

Cajal, S. R., 212
Campbell, C. B. G., 3, 395, (405)
Capo, T., 103, 222
Carazzi, D., 217, 225, *227,* 230
Cardot, H., 119
Carefoot, T. H., 43, *44,* 44, 48, 88n,
    287, 299, 301, *302,* 302, *303,* 304,
    322, 324
Carew, T. J., 42, *42,* 43, 44n, 69, 71,
    72, *74,* 75, *75,* 76, 129, 131, 136,
    187n, 188n, 190, *191,* 195, *197,*
    197, 198, 210–211, 214, 269, *270,*
    272, 273, 274, *275,* 275, 282, 283,
    296, 299, *299,* 301, 322, *324, 336,*
    337, 346, 354, *355,* 356, 368, 400
Carlson, A. J., 136, *288*
Carthy, J. D., *280*
Castellucci, V. F., 43, 65, 66, 136,
    160, 171, 175, 179, 187n, 188n,
    189, 190, 191, *192,* 193, *193,* 195,
    195, 196, 196n, *197, 198,* 198,
    199, 203, 204, 211, 217n, *219,*

Castellucci (*continued*)
    238, 250, *251, 253,* 269, *335,* 337,
    338, *339, 340, 343,* 343, 345, *347,*
    356, 358
Cather, J. N., 224, 225, 225n, (263)
Cedar, H., 337, *339*
Chace, L., 323
Chalazonitis, N., 163, 163n, 165
Chan-Palay, V., 206
Chang, J. J., 321, 382
Chapman, G., *280*
Chapple, W. D., 134
Chiaje, S., delle, 29
Christensen, A. M., 295
Civil, G. W., 82
Clark, R. B., *280*
Clarke, A. H., Jr., 10
Clarke, B., 413
Clement, A. C., 224, 225n, 255
Cobbs, J. S., 308
Coggeshall, R. E., 61, *62,* 65, 97, 103,
    150, *153,* 153, *154, 155,* 156, 157,
    158, 160, 161, *162,* 165, 175, 179,
    179, 181, 214, 243, 248, 324, 391,
    392, 395
Cohen, J., 104, 124, 127, 195, 211,
    308–310, *309,* 310n, *349, 350,*
    351, 356, 358, 360
Cohen, M. J., 165, 171
Collins, S. D., 379, *380*
Conklin, E. G., 226, 242, 248
Copeland, M., 61
Cottrell, G. A., 395
Crampton, H. E., Jr., 224
Creek, G. A., 22
Crick, F., 399
Crofts, D. R., 20, *21,* 21, 22, 243, 248
Crow, T., 377
Crozier, W. J., 272n
Cuvier, G., (5), 27, *28, 95*

Dainton, B. H., 287
Dan, K., 224, (264)
Davidson, E. H., 224, 225, 255, (263)
Davis, W. J., 170, 314, 315, 316, *317,*
    323, 326, *327,* 327, 364, 368, *369,*
    379, 395
Dawson, E. Y., 44n, 48, 300
Dennis, M. J., 153n
DeRoon, A. C., 249
Dessi Fulgheri, F., 104
Dethier, V. G., 360n, 362n, 370, (358)
Detwiler, P. B., 64
Dewhurst, S. A., 351, 395
Dieringer, N., 82, 83, 272n, 273, 275,
    276, 296, 354, *355,* 365
Dijkgraaf, S., 61, 65

Divaris, G. A., 79, 80
Dobzhansky, T. G., 400, 402
Dogiel, J., 136
Dorsett, D. A., 131, 295, 395
Doty, R. W., 268
Dower, W. J., 161, *162,* 163
Downey, P., 373, 374
Drew, G. A., 73
Dudek, F. E., 324
Duffy, E., 352

Eales, N. B., 18, 24, 27, 29, 32, *34,*
    35, 37, 38, *38, 39,* 43, (45), 68,
    76, 80, 85, 91, 92n, *93,* 94, 96,
    97, 103, 119n, *122,* 123n, 125,
    129, 139n, 292, 299, 301,
    321, 322
Edwards, J. S., 245
Eisenstadt, M., 126, 351, 395
Emery, D. G., 307
Enriques, P., 82, 120, 124, 127, 130,
    133, 136
Erlanger, R. V., 248
Eskin, A., 52, 53, *54,* 55-56
Everhart, T. E., *181*
Eyman, E. D., 318n

Farmer, W. M., 292
Feder, H., 295
Feinstein, R. M., 82, 83
Filatova, Z. A., 10
Florkin, M., 88n
Flury, F., 274
Forrest, J. E., 89n
Fraser Rowell, C. H., 318n, 319,
    320, 382
Frazier, W. T., 61, *154,* 160, 179, *179,*
    243, 324
Fredman, S. M., 210, 211, 282
French, J. D., 353
Fretter, V., 6, 8, 12, 20, 22, 23, (26),
    31, 69, 89n, 97, 112, (263),
    388, 391
Friedrich, H., 64
Frings, C., 50, 68, 308
Frings, H., 50, 68, 308
Fröhlich, F. W., 69, 123n, 129, 130,
    130n, 210
Furshpan, E. J., 166

Gainer, H., 157, 324
Gallin, E. K., 64
Garcia, J., 2, 382
Gardner, D., 127
Garstang, W., 8, 20, 22, 43, 299
Gascoigne, T., 89n
Gelperin, A., 66, 287, 321, 362n, 382,
    383, 395, 396

Gerhardt, U., 323
Geronimo, J., 50, *51,* 42, 56n
Gersch, M., *222*
Gerschenfeld, H. M., 153n, 181
Getting, P. A., 66, 292, 296, *297*
Ghiretti, F., 83
Ghiselin, M. T., 22, 97, *98,* 388
Giller, E., Jr., 83, 133
Gillette, R., 314, 316, 317, 395
Gilula, N. B., 214
Globus, A., 165, 166, 171, 171n
Gold, P., 266, 272, 274
Goldman, J. E., 79, 126, 165, 277,
    351, 395
Gooden, B. A., 82, 83
Goodrich, E. S., 84, 85
Graham, A., 6, 8, 12, 22, (26), 69,
    89n, 97, 112, (263), 388, 391
Granzow, B., *320,* 321
Grasse, P. P., 419
Graubard, K., 175, 179
Gray, J., 66, *271, 288*
Green, K. F., 382
Grether, W. F., 354
Grillner, S., 283
Groves, P. M., 351
Guiart, J., 29, 32n, *36, 49, 71, 121,*
    292, 390

Haas, G., 3
Hadfield, M. G., 217n, *220,* 238,
    254n, 260
Haeckel, W., *115*
Hager, J. L., 373, 382
Hamilton, P. V., 290, 292
Hamilton, W. J., III, 295
Hankins, W. G., 2
Hardy, A., 42
Harris, A. J., 153n
Harris, L. G., 256n
Harrison, F. M., 84, 85, 86
Hawkins, R. D., 2, 345, 356
Hebb, D. O., 353, 359
Hegner, R. W., 420
Heinroth, O., 394
Hening, W., 69, 129, 131, 210–211,
    272, 275, 282, 283, 296, 337
Henkart, M., 165
Herter, K., 210
Hessels, H. G. A., 61, 65
Heyer, C. B., 211
Hill, R. B., 79
Hinde, R. A., 295, 382, 402
Hodos, W., 3, 7, 266, 387, 395, (405)
Hoffman, F. B., 188–210
Hoffman, H., 108n, 119n, *121,* 123n,
    126n, 139, 140, *141,* (147), *389*

Hogan, M. J., 163
Holtzman, E., 160, 163
Horridge, G. A., 108n, *110, 111, 113, 115, 119,* 123n, (147), 150, 187n, 210, 214
Howells, H. H., 89n, 91, *93,* 94, *95,* 96, 97, 308
Howieson, D. B., 198
Hoy, R. R., 157, *401,* 403
Huber, F., 403
Hudson, D. J., 53, 56, 284, 286, *286,* 287
Hughes, G. M., 133, *134,* 390
Hughes, H. P. I., 50, 52n, 131
Hunsak, D., II., 94
Hunter, P. J., 323
Hurst, A., 89n
Hyman, L. H., 5n, 12, (26), *30,* 84, 85, 94, 96, 248, (263)

Iles, J. F., 166
Impelman, D., 296
Inaba, A., 223

Jacklet, J. W., 50, *51,* 51, 52, *53,* 53, *55,* 56n, 136, 187, 187n, 188n, 190, 198, 199, 203, 211, (215), 284, 285, 286, 324, 325
Jacobson, M., 250, (264)
Jahan-Parwar, B., 49, 50, 57, 61, 123, 126, 210, 211, 282, 306, 310, 373, 374
Jennings, H. S., 265
Jones, H. D., 66, *81,* 81, 82
Jordan, H. J., 50, 68, 123, 123n, 129, 130, 187n, 210, 304
Jourdan, F., 175
Jouvet, M., 353

Kandel, E. R., 18, 43, 61, 65, 66, 69, 71, 72, *74,* 75, *75,* 76, *78,* 79, 82, 86, 103, 104, 126, 127, 129, 131, 136, *153,* 153n, *154, 160,* 160, 165, 166, *167, 168, 169,* 170, 171, *174,* 175, 179, *179,* 183, 187, 187n, 188n, *189,* 190, *191,* 191, *192,* 193, *193,* 195, *195, 196,* 196n, *197,* 197, *198,* 198, 199, 203, 204, 210–211, 214, 217n, *219,* 238, 243, 244, 249, 250, *251, 253,* 269, *270,* 272, 272n, 273, 275, 277, *279, 281,* 282, 283, 296, *298,* 305, *319,* 321, 324, *334, 335, 336,* 337, *338, 339,* 340, *343,* 343, 345, *347,* 351, *355,* 356, 358, 368, 389, *392,* 395, *396,* 400
Kendel, P., 222
Kaneko, C. R. S., 318, 318n, *320*

Karlson, R. H., 254n
Karsson, U. L., 211
Karsznia, R., 230
Kater, S. B., 104, (185), 211, 318, 318n, *319,* 319, *320,* 321, 382
Kay, E. A., 43
Kennedy, D., 157, 170, (185)
Kim, K. S., 83
King, D. G., 165, 170, 175, (185)
Kirschner, L. B., 84
Klein, M., 340, 343
Knight, J. B., 6
Koester, J. D., *78,* 79, 82, 83, 86, 104, 126, 136, *160,* 187n, 199, 211, 272n, 273, 275, *276,* 276, 277, *278,* 351, 354, *355,* 365, 395
Kohn, A. J., 295
Koike, H., 126, 351, 395
Koningsor, R. L., Jr., 94
Kovac, M. P., 314, 315, 316, 326, 237, 295
Krasne, F. B., 170
Krauhs, J. M., 163
Kravitz, E. A., 165, 166, (185)
Kriegstein, A. R., 43, 45, 82, 217n, *219, 233,* 235, *236,* 238, *241,* 243, *246,* 246, 251, *253,* 390
Krijgsman, B. J., 79, 80
Krull, H., 424
Kuffler, S. W., 153n, 156
Kumarasiri, M. H., 83
Kumé, M., 224, (264)
Kunze, H., 126, 395
Kupfermann, I., 42, *42,* 43, 44n, 50, 57, 58, 61, 68, 75, *78,* 79, *87,* 88, 91, 94, 104, 124, 126, 127, 130, 136, *154,* 160, 179, *179,* 187n, 188n, *189,* 190, 195, *195, 196,* 196n, *197, 198,* 198, 211, 243, 249, 269, 272, 274, *275,* 275, 277, 284, 285, 299, *299,* 301, 304, 305, 308, *309,* 310, 310n, 322, 324, *335,* 345, *346, 347, 349, 350,* 351, 354, *355,* 356, 357, 358, 360, 360n, *361, 362, 363,* 364, 377, (385), 395, *396*

de Lacaze-Duthiers, H., 29
Lang, A., 22
Lang, F., 22
Lankester, E. R., (5), 6
Lasek, R. J., 161, *162,* 163
Laverack, M. S., 61, 310
Lee, C. K., 161, 304
Lee, R. M., 49, 50, 56, *57,* 61, *142, 306,* 306, *307,* 314, *316,* 364, *366,* 372, *373,* 373, 374

Lehrman, D. S., 143
Lemche, H., 8, *10,* 10
Lent, C. M., 397, *398*
Levenstein, R. Y., 10
Levi-Montalcini, R., 244
Levitan, H., 313
Levitan, I. B., 337
Lewis, E. R., *181*
Lewis, J. R., 41n,
Lewis, P. R., 88n
Lewis, R. D., 287
Lickey, M. E., 53, 56, 284, 286, *286,* 287, 308, 377, *378*
Liebeswar, G., *78,* 79, 82, 104, 136, 211, 272n, 277, 351, 395
Liegeois, R. J., *142*
Linneaus, C., 27
Lissmann, H. W., 66, 287, *288*
Little, C., 84, 85
Lloyd, D. P. C., 268
Lockard, R. B., 266
Loeb, J., 265
Loh, Y. P., 324
Luborsky-Moore, J. L., 51
Lukowiak, K. D., 187, 187n, 188n, 190, 196n, 198, 199, 203, 211, 214, (215)
Van Lummel, L. E. Ae., *110*
Lupo di Prisco, C., 104
Lux, H. D., 165, 166, 171, 171n

MacFarland, F. M., 91, 119n, 123n, *137,* 138, *139,* 140
MacGinite, G. E., 44n, 45, 48, 222, 322, 323
MacGinitie, N., 45
Magoun, H. W., 353
Malmo, R. B., 353
Mandelbaum, D., *350,* 351, 358, 360
Manning, A., 2, 402, 403, 404, (405)
Manwell, C., 96
Marcus, E., *36,* 96, 259
Margoliash, E., 399
Markman, B., 255
Marler, P., 295
Martin, A. W., 84, 85, 86
Martin, R., 430
Mayeri, E., *78,* 79, 82, 104, 136, 211, 272n, 277, 325, 351, 395
Mayr, E., 269, 394, 400, 402, 404, (405)
Mazzarelli, G., 29, 91, 119, 123n, 125, 138
McCaman, R. E., 351, 395
McGowan, B. K., 384
McKee, A. E., 61
McLean, N., 94
McMahan, U. J., 165, 171, 175

Meglitsch, P. S., *7,* (26), 85, 91
Menzies, R. J., 10
Merickel, M. B., 318, 318n, *320*
Merton, H., 58, 58n
Miller, M. C., 43, 44
Miller, N. E., 287n
Milne-Edwards, H., 29, 76
Moller, R., 197
Morgan, T. H., 255
Moritz, C. E., 242
Morris, D., 268
Morris, M. C., 69
Morton, J. E., 8, *8,* 11, 12, 22, (26), *30,* 31, 32, 32n, 66, 66n, 69, 71, 94, 295
Moruzzi, G., 353
Mpitsos, G. J., 314, 315, *317,* 323, 326, *327,* 364, 368, *369,* 379, 380, 395
Muller, K. J., 165, 171, 175
Mulloney, B., 166
Murray, M. J., *142,* 310, *311,* 311, 313, 326, 379

Naef, A., 8, 22, *109*
Nauman, A. F., 163
Neale, E. A., 157
Neilson, D. R., Jr., 212
Nelson, G. J., 7
Neu, W., 69, 290, *291*
Newby, N. A., 187n, 199, 322
Newell, R. C., *41*
Nicaise, G., 175
Nicholls, J. G., 156
Nichols, D., *280*
Nicholson, C., (185)
Noble, R. G., 61
Nordlander, R. H., 245
Novikoff, A. B., 160, 163

Odhner, N. H., *30*
Orkand, P. M., 104, 211
Orkand, R. K., 104, 156, 211
Ortmann, R., 143
Osborne, N. N., 395
Owen, G., *89,* 89n, 90, 91

Paine, R. T., 310, *311,* 311, 313, 322
Palay, S. L., 160, 175, 195, 206
Palovcik, R. A., 314, *316,* 364
Parker, G. H., 212, 215, 277, 282
Pavlov, I. P., 435
Peake, J., (26)
Pearson, K. G., 283
Peeke, H. V. S., 89n
Pelseneer, P., 6, 8, 15, 23, 24, 29, *30,* 32n, *117*

Peltrera, A., 254, 255
Pentreath, V.W., 165, 171
Perdeck, A. C., 402
Peretz, B., 76, 136, 187n, 188n, 190, 195, 196, 196n, 197, 198, 211, 214, 272, 275
Perlman, A. J., 187n, 188n, 199, *201, 202*
Peters, A., 175
Phillips, D. W., 295
Pinneo, J. M., 326, *327,* 368, *369*
Pinsker, H. M., 72, *74,* 75, *75,* 76, 82, 83, 136, 187n, 188n, *189,* 190, *191,* 195, *195, 196,* 196n, *197,* 197, *198,* 198, 211, 214, 269, 272, 308, *334, 335, 336,* 337, 345, *347,* 356, 357, 377
Pitman, R. M., 165, 171
Pojeta, J., Jr., 11, 14
Poole, M., 65, 217n, 225, 229, *230,* 230, 231, *233,* 233, *235*
Potter, D. D., 166
Potts, W. T., *16,* 84, 85
Power, M. E., 249
Preston, R. J., 49, 50, 56, *57,* 61, 304, *306,* 306, *307,* 314, 364, *366*
Prior, D. J., 66, 209, 211, (215), 287, 321
Prosser, C.L., 187n
Przybylski, R. J., 161
Purchon, R. D., 30, 31, 74, 276
Purves, D., 171

Ram, J. L., 368, *369*
Ramsay, J. A., 80
Rang, S., 28, *28*
Ranzi, S., 249, 255
Raven, C. P., 224, 225n, 228, 229, 242, 244, 248, 249, 255, (264)
Read, K. R. H., 95, 96
Redi, F., 27
Reingold, S. C., 321, 382
Remler, M. P., (185)
Riegel, J. A., 84
Ries, E., *222*
Robbins, M. R., 314, *316,* 364
Rogers, T. D., 163, 165
Rokop, F. J., 10
Rondelet, G., 27
Rose, R. M., 89n, 310
Rosenbluth, J., 150, 153, 175
Rossi-Fanelli, A., 94, 96
Rossner, K. L., 126
Rosza, K. S., 211
Rubinson, K., 72, *74,* 75, *75,* 76, 136, *191,* 195, *197,* 197, 198, 214, 272
Rudman, W. B., 89n

Runham, N. W., *90,* 323
Runnegar, B., 11, 14, (26)
Runnström, J., 255
Rusiniak, K. W., 2
Russell-Hunter, W. D., 69, *71, 73*

Sakharov, D. A., 126, 395
Sarne, Y., 157, 324
Satir, P., 214
Saunders, A. M. C., 65, 217n, 255, 229, *230,* 230, 231, *235*
Saunders, J. W., Jr., 224, 230, (264)
Schacher, S., 153n, 243, 244
Scharrer, B., 143
Scharrer, E., 143
Scheltema, R. S., 259
Schmale, M., 82, 83
Schoenlein, K., 82, 136
Schonberg, M., *350,* 351, 358, 360
Schrader, K., 245
Schubert, P., 165, 166, 171, 171n
Schwartz, J. H., 83, 126, 133, *160,* 165, 166, 171, *174,* 175, 183, 331, *339,* 351, 395
Segundo, J P., 313
Selden, E. B., 52n
Seligman, M. E. P., 372, 382
Selverston, A. I., 166, (185)
Sener, R., 56n
Senseman, D., 395, *396*
Sherrington, C. S., 268, 326
Shimahara, T., 343
Siegler, M. V. S., 314, 315, *317,* 395
Simon, S., 325
Simpson, G. G., 3, 6n, 384
von Skramlik, E., 79
Slinn, D. J., 293, *294,* 318, 318n
Smith, F. E., ·*77, 223*
Smith, F. G. W., 20, 242
Smith, J. E., *280*
Smith, J. T., 53, 56, 284, 286, *286,* 287
Smith, M., 61
Smith, S. T., *257*
Smock, T., 136, 325
Sokolov, P. G., 287
Sokolova, M. N., 10
Sordahl, L. A., 163
Sparrow, A. H., 163
Spencer, W. A., 187, 187n, 199, 211, 212
Spengel, J. W., 18, *19, 113, 114, 115*
Spielmann, W., 230
Spira, M. E., 211, 313
Spitznas, M., 163
Spray, D. C., 211, 313
Stadhouders, A. M., 249

Starmühlner, F., 90
Stasek, C. R., 11, 12, 14
Stempell, W., *111*
Stephenson, A., 40, 40n, 41n
Stephenson, T. A., 40, 40n, 41n
Stevenshon-Hinde, J., 382
Steward, O., 85
Stinnakre, J., 58, 58n, *60*, 88
Stirling, Ch. A., 170
Stolzberg, S., 95n, 165
Stone, B. A., 94
Straub, W., 83
Stretton, A. O. W., 165, 166, (185)
Strumwasser, F., 56n, 120, 136, 284,
    285, 285, 324, 325
Sulston, J. E., 245
Susswein, A J., 91, 94, 354, *355*, 360,
    362, 362, *363*, 364
Swartz, F. J., 161, *162*
Switzer-Dunlop, M., 217, *220*, 238

Tardy, J., 256, 259
Tauc, L., 58, 58n, *60*, 88, 126, 133,
    153n, 181, 313, 343, 390,
    395, *396*
Ten Cate, J., 123n, 130
Thiele, J., *30*
Thompson, E. B., *30*, 31, 32n, *160*,
    165, 166, 171, *174*, 175, 179, 183,
    204, 292
Thompson, E. L., 372
Thompson, R. F., 212, 351
Thompson, S. H., 292
Thompson, T. E., 14, 22, 24, (26), 45,
    82, 89n, 97, *102*, 103, (106), 220,
    221, 256; *257, 258*, 259, *260*, 273,
    293, *294*, 304, 318, 318n, 322,
    323, 324
Thorndike, E. L., 446

Thorpe, W. H., 212, 332
Thorson, G., 256
Tiffany, W. J., III, 84
Tilton, B. E., 273
Tinbergen, N., 326
Tobach, E., 266, 272, 274
Toevs, L. A. S., 324
Tomasucci, M., 104
Trappmann, W., 447
Trinkhaus, J., 230
Tuley, F. H., Jr., 165
Twarog, B. M., 104
Tweedle, C. D., 165, 171

Underwood, A. J., 22
Ungersted, U., 353

Valenstein, F., 95n, 165
Verworn, M., 265
Villee, C. A., *77, 223*
Vles, F., 227
Von Baumgarten, R. J., 61
Vorwohl, G., 84–85

Wachtel, H., 296
Wald, F., 52n
Walker, W. F., Jr., *77, 223*
Watson, J. D., (264)
Waziri, R., 61, *154*, 160, 179, *179*,
    243, 324
Webster, H. deF., 175, 195
Weel, P. B., van, 86, 86n, *87*
Weevers, R. de G., 131, 292
Weinreich, D., 126, 395
Weiss, K. R., 58, 83, 124, 126, 127
    211, 273, 276, *309*, 310, 310n,
    348, *349, 350*, 351, 354, *355*, 356,
    360, 362, *363*, 364, 365, 395, *396*
Weiss, P., 82, *87*, 88

Wells, J., 143, *144*
Wells, M. J., *13*, 66, 143, *144*, (147),
    226, 384
Welsh, J. H., 79
Wenz, W., 8n
Wheeler, W. M., 245
Whitman, C. O., 394
Wiederhold, M. L., 61, 64
Wilbur, K. M., 6, (26), (106)
Willigen, C. A. van der, *111*
Willows, A. O. D., *114*, 272n, 282,
    293, 296, *297*, (329), 348, (385)
Wilmoth, J. H., *21, 228*
Wilson, E. B., 224
Wilson, E. O., 2
Winckworth, R., 6
Wine, J. J., 157
Wingstrand, K. G., 8, *10*
Winkler, L. R., 29, 44n, 48, 49, 68,
    76, 83, 92n, *102*, 273, 308
Winlow, W., 166, *167, 168, 169*
Wiren, A., *110*
Wittenberg, B. A., 95n, 165
Wittenberg, J. B., 95, 95n, 165
Wolff, H. G., 61, *63*, 64
Wölper, C., 61
Wright, D. L., 76, 80, 82

Yaksta, B. A., 161, *162*
Yochelson, E. L., 6
Yonge, C. M., 6, 8, 11, 12, (26), *30*,
    32n, 42, 69, 70, 72, 73,
    (106), *294*
Young, J. Z., 118, *119*, 143

Zeevi, Y. Y., *181*
Ziegler, A., 266, 272, 274
Zs-Nagy, I., 395
Zulch, K. J., Mihaud, 32n

Subject Index

In this index all references are to *Aplysia* unless otherwise noted.

Abdominal
  aorta, 80, 137
  ganglion. *See* Ganglia
  nerve (A6), 138
  sinus, common ventral, 77
Acephaly (cephalopods), 255
Achatina, 85
Acochlidiacea, 31
Aceola, 31
Acteon, 32, 273
Acteonidae, 391
Action potential
  in optic nerve, 52
  in osmoreception, 58
  in statocyst receptor neurons, 64
Adalaria, 259
Adaptation, in opisthobranch biology, 4
Agriolimax, 66
Akeridae, 32, 140, 390–391
Albumen gland, 97, 100, 103
Amino acids, osmoregulatory role, 88n
Ammonia, urinary (molluscs), 85
Ammonoidea, 15
Amphineura, 14
Ampulla, 97
Anaspidea, 31–32, 70, 140, 276, 295, 390
Annelida, 8
Anodonta, 84
Anophthalmia (cephalopods), 255
Ansel's membrane, 97
Anus, 71, 97
  anal cells (embryo), 231, 234
  development, 235, 237
  innervation, 138
  in molluscs, 69
  oral-anal flexion, 18, 231, 245

Aorta, 80, 137
Aplacophora, 14, 108–109
Aplysiidae, 32, 140, 142
Aplysiomorpha, 31
Appendages, hemal meshwork, 69
Appetitive response. *See* Feeding
Archenteron, 231
Arion, 323
Arminacea, 31
Arousal, 332, 352
  as motivation component, 351, 359–360, 364–366,
    384
  as nonunitary process, 351, 353–354, 356, 384
  sensitizing and nonsensitizing components,
    365–366
  states of, 364–365
  systems, positive vs. negative, 354–356
Arteries. *See* Circulatory system

Bag cells, 143, 242–243
  electrical coupling, 324
  hormone, 136, 249, 324–325
  numeric increase postmetamorphosis, 248–249
  processes, 150, 160
Behavior. *See also* Arousal; Conditioning; Fixed-
    action patterns; Fixed acts; Habituation;
    Learning; Motivational state; Reflexes; Sensi-
    tization; and *specific behaviors*
  adaptive function, 399–400
  all-or-none responses, 268
  alteration by arousal stimuli, 351–353, 356,
    360
  autonomic regulation of, 274–277
  causation, proximal vs. ultimate, 2
  choice, and response hierarchies, 325–327
  classification, 267–269

Behavior (*continued*)
  closed vs. open program, 269
  comparative or evolutionary perspective, 2
  comparative studies, 265–267, 328–329, 333,
      387–388, 394–395
  complex responses, 268
  coordination of responses, 326
  cue-directedness, 359, 366
  defensive behavior (*q.v.*)
  determinants, structural vs. functional, 328
  developmental sequence, 217, 253
  differences, neuronal origins, 397
  effector system
      differential utilization, neural control, 272
      innervation patterns, 104–105, 188
      stimulus intensity discrimination, mecha-
          nism, 209
  elementary responses, 268
  environmental effects on, 274, 328
  escape behavior (*q.v.*)
  functional perspective, 2
  goal-directedness, 359, 366
  goal-oriented, stages, 360
  graded, 268
  hierarchy, and motivational state, 368
  higher-order, 321–325
  and homeostatic regulation, 359, 370–371
  homologies, 267, 387–388, 394–395
      comparative neural circuitry, 397–399
      in opisthobranchs, 394–397
      structural evidence for, 387
      studied with identified cells, 3, 387–388
  hormonal modulation, 325
  inherited vs. acquired, 268–269
  intervening variables, 331, 371
  modification. *See also* Conditioning; Habitua-
      tion; Learning; Sensitization
      initiation by nervous system, and specia-
          tion effects, 405
      in protozoa, 212
  neural control
      higher-order phenomena, 267
      of homologous behaviors, 395
      redundancy, 212
      relationship to behavior, 397–399
  patterns, evolutionary significance, 400–405
  predator-prey recognition, 295–296
  priming, 360
  quiescence, 360, 364, 366
  repertory, sequential development, 217, 253
  response hierarchies and behavioral choice,
      325–327
  responsiveness changes, intervening variables,
      331
  sexual behavior (*q.v.*)
  sign stimuli for, 295
  and speciation, 400–403
  stereotypic, 268

  as taxonomic tool, 394
  thresholds for homologous responses, and
      speciation, 404
Benthic phase (larva), 238
Bivalvia, 14–15, 109–112
Blastocoel, 76n, 223, 231n
Blastomeres, 224–225, 261–262
Blastula, 223–225
Blochmann's gland, 71, 272–273
Blood, 83, 95
Body cavity, 8n, 77
Body plan, 48, 105
  in detorsion, 23–24, 26, 29
  growth axes, 47–48
  of molluscs, 5–7, 24, 47–48
  symmetry, 23, 47–48
  torsion, 231–232
Bohadsch's gland, 71, 274
Buccal cavity, 91
Buccal mass, 91
  in feeding, 91, 308, 318–319 (pulmonates)
  innervation, 127–129, 211, 318–319 (pulmonates)
  muscles, 94, 318 (pulmonates)
Bullomorpha, 29–30
Burrowing, 277, 288–290
Bursa copulatrix, 136

Calcium
  in phototransduction, 163–165
  in presynaptic facilitation, 340, 343
  in transmitter mobilization and release, 340, 351
cAMP. *See* Cyclic nucleotides
Cardiorespiratory system. *See* Circulatory system;
      Heart; Respiratory system
Cecum, 96
Cephalaspidea, 30, 32, 90, 310–313, 391
Cephalization, 108, 114, 145
Cephalopoda, 15
  developmental deformities, 255
  eye, 119, 249, 255
  inking response, 273
  learning capabilities, 374
  nervous system, 118
  optic gland, 143
  salt resorption by kidney, 84
  segmentation, 11
Chemosensation, 56–57, 61
Chiastoneury, 112
Chilina, 32
Chitons, 11
Chromosomes, 223
Cilia, velar, 230, 238
Circulatory system, 76–77, 105–106 (*see also* Heart;
      Hemodynamics; Respiratory system)
  aorta, 80, 137
  behavioral adjustments, 276–277
  circulation, functions, 76–79
  crista aortae, 80

gastroeseophageal artery, 80, 137
gill veins, 72, 75–76, 78–79
heart rate changes in behavioral responses, 276–277
hydraulic skeleton (appendages), 66, 69
neural control, 135–137, 277
pressure gradients, 80–82
renal portal system, 79
venous sinuses, 77–79
Cleavage. *See* Egg
Clocks, biological. *See* Oscillators
Coleoidea, 15
Color variations, 240, 301
Comparative studies, 265–267, 328–329, 387–388, 394–395
  experimental advantages of *Aplysia,* 119–120, 266
Conditioning, 265–266
  classical, 332, 372 (gastropods)
  of gastropods, 372, 379–382
  instrumental, 332
  operant, 332, 372–374
  vs. sensitization, 333
Consummatory response. *See* Feeding; Motivational state
Copulation. *See* Sexual behavior
Cordon, 103, 222
Corollary discharge, of motor activity, 315
Coupling chain, 321–322
Crawling. *See* Locomotion
Ctenidia, 7, 72
Cyclic nucleotides
  in habituation, and sensitization, 337, 343, 351, 383
Cyclopia (cephalopods), 255
Cylindrobulla, 32

Defensive behavior
  of adult, 70, 253
  alterations by arousal stimuli, 356, 360
  development, 252–253
  fixed acts, 272–274
  habituation, 333, 337, 345
  larval, 252–253
  neural control, 135–136
  reflex vs. fixed act, neural control, 273
  reflex threshold, 274
  reflexes, 269–272
  sensitization, 333, 337, 356
  suppression, food-induced, 326–327, 354–356
Dendronotacea, 31
Determinants of cleavage, 224–225, 255
Detorsion, 23–24, 26, 29, 112, 120, 245–247
Development. *See also* Egg; Larva; Metamorphosis
  behavioral triggering step, 254
  determinants, 224–225, 255
  determinate, 224

direct vs. indirect, 256
embryonic, 225–234
juvenile (postmetamorphic), 218, 239–240, 250–253, 262
larval, 218, 235–238, 262
malformations, 254–255
metamorphic, 218, 250–253, 262
morphogenetic substances, 224
regulative, 224n
types (opisthobranchs), 256–259
Diet. *See* Feeding
Dibranchia, 15
Digestive system, 91 (*see also* Feeding)
  anal cells (embryo), 231, 234
  anus, 69 (molluscs), 71, 97, 235, 237
  archenteron, 231
  buccal mass, 89–90 (gastropods), 91, 94
  cecum, 96
  development, 231, 237, 245–246
  digestion, 96
  digestive gland, 91 (gastropods), 96
  esophagus, 94, 96
  fecal string, 96–97
  foregut, 89–90 (gastropods), 91–94
  in gastropods, 89–91
  gizzards, 96
  hepatopancreas, 96
  hindgut, 91, 96
  intestine, 96
  midgut, 91 (gastropods), 96, 230
  midgut gland (gastropods), 91
  mouth, 91, 231, 234
  neural control, 127, 136
  odontophore, 90 (gastropods), 91–94, 308
  oral-anal flexion, 18, 231, 245
  radula, 48, 91–92, 238, 308
  rectum, 91
  red spots (perivisceral membrane), 237, 239
  rhochidion, 91
  salivary glands, 94
  stomach, 91 (gastropods), 96
  torsion and detorsion, 23–24, 245–246
Distribution of *Aplysia,* 38–45
DNA of cell nucleus, 161
Dolabellinae, 140
Dolabriferinae, 140
Doridacea, 31
Doridomorpha, 29, 31
Drive, 331, 359
Drive state. *See* Motivational state

Ectoderm, 229–230, 240, 242–245, 255
Efference copy, of motor activity, 315
Egg, 97, 103–104, 220
  animal pole, 226, 229
  capsules, 103, 222
  centrolecithal, 220n
  cleavage, 223, 260

Egg (*continued*)
  cordon, 103, 222
  cortex, 224–225
  determinate, 224, 261
  development, 223–229, 260–262
  indeterminate, 224
  macromeres, 224–229
  meridionality, 223, 225, 260
  micromeres, 224–229
  mitotic spindles, 220n
  morphogenetic substances, 224, 261–262
  mosaic eggs, 224n
  oblique cleavage, 223, 260
  polar lobe, 224
  regulative eggs, 224n
  release and passage, 97, 103–104
  segmentation, 223
  spawn production, 323–324
  spiral cleavage, 223, 226–228, 260
  strand, 103, 222
  symmetry, 225, 228
  telolecithal, 220n
  vegetal pole, 220n, 224–225, 230–231
Egg-laying, 101, 323–324
  as fixed act, 324–325
  hormonal control, 136, 249, 324–325
Elysiomorpha, 29
Embryo
  archenteron, 231
  blastocoel, 76n, 223, 231n
  blastomeres, 224–225, 261–262
  blastula, 223–225
  cleavage stages, 218, 223–229
  coelomic cavity, 225
  development, 218n, 223–233, 262
  epiboly, 230
  gastrulation, 223, 225, 230–231
  germ layers, development, 229
  lithium-treated (pulmonates), 255
  oral-anal flexion, 231, 245
  prototroch, 230, 233–234
  rotation within egg capsule, 230–231
  segmentation cavity, 231
  torsion, 231–232
  trochophore, 225, 230, 240
Endoderm, 225, 229, 255
Environment
  and dietary preference, 299
  and habituation, 346–348
Eolidacea, 31, 292
Eolidomorpha, 29, 31
Epiboly, 230
Equilibrium reception, 61–65
Erythrocruorin, 83
Escape behavior, 295–296
  as fixed-action pattern, 295–296
  habituation (opisthobranchs), 348

  neural control (nudibranchs), 296
Esophagus, 94, 96
  crop, 94, 96
  in gastropods, 91
  innervation, 127
  suction (sacoglossans), 90
Euthyneury, 17, 23, 112
Evolution
  behavioral significance for, 404
  convergence and divergence, 6n
  parallelism, 6, 118
  programmed, 32
  selection pressures for, 388, 404
  significance for behavior, 387–388
  transitional position of *Aplysia*, 120
Eye, 50–52
  cephalopod vs. vertebrate, 119
  circadian periodicity, 53–56, 285
  development, 237, 239, 242
  developmental deformities (cephalopods), 255
  entrainment, 53–56
  and locomotor modulation, 285–286, 375
    (nudibranchs)
  neural control, 52
  ocular clock, 53–56
  and optic ganglion, 249 (cephalopods), 375
    (nudibranchs)
  receptor cells, 50–51, 375 (nudibranchs)
  responses to light, 52, 375–376 (nudibranchs)
  retinal neuropil, 51
Exteroceptive stimuli, 365

Feeding, 304 (*See also* Digestive system)
  adult behavior, development, 250–252
  appetitive response, 252, 304–307, 314 (opistho-
    branchs)
    alterations by arousal stimuli, 354
    chemosensation in, 304–307
    development, 252, 254
    neural control, 124
  avoidance training (gastropods) 379–382
  behavior sequence, 298
  biting response (*see* consummatory response)
  central program (pulmonates), 318–319
  circadian rhythm, 360n
  components, 252, 254
  conditioning (gastropods), 379–382
  consummatory response, 252, 308, 311 (opistho-
    branchs)
    biting response, 92–94, 308
    development, 252, 254
    entrainment, 377
    excitation by chemical stimuli, 363
    inhibition by food bulk, 360–363
    neural control, 123–124, 308–309, 313–316
      (opisthobranchs), 318–320 (pulmonates)
    sensitization, 348–351, 356

suppression by noxious stimuli, 356
swallowing, 94, 308
diet, 299
preferences, 301; environmental variations, 299; and morphological adaptation, 303–304; variations with age, 218, 301
and spawn production, 324
distance chemoreception, 304–307, 314
food
avoidance learning (gastropods), 379–382
detection (opisthobranchs), 310–311, 314
-gathering organ, 48, (larva) 234
pigments, 301
-seeking component (*see* appetitive response)
trail chemoreception (cephalaspids), 310–311
foregut role, 91–94, (opisthobranchs) 90
and growth, 301–302
habituation (opisthobranchs), 379
head movements, neural control, 129–130, 308
and heart rate, 351, 354
in larva, 250–252
at metamorphosis, 238–239
modification of, by training, 377, 379–382 (gastropods)
molluscan patterns, 88
neural control, 123–124, 127, 313–316, 348–351 (opisthobranchs), 318–321 (pulmonates)
nutrition, 301–302
in opisthobranchs, 310–318
orienting response, 50, 304–305, 359
neural control, 308
pharyngeal response (*see* consummatory response)
phases (opisthobranchs), 313–314
as preferred behavior, 326–327
in pulmonates, 318–321
reverse peristalsis, 308
satiation, signal for, 360, 363–364
and sexual behavior (opisthobranchs), 322–323, 326
suspension feeding (prosobranchs), 90
swallowing, 94, 308
tracking response (*see* appetitive response)
Fertilization, 97, 101–103
Fertilization chamber, 103
Fixed-action patterns
definition, 268
in circulatory adjustments, 277
escape behavior, 295–296
in feeding, 298, 308
in locomotion, 277–282
respiratory pumping, 274–275
sign stimuli, 295
Fixed acts
definition, 268
egg laying, 324–325
escape swimming (opisthobranchs), 348

inking, 272–273
vs. reflex acts, 268
respiratory pumping, 272
Foot, 68–69, 105 (*see also* Locomotion)
crawling waves, 277
development, 233–235, 237
in gastropods, 66–68
innervation, 129, 131–132, 146, 210
larval, 235
in molluscs, 7, 66
movements, neural control, 210
muscles (gastropods), 66–68
pedal ganglia. *See* Ganglia

Gametolytic gland, 101
Ganglia. *See also* Nerves; Neurons
abdominal ganglion, 134–138, 146
in anaspids, 140
bag cells, 136, 143, 150, 160, 242–243, 248–249, 324–325
connectives, 133, 242–243
in defensive withdrawal, 136, 190–193, 269
development, 242–244, 263, 390–391
in digestive system control, 136
hemiganglia, 133–135, 146, 243, 390
in inking, 71, 136
nerve trunks from (A1–A6), 136–138
neuroendocrine functions, 135
in osmoreception, 58
in primitive vs. advanced forms, 120, 145–146, 390–391
reproductive functions, 136
respiratory function, 136, 190
respiratory pumping command cells, 275
siphon innervation, 191, 199
structure, 150
torsion in, 120, 138, 248
trophospongium, 160
variations, 138
visceral functions, 135
white cells, 150, 160
branchial ganglion, 136
buccal ganglia, 127
connectives and commissure, 123, 127
consummatory response control, 124, 308
development, 242
in esophageal movements, 127
nerve trunks from (B1–B6), 127–129
odontophore control, 127
central ganglia, homologies (opisthobranchs), 388–394
cephalization, 108, 114, 145
cerebral ganglia, 123
appetitive response control, 124
in chemosensation, 57
connectives and commissure, 123, 127, 129, 132, 242

Ganglia (*continued*)
  consummatory response control, 124, 308
  in copulation, 124, 130
  development, 242
  in feeding, 123–124
  in locomotor control, 123, 282
  metacerebral cells (*see* Neurons)
  nerve trunks from (C1–C6), 126–127
  in ocular control, 52
circumesophageal ring, 114, 123, 138; in opistho-
    branchs, 140, 142, 145, 388
connective tissue sheath, 150
connectives and commissures, 18–20, 129, 132–
    133, 156–157, 242, 242–243 (prosobranchs),
    248
development, 240, 242–243, 245–249, 262–263
fusion, 133, 138; in anaspids, 390; in gastro-
    pods, 108, 114, 116–118, 145
genital ganglion, 137, 242
glial cells, 155–157
head ganglia, 123
homologies, study of, 388–394 (opisthobranchs),
    397 (other invertebrates)
identified cells as markers, 140, 391–394,
    395–397
infraesophageal ganglion, 248
intestinal ganglion, 135n
invertebrate types, 149–150
labels for electron-microscopy, 170–175
metacerebral cells (*see* Neurons)
neurohemal organ, 150
neuronal markers, 391–394, 395–397
neuropil region, 165–166, 170–181
optic ganglion, 124–125, 242, 249, 375 (nudi-
    branchs)
osphradial ganglion, 58, 242
pallial ganglia, larval, 133–135
parietal ganglion, 135n, 246–248
parietovisceral ganglion, 134–135
pedal ganglia, 129, 138
  connectives and commissures, 123, 129, 132,
      242
  in copulation, 124, 130
  development, 242, 249
  in escape response (opisthobranchs), 296
  in head movement control, 130, 308
  in locomotor control, 129–130, 282
  nerve trunks from (P1–P10), 131–132
  in penile control, 129–131
pleural ganglia, 132, 134
  connectives, 129, 132–133, 242
  development, 242
  in escape response (opisthobranchs), 296
  functions, 133
  nerve trunks from (PL1–PL2), 133
regions innervated by, 390
ring ganglia, in locomotor control, 283
structure, 150 (molluscs), 183 (invertebrates)

subesophageal ganglion, 248
subintestinal ganglion, 120, 135, 243
supraesophageal ganglion, 242, 246–248
supraintestinal ganglion, 120, 135, 243
torsion and detorsion, 248, 390; in gastropods,
    246–248; in opisthobranchs, 391
visceral ganglion, 120, 135, 243, 248
visceral loop (gastropods), 116, 118, 140, 145
Gastropoda, 16–17
  adaptive divergence, 25
  cephalization, 114, 145
  classification, 17
  conditioning, 372
  digestive system, 89–91
  experimental advantages, 119–120
  feeding behavior, 88–89
    modification by training, 372, 379–382
  foot, 66–68
  nervous system, 25, 112–118, 145
  neural concentration, 114
  respiratory pumping, 275–276
  torsion and detorsion in, 16–20, 25–26, 112, 116,
      145, 246–248
Gastrulation, 223, 225, 230–231
Genetic program
  closed vs. open, 269
  in lithium-treated cells, 255
  research, advantages of *Aplysia*, 263
Genital
  aperture, 50, 70, 104, 322
  ducts, 97, 100, 104, 136–138
  ganglion, 137, 242
  gland, 136–138
  groove, 104, 240
  nerve (A5), 136–137
Gill, 72
  development, 239
  in gastropods, 17
  innervation, 136, 138
  in molluscs, 11–12, 72–74
  movements, neural control, 133, 136, 190, 395–
      396 (nudibranchs)
  muscle groups, 75–76
  muscular contractions, 75
  in opisthobranchs, 29
  pinnule response, 196–198, 272
  pinnules, 72
  respiratory pumping, 73–74, 272
  veins, 72, 75–76, 78–79
  withdrawal reflex, 188–190, 269
    habituation, cellular analysis, 345
    neural control, 136, 189–195, 269, 356
    as relayed reflex, 188–190
    sensitization, 337, 351, 356
    suppression, food-induced (opisthobranchs),
        326–327, 356
Gizzards, anterior and posterior, 96
Glial cells, 155–157

Golgi complex, 157–160
Gonad, 97
Gonoduct, 97
Gut. *See* Digestive system
Gymnosomata, 31, 295

Habituation, 332–333
  definition, 333
  cellular analysis, 337
  in central and peripheral pathways, 199, 206,
      211–213
  centralization, 212–213
  environmental effects on, 346–348
  evolution, 212–213
  from homosynaptic depression, 212, 337
  kinetics, neural basis, 199, 206
  long-term, cellular analysis, 345
  in metazoa, 212
  molecular model, 343
  in opisthobranchs, 348
  in protozoa, 212
  of receptor cells, 212
  from sensory adaptation, 212
  short-term, 333, 337, 343
  stimulation pattern effect, 333
  time course, 333–337
Hair cells, 61–64
Head, 48–50, 105
  developmental deformities (cephalopods, pulmo-
      nates), 255
  innervation, 123, 129
  in molluscs, 6
  waving, 48, 56, 252, 305, 307
    neural control, 129–130, 308
Heart. *See also* Circulatory system; Hemo-
      dynamics; Pericardium
  cardiac output regulation, 82–83
  development, 237–240
  function, 79
  in molluscs, 12
  myogenicity, 79, 106, 211
  neural control, 79, 137, 211, 277
  rate, alterations by arousal, 276–277, 351–352,
      354, 365
  structure, 79, 105–106
Helisoma, 104, 318–321, 382
Helix, 82, 84–85, 287
Hemocoel, 77
Hemocyanin, 83, 95
Hemodynamics
  cardiac and arterial filling, 80–82
  cardiac output regulation, 82–83
  constant volume hypothesis, 80–82
  neural control, 136
Hemolymph, 66, 69, 83, 136
Hermaphroditic ducts, 97, 100, 104, 136–137
Hermaphroditic gland, 97, 137
Hermaphroditism, 97, 106, 321

Hermissenda, 64, 374–375
Homeostasis and behavior, 359, 370
Hormones
  bag-cell hormone, 136, 249, 324–325
  of optic gland (cephalopods), 143
  in osmoreception, 58, 61, 88, 106
  of ovotestis, 104
  in reproductive activity, 136, 324–325
Hybrids, 400–404
Hydraulic skeleton, 66, 69
Hypobranchial gland, 71

Ink vesicles, 237, 239
Inking
  as fixed act, 272–273
  function, 273, 400, 404
  neural control, 135, 273, 395, 400
  response threshold, 273–274, 404
Insects, 245
  blowfly, 360n
  cricket, 403–404
  fruit fly *(Drosophila),* 249
Interoceptive stimuli, 365
Intertidal zone, 40
Intestine. *See* Digestive system
Ionic conductance
  Ca^{++} influx across membrane, in sensitization,
      340
  K^+, in phototransduction, 163–165

Juvenile
  development (postmetamorphic), 218, 239–240,
      262
  heart rate, 82, 239–240

Kidney, 70, 79, 85
  development, 231, 235, 237
  function (molluscs), 84–85
  innervation, 85, 136–137
  in molluscs, 83–85
  renal
    cavity, 79, 85
    pore, 85
    sac, 85
    vein, 79, 85
  in vertebrates, 83–85

Labial nerves (C1, C3), 126–127
Lamellibranchia, 15
Larva, 218, 235
  behavioral repertory, 250
  defensive response, 252–253
  development, 218, 235–239
  developmental types (opisthobranchs), 256–259
  double monsters, 254–255
  feeding, 250–252
  locomotion, 231, 234, 237–238, 250
  metamorphosis *(q.v.)*

Larva (*continued*)
  in opisthobranchs, 256–259
  organs, 235–239
  settlement, 238
  shell, 70, 237
  trochophore, 225, 230, 240
  veliger stage, 218, 225, 234–237, 250–253, 262
  velum, 231, 234–235, 238–239
Laurencia, 43, 238, 252, 254, 299
Learning. *See also* Conditioning; Habituation;
    Sensitization
  definition, 331
  associative, 1, 265, 332, 371–372, 374, 382, 384
  classification, 332
  complexity, 332
  nonassociative, 332–333, 374, 383
  from postural cues (molluscs), 374
  short-term vs. long-term, 332
Leeches, 397
Life cycle phases, 218
Light
  ocular clock, entrainment of, 53–56
  phototactic response, inhibition by rotation
      (nudibranchs), 374–377
  phototransduction, 163
  receptors, ocular and extraocular, 286
Limax, 287, 321, 323, 382
Lipochondria, 163–165
Lithium, and developmental deformities, 255
Littoral zones, 40
Locomotion, 277 (*see also* Foot)
  adult behavior, development, 250
  arousal states, 364–365
  benthic phase (larva), 238
  burrowing, 277, 288–290
  crawling
    circadian periodicity, 284–286, 287 (pulmo-
        nates)
    entrainment of, 284
    pedal waves, 66 (molluscs), 277–282
    phase in larval development, 237–238
    in pulmonates, 287
    timing signals, 286–287
  development, 238, 250
  escape behavior, 295–296, 348
  as fixed-action pattern, 277–282
  hydraulic function of hemolymph (molluscs),
      66, 69
  in larva, 231, 234, 237–238, 250
  neural control, 123, 129–132, 210–211, 282–284
  operant conditioning, 372–374
  pelagic phase (larva), 238
  rhythm, modulation by biological clocks,
      285–287
  sensory-sensory interactions (nudibranchs),
      374–377
  shifting patterns, evolutionary importance, 405

  swimming, 290–292
    escape behavior (opisthobranchs), 292, 348
    in opisthobranchs, 292–295
    parapodial activity in, 210, 290–292, 293
        (opisthobranchs)
    and respiratory pumping, 276
    transition from crawling, 292
    transition to crawling, 238, 250
  variations with age, 218
Loligo, 255
Lymnaea, 255
Lysosomes, 157, 163

Mantle
  cavity, 70–71, 105
    in detorsion, 23
    in molluscs, 7, 11, 69
    in torsion, 18
  development, 237, 239–240
  external organs, neural control, 135–136
  in molluscs, 7
  shelf, 70
    innervation, 131–132, 269
    mechanoreceptors, 66
Maritime zone, 40–41n
Mating patterns. *See* Sexual behavior
Mechanoreception, 65–66
Membrane
  Ca^{++} influx across, in sensitization, 340
  K^+ conductance in phototransduction, 163–165
Mesoderm, 225, 229, 255
Mesogastropoda, 29, 246–248
Metamorphosis, 218, 238–239
  behavioral triggering step, 254
  defensive behavior changes, 252–253
  feeding behavior changes, 250–252
  locomotor changes, 250
  in opisthobranchs, 259–260
  postmetamorphic (juvenile) development, 218,
      239–240, 245–246, 259–260 (opisthobranchs), 262
  torsion and detorsion, 245
  triggering substance, 43, 238, 252, 254, 256–257,
      259–260 (opisthobranchs)
  velar cilia shedding, 238
  velar lobe fusion, 239
Metapodial glands, 237
Metapodium, 68–69, 239
Microscopy
  electron-microscopic labels, 170–175, 184
  staining media, 170–171
Migration between shore zones, 43–44
Mollusca
  body plan, 5–7, 24, 47–48
  central nervous system, 108–119
  classification, 12, 25
  ctenidia, 7
  digestive tract, 89

egg, development of, 260–262
equilibrium reception, 64–65
evolution, 8
feeding patterns, 88–89
foot, 7, 66
ganglionic structure, 150
growth axes, 47
kidney, 83–85
learning capabilities, 374, 384
mantle cavity, 7, 11, 69
morphotype, 6–7, 24
nervous system, 108–119, 145
neuroendocrine systems, 142–143, 146
primitive, characteristics, 11–12, 108
proto-mollusc, 11–12
segmentation, 10–11
shell, 11–12
skin, 48
urine formation, 83–86
Monoplacophora, 12–13, 108–109
Morphotype, 6–7
Mosaicism, 224n
Motivation, 331, 359
Motivational state, 359
     and arousal, 359–360, 364–365
     and behavioral hierarchy, 368
     cellular-level examination of, 370–371
     functional components, 366–367
     goal-specific, 359, 366
     and homeostasis, 359, 370
     as intervening variable, 359
     and reinforcing value of stimulus, 366–367
     and sensitization, relationships, 365–366
Motor neurons. For individual cells see Neurons
     in circulation control, 277, 351, 356
     corollary discharge, 315
     discharge activity, efference copy of, 315
     in feeding, 124, 127, 308, 351
          opisthobranchs, 313–315
          pulmonates, 318–321
     in gill movements, 190, 269, 295–297 (nudi-
          branchs)
     in inking, 273, 400
     in locomotion, 282–284
     in mating song (cricket), 403–404
     peripheral vs. central (Spisula), 209
     in polyneuronal innervation, 105
     in respiratory pumping, 272
     in speciation, 404
Mouth, 91, 231, 234
     oral-anal flexion, 18, 231, 245
     oral veil, 49, 126, 239, 304, 307, 314
Mucous gland, 97, 100, 103
Mucus, 48
Muscles
     buccal mass, 94
          neural control, 212

     in pulmonates, 318
     columellar, 68
     effector organs, innervation, 104–105
     of foot (gastropods), 66–68
     of gill, 75–76
     polyneuronal innervation, 105
     of shell, 68
     velum retractors (larva), 233
Myoglobin, 94–96, 165
Mytilus, 104

Nautiloidea, 15
Navanax, 90, 142, 310–313, 322–323, 326, 364, 379
Neck, 48
Neopilina, 8–12
Nerves. See also Ganglia; Torsion
     *, 132n
     A1–A6, 136–138
     A1a (vulvar nerve), 132, 136
     A6c (siphon nerve), 132, 138
     abdominal nerve (A6), 138
     anal nerve (A6), 138
     B1–B6, 127–129
     branchial nerve (A2), 136
     buccal nerves (B4–B6), 129
     bursa copulatrix nerve (A3), 136
     C1–C6, 126–127
     circulatory supply, 150
     circumesophageal ring, 114, 123, 138, 140;
          in opisthobranchs, 145, 388
     connectives, glial cells of, 156–157
     esophageal nerve (B2), 127
     genital nerve (A5), 136–137, 242
     labial nerves (C1, C3), 126–127
     optic nerve (C5), 52, 127
     osphradial nerve (A2), 136
     P1–P10, 131–132
     P5 variations, 138
     P5a, 132–133, 138
     P5a*A6c, 132, 138
     P5b*A1a, 132, 136, 138
     parapedal commissure nerve (P10), 132
     parapodial nerves (P6, P7), 132
     pedal nerves (P1, P8, P9), 131–132, 138
     pericardial nerve (A4), 137
     periesophageal ring, 140, 142
     peripheral nerves
          distribution variations, 138–140
          glial cells of, 156–157
     pharyngeal nerves (B4, B5), 129
     PL1, 131, 133
     pleural mantle nerve (PL2), 133
     radular nerve (B1), 127
     salivary gland nerve (B3), 129
     siphon nerve (A6), 132, 138
     spermathecal nerve (A3), 136
     statocyst nerve (C6), 127

Nerves (*continued*)
  tegumentary nerves (P2–P5), 131–132, 136
    opaline branch (P5a), 133
    pleural root (PL1), 131, 133
  tentacular nerves (C2, C4), 126–127, 131
  visceral loop, 133, 140
  vulvar nerve (A1), 132, 136, 138
Nervous system. *See also* Ganglia; Glial cells;
    Habituation; Sensitization
  in anaspids, 140–142
  archetype (gastropods), 18–20, 25, 145
  autonomic regulation of complex behavior,
    274–277
  autonomic responsiveness, alteration by arousal
    stimuli, 352
  axon reflexes, 208
  behavioral change initiation, and speciation, 405
  in bivalves, 109–112
  brain, 118
  central innervation without peripheral motor
    cells, 211
  central vs. peripheral control, 187–188, 208–209
    (*Spisula*), 209–211, 214
  cephalization, 108, 114, 145
  in cephalopods, 118
  chiastoneury, 112
  circulatory supply, 150
  concentration, 108, 114, 145
  detorsion, 23, 112, 120, 246, 390
  developmental stages, 240–245, 262–263
  equilibrating functions after statocyst removal,
    64–65
  euthyneuran, 17, 23, 112
  evolution, 212, 215
  experimental advantages of *Aplysia*, 119–120
  in gastropods, 25, 112–118, 145
  innervation patterns, 188–189
  integrative action, from response hierarchy and
    choice, 326
  metamorphic development, 245
  in molluscs, 108–119, 145
  neural circuitry
    in different behaviors, 397
    in homologous behaviors, 399
    variations in related organisms, 395–397
  neuron doctrine, 206
  of opisthobranchs, 116, 120, 145–146, 388–390
  orthoneuran, 112
  peripheral pathways, 214–215
  postmetamorphic (juvenile) development, 245–
    246, 248
  primitive, 108–109
  in prosobranchs, 112–114
  in pulmonates, 112, 116
  in speciation, 405
  streptoneuran, 17, 23, 112, 248
  torsion, 16, 18, 120, 245, 390

  gastropods, 16–20, 25–26, 112, 116, 145,
    246–248
  transitional phylogenetic position of *Aplysia*, 120
  variations, 138–140, 140–142 (anaspids)
  zygosis (gastropods), 116
Neuroblast, 245
Neuroendocrine
  processes, neural control, 135
  system (molluscs), 142–143, 146
Neurohemal organ, 150
Neuron doctrine, 206
Neurons. *See also* Motor neurons; Sensory neu-
    rons; Synapse; *and individual organisms*
  axon formation, 244
  axonal relationships, 153, 166–170, 181–183, 244
  bag cells, 136, 143, 150, 160, 242–243, 248–249,
    324–325
  cell body, 120, 153–155, 183–184
    and axon terminals, relationships, 153,
      166–170, 181–183
    glial cells associated with, 157
  cell lineage studies, 244–245
  cell size increase postmetamorphosis, 249
  central, evolution from primitive cell, 212
  command cells
    in circulation control, 277
    in feeding, 310, 316–318 (opisthobranchs),
      318–321 (pulmonates)
    in locomotion, 283
    in respiratory pumping, 272, 275
  comparative studies, 387–388, 394–395
  conductance to K$^+$, 163–165
  cyberchron network (pulmonates), 319
  cytoplasm, 157–160, 181–183
  dendritic trees, 153–155, 166
  differentiation, and peripheral organ develop-
    ment, 249
  DNA content of nucleus, 161–163
  endogenous burst activity, 277
  endoplasmic reticulum, 157
  fine structure, 157–160
  ganglionic markers, 140, 391–394, 395–397
  glial indentations, 160
  granules, 160
  growth factor, 244
  identified cells
    in ganglionic development, 243–244
    as labels for synaptic regions, 170–171
    lineage studies, 244–245
    in study of behavioral homologies, 4, 394
    in study of ganglionic homologies, 391–394
    in study of speciation, 405
  invertebrate vs. vertebrate, 153–155, 161–162,
    177, 187
  L2–L6, 160, 166
  L3, 171n, 175
  L7, 166, 171n, 175, 345

L10, 166, 171n, 175, 183, 277, 395
L14A, 395
L23–L29, 356
labeling for microscopy, 170–175, 184
LD$_{H11-H13}$, 211
LD$_{HE}$, 211
lipochondria, 163–165
LP1, 133–134, 244–245
metacerebral cells, 125
    in feeding, 124, 310, 316 (opisthobranchs),
        319–321 (pulmonates), 356
    in opisthobranchs and pulmonates, 395
    in sensitization, 348–351, 356
myoglobin content, 165
neuroblasts, 245
neurogenesis, 217, 249
neuropil region, 153, 157, 165–170, 175–181
nucleus, 160–163
numeric increase postmetamorphosis, 248–249
organelles, 157, 181–183
PL-1 (nudibranchs), 397
paracerebral neurons, in feeding (opistho-
    branchs), 316
peripheral, types, 203–204, 214–215
pigment granules, 157, 163–165
plasticity, in habituation, 212–213
polarity, invertebrate vs. vertebrate, 153
polyploid, 161
presynaptic terminals, 175–179
protein synthesis in, 156
R2, 133–134, 161, 163, 171n, 183, 243–245, 249
R3–R14, 150, 160
R15, 58, 61, 88, 106, 160, 166, 181
RB$_{HE}$, 166, 211, 277, 356
receptive and transmitting poles, 155
Retzius cells (leeches), 397
ribosomes, 157
serotonergic, in sensitization, 340, 345,
    351, 383
and speciation, 403–404
spines, 177
staining of individual cells, 165–166
of statocyst, development, 65
trophospongium, 160
varicosities, synaptic, 177
vesicles, 175, 177, 179, 184
white cells, 150, 160, 166
    in withdrawal reflexes, 269–272
Neuropil, 153, 157
    axonal relationships, 166–170
    structure, 165–166, 175–181
Neurosecretory cells, 143 (gastropods), 150, 160,
    166, 203–204, 244, 324
Nose bearers, 50
Notarchinae, 140
Notarchus, 70, 73, 140
Notaspidea, 31–32, 293, 314–318

Nucleus, 160,163
Nudibranchia, 29, 31–32, 45, 64, 259–260, 292, 296,
    304, 323, 348, 374–375, 395–397
Nutrition, 301–302

Octopus, 84
Odontophore, 91–92
    in feeding, 92–94, 308
    in gastropods, 90
    neural control, 127
    in pulmonates, 318
Oocytes, 101, 103
Opaline, 273–274
Opaline gland, 71, 274
    innervation, 132, 136, 138, 140
Operculum, 21, 234–235, 239
Opisthobranchia, 17, 29–37
    adaptive behavior, study of, 4
    annectant form, 32
    associative learning, 379–381
    behavioral homologies, 394–397
    classification, 17, 29–31
    in comparative studies, 4, 266
    conditioning, 379–381
    development, types of, 256–259
    escape behavior, 295–296, 348
    evolutionary relationships, 31–32, 388, 391
    experimental advantages, 119–120, 266
    feeding behavior, 90, 310–318
    gill, 29
    learning capabilities, 332–333, 348
    locomotor patterns, 277, 288, 292–295
    metamorphic and postmetamorphic develop-
        ment, 256–257
    nervous system, 116, 120, 388–390
        central ganglia, homologies, 388–394
        primitive, 145–146
    phototaxis, inhibition by rotation (nudibranchs),
        374–377
    phylogeny, transitional position of *Aplysia*, 120
    respiratory pumping, 275–276
    sexual behavior, 322–323
    transitional group in, 391
Optic
    ganglion, 124–125, 242, 249, 375 (nudibranchs)
    gland (cephalopods), 143
    nerve (C5), 52, 127
Oral
    -anal flexion, 18, 231, 245
    tube, 68
    veil, 49, 126, 239, 304, 307, 314
Organs
    development, 231, 237, 239, 249
    effector systems, neural control, 104–105, 188,
        209, 272
    in torsion, 18
Orientation indicators, 61

Orienting response. *See* Feeding; Motivational
    state
Orthoneury, 112
Oscillators, 285–286
    extraocular clocks, 286
    in feeding movements (opisthobranchs),
        316–318
    in locomotor rhythm, 286
    ocular clock, 53–56, 286–287
    synchronization, 286–287
Osmoreception, 58–61
Osmotic regulation, 58–61, 86–88, 135
Osphradium, 11, 58, 61, 70, 88
    chemoreceptor function, 61
    ganglion, 58, 242
    hormonal activity, 58, 61
    in molluscs, 7
    nerve (A2), 136
Oviductal channel, 104
Ovotestis, 97, 103–104, 137
Ovum. *See* Egg

Pallium (molluscs), 7
Parapodia, 69–71
    defensive withdrawal, 130
    development, 239–240
    functions, 70
    innervation, 129–132, 209–210, 244
    in swimming, 290–292
Patella, 81
Pelagic phase (larva), 238
Pelecypoda, 14–15
Penis, 101, 322
    neural control, 124, 127, 129, 131
Pericardium
    cavity, 79
    constant volume hypothesis, 80–82
    filtration into, 86, 136
    innervation, 136–137
Periesophageal ring, 140, 142
Phalloform organs (opisthobranchs), 313
Pharyngeal nerves (B4, B5), 129
Pheromones, 290, 313, 404
Phyllaplysia, 97
Physa, 372
Pigment granules, in neuronal cell body, 157,
    163–165
Pink color, in juvenile development, 239
Pinnules (of gill), 72
Pleurobranchaea, 32, 142, 292, 314–318, 323, 326,
    364, 368–369, 379
Pleurobranchomorpha, 29, 31
Polyplacophora, 14, 108–109
Potassium, in phototransduction, 163–165
Potentials
    depolarizing, 52, 64
    excitatory postsynaptic potential (EPSP), 206
    in glial cells, 156

hyperpolarizing, 52, 64
ocular polarization response to light, 52
in sensory hair cells (nudibranchs), 64
in statocyst neurons, 64
synaptic
    in feeding behavior modulation
        (opisthobranchs), 318
    in habituation, 337, 345
    in sensitization, 337
Prevelum, 230–231
Proboscis (opisthobranchs), 314
Propodium, 237
Prosobranchia, 17, 64, 90, 112–114, 231–232, 242–
    243, 246, 388
Prostate gland, 101
Protein synthesis
    in glial cells, 156
    in habituation and sensitization, 337
    in neurons, 156
Protobranchia, 15
Prototroch, 230, 233–234
Pseudosiphon, 71
Pterotrachea, 64
Pulmonata
    classification, 17
    developmental deformities, 255
    evolutionary relationships, 32
    feeding behavior, 318–321
    learning capabilities, 332–333
    locomotor patterns, 287
    nervous system, 112, 116
    sexual behavior, 323
    taste-avoidance learning, 382
Purple (ink) gland, 71, 272–273
    innervation, 71, 136, 138
    mechanoreceptors, 66

Radula, 48, 91–92
    in feeding, 308
    in gastropods, 90
    neural control, 127, 308–309
    in pulmonates, 318
Rectum, 91, 137
Red spots
    on epidermis, 240
    on perivisceral membrane, 237, 239
Reflexes
    axon reflexes, 208
    central vs. peripheral mediation, 195, 208
    conjoined responses, 195, 199, 214
    defensive withdrawal, 188–189, 199, 269
        neural control, 136, 189–195, 269–272
    feeding, reflex patterns in, 304–305
    local responses, 188, 196, 209, 214, 333n
    reflex acts
        definition, 268
        vs. fixed acts, 268–269
    reflex patterns, definition, 268

relayed reflexes, 188, 214
remote reflexes, 188
Reproductive activity. *See also* Egg-laying; Sexual
  behavior
  hormonal control, 136
  neural control, 135–136
  species-isolating mechanisms, 400–403
Reproductive maturity, 240
Reproductive system. *See also* Egg; Genital;
    Sexual behavior; Sperm
  albumen gland, 97, 100, 103
  ampulla, 97, 100
  autospermal grooves, 100–101
  egg (*q.v.*)
  fertilization, 97, 101–103
  fertilization chamber, 103
  follicles, 97
  gametolytic gland, 101
  genital (*q.v.*)
  gonad, 97
  gonoduct, 97
  hermaphroditic ducts, 97, 100, 104
  membrane gland, 103
  mucous gland, 97, 100, 103
  oocytes, 101, 103
  oviductal channel, 104
  ovotestis, 97, 103–104
  ovum (*see* Egg)
  penis (*q.v.*)
  prostate gland, 101
  seminal
    groove, 70
    receptacle, 97, 101
    vesicles, 97, 103
  sperm (*q.v.*)
  vagina, 97, 104, 136
  white mucous gland, 103
  winding gland, 103
Research advantages of *Aplysia,* 119–120, 263, 266
Respiratory
  activity, neural control, 135
  cavity (molluscs), 69
  organ (*see* Gill)
  pigments, 83
  pumping
    as fixed act, 272
    as fixed-action pattern, 274–275
    in gastropods, 275–276
    neural control, 272, 275
    and swimming, 276
  space, 70–71; in molluscs, 7, 69
  system (*see also* Circulation; Gill)
    neural control, 133, 136, 190
    pleural mantle nerve (PL2), 133
Response hierarchies and behavioral choice, 326
Rhabdomere, 51
Rhinophores, 50, 240
Rhochidion, 91

Sacoglossa, 31–32, 90
Salivary
  glands, 94, 127
  nerve (B3), 129
Saponins, 296
Satiety, 360
Scaphopoda, 14
Seaweeds, 299–300 (*see also* Feeding; Meta-
    morphosis)
Segmentation. *See* Egg
Seminal
  groove, 70
  receptacle, 97, 101
  vesicle, 97, 103
Sense organs, 50, 105
  anterior, innervation, 123
  eye, 50–52, 375 (nudibranchs)
  mechanoreceptors, 65–66
  osphradium, 58
  sensory interactions (nudibranchs), 375
  statocysts, 61–65, 235, 240; in
      nudibranchs, 375
  tentacles, 49–50, 56–57, 239, 304, 314
  withdrawal reflexes, 269
Sensitization, 332–333
  definition, 333, 353
  as arousal component, 351, 353, 365–366, 384
  cellular analysis, 337–345, 383
  vs. conditioning, 333
  of fixed-action patterns, 348–351
  molecular model, 343, 358, 383
  and motivation, relationships, 365–368
  negative and positive systems, 356–358
  by noxious stimuli, 356
  sensory interactions (nudibranchs), 375
  short-term vs. long-term, 333, 337
  stimulation pattern effect, 333
  time course, 333, 337
Sensory neurons
  in feeding, 313 (cephalaspids), 319 (pulmonates)
  and peripheral motor cells, 209
  in sensitization, 340
Septibranchia, 15
Set point, in homeostatic regulation, 359
Sexual behavior. *See also* Reproductive activity
  behavioral differences and sexual isolation, 402
  copulation, 101–102, 321
    and burrowing, 288–290
    circadian activity, 322
    coupling chain, 321–322
    endocrine control, 322
    neural control, 124, 136, 322
  and feeding behavior (opisthobranchs), 322–323,
      326
  mate selection, restrictive mechanisms, 400–402
  mating patterns, 321–322, 322–323 (opistho-
      branchs), 400–402
  mating signals between conspecifics, 404

Sexual behavior (*continued*)
  singing neural circuitry for (cricket), 403–404
Sexual dimorphism, 97
Shell, 70
  in detorsion, 23
  development, 231, 234, 237, 239
  function in larva, 70
  in molluscs, 11–12
  muscles, 68
Shore zones, 40–42, 45
  and *Aplysia* populations, 43–45
  and dietary preferences, 299
  and habituation, 346–348
Sign stimuli, 295
Siphon, 70–71
  axon reflexes, 208
  development, 237, 239–240
  innervation, 65, 138, 191, 199, 203–206, 209,
    (*Spisula*), 269
  mechanoreception, 65
  peripheral-central interactions, 199, 208–209
    (*Spisula*)
  pseudosiphon, 71
  withdrawal reflex, 136, 269
    vs. gill withdrawal reflex, 199
    and inking response, 272–273
    neural control, 199, 204–206, 269, 356
Skin, 48, 240
Somatic motor apparatus, 66–69
Spawn production, 323–324
Speciation, 400–405
Sperm, 97–100, 220
  allosperm, 97, 101–103
  autosperm, 97–103
  autospermal grooves, 100–101
Spermatheca, 101, 103, 136
Spermatic duct, 100
Spermatocyst, 101
Spisula, 208–209
Staining media, 170–171
Statoconia, 61, 64
Statocysts, 61–65, 235
  angular range, 64
  development, 235, 240
  hair cells, 61–64, 375 (nudibranchs)
  innervation, 64, 127, 129
  in nudibranchs, 64, 375
  and pedal ganglia, 249
Statolith, 240
Statolymph, 61
Stomach, 91 (gastropods), 96
Streptoneury, 17, 23, 112, 248
Suprapallium, 70
Swallowing, 94, 308
Swimming. *See* Locomotion
Synapse
  chemical mediation of peripheral neural path-
    ways, 197
  classes, ultrastructures, 175

en passant synapses, 179
  excitatory postsynaptic potential, 206
  in habituation, 212, 337, 345
  homosynaptic depression, 212
  inhibition, in sensory interactions (nudibranchs),
    375–376
  postsynaptic
    contacts, multiple, 170, 177–179
    inputs, multiple, 179–181
    receptive surface, 177
    sites, presynaptic terminals as, 177
  potentials, in feeding behavior modulation
    (gastropods), 316, 318
  presynaptic
    facilitation, 212, 340
    terminals, 170, 175–177, 184
  in sensitization, 337, 340, 351
  spines, 177
  terminals, vesicles, 175, 177, 179, 184
  transmitter content, 175–177
  varicosities, 177
Synophthalmia (cephalopods), 255
Syphonota, 142

Tail (metapodium), 68, 239
Tectibranchia, 29
Teeth (of radula), 90 (gastropods), 91–92
Tegumentary nerves (P2–P5), 131–132, 136
Tentacles
  anterior, 49–50, 239, 304, 314
  chemosensation, 56–57, 314
  innervation, 126–127
  posterior (rhinophores), 50, 240
  tentacular groove, 57
  tentacular nerves (C2, C4), 126–127, 131
Tetrabranchia, 11, 15
Thecosomata, 31–32, 295
Torsion, 16, 18, 120, 231–232, 245
  adaptive advantage, 22–23, 248
  detorsion, 23–24, 26, 29, 120, 245–246, 390
  in gastropods, 16–20, 25–26, 112, 116, 246–248
  ontogenetic origins, 20–22, 26
  phylogenetic origins, 22–23
Transmitter
  mobilization and release, in sensitization,
    340–343
  synaptic content of, 175–177
Tritonia, 45, 260, 292, 296, 323, 348, 395–397
Tritoniomorpha, 29, 31
Trochophore, 225, 230, 240
Trophospongium, 160
Turbellaria, 8–10

Urinary system, 83–88, 106
Urine formation, 83–86

Vagina, 97, 104, 136
Varicosities, synaptic, 177
Veins. *See* Circulatory system

Veliger. *See* Larva
Velum, 231, 234–235, 238–239
Vertebrates
    eye (vs. cephalopod eye), 119
    habituation, 212
    kidney (vs. molluscan kidney), 83–85
    learning capabilities (vs. molluscs), 384
    neurons (vs. invertebrate neurons), 153–155,
        161–162, 177, 187
    pituitary system (vs. cephalopod optic gland),
        143
    reticular system in arousal, 353
Vesicles, neuronal, 175, 177, 179, 184
Visceral
    functions, neural control, 135–138

    loop (gastropods), 116, 118, 140, 145
    mass, 7 (molluscs), 70, 105
    torsion and detorsion, 18, 231, 245
Viviparus, 84–85
Vulvar nerve (A1), 136, 138

Walking. *See* Crawling
Water balance, 83–86
White cells, 150, 160, 203–204
White spots, in juvenile development, 239–240
Winding gland, 103
Withdrawal reflex. *See* Defensive behavior; Gill

Zygosis (gastropods), 116